Integrated Reservoir Asset Management

Integrated Reservoir Asset Management
Principles and Best Practices

John R. Fanchi

ELSEVIER

AMSTERDAM • BOSTON • HEIDELBERG • LONDON
NEW YORK • OXFORD • PARIS • SAN DIEGO
SAN FRANCISCO • SINGAPORE • SYDNEY • TOKYO

Gulf Professional Publishing is an imprint of Elsevier

Gulf Professional Publishing is an imprint of Elsevier
30 Corporate Drive, Suite 400
Burlington, MA 01803, USA

The Boulevard, Langford Lane
Kidlington, Oxford, OX5 1GB, UK

Notices
Knowledge and best practice in this field are constantly changing. As new research and experience broaden our understanding, changes in research methods, professional practices, or medical treatment may become necessary.

Practitioners and researchers must always rely on their own experience and knowledge in evaluating and using any information, methods, compounds, or experiments described herein. In using such information or methods they should be mindful of their own safety and the safety of others, including parties for whom they have a professional responsibility.

To the fullest extent of the law, neither the Publisher nor the authors, contributors, or editors, assume any liability for any injury and/or damage to persons or property as a matter of products liability, negligence or otherwise, or from any use or operation of any methods, products, instructions, or ideas contained in the material herein.

Library of Congress Cataloging-in-Publication Data
Fanchi, John R.
    Integrated reservoir asset management : principles and best practices / John Fanchi.
        p. cm.
    Includes bibliographical references and index.
    ISBN 978-0-12-810200-8
    1. Oil reservoir engineering. 2. Petroleum reserves. I. Title.
    TN870.57.F36 2010
    622'.3382–dc22   2010006757

British Library Cataloguing-in-Publication Data
A catalogue record for this book is available from the British Library.

For information on all Gulf Professional Publishing
publications visit our Web site at *www.elsevierdirect.com*

10 11 12 13 14   10 9 8 7 6 5 4 3 2 1
Printed in the United States

Contents

Preface ix

About the Author xi

1 Introduction 1
1.1 Life Cycle of a Reservoir 1
1.2 Reservoir Management 5
1.3 Recovery Efficiency 7
1.4 Reservoir Management and Economics 8
1.5 Reservoir Management and the Environment 11
CS.1 Valley Fill Case Study: Introduction 14
Exercises 15

2 Fluid Properties 17
2.1 The Origin of Fossil Fuels 17
2.2 Description of Fluid Properties 18
2.3 Classification of Petroleum Fluids 23
2.4 Representation of Fluid Properties 27
CS.2 Valley Fill Case Study: Fluid Properties 31
Exercises 32

3 Geology 33
3.1 The Geologic History of the Earth 33
3.2 Rock Formations and Facies 36
3.3 Structures and Traps 39
3.4 Petroleum Occurrence 39
3.5 Geochemistry 40
3.6 Basin Modeling 42
3.7 Porous Media 43
3.8 Volumetric Analysis 46
CS.3 Valley Fill Case Study: Geologic Model 47
Exercises 47

4 Porosity and Permeability 49
4.1 Bulk Volume and Net Volume 49
4.2 Porosity and Grain Volume 50
4.3 Effective Pore Volume 50
4.4 Porosity Compressibility 51
4.5 Darcy's Law and Permeability 51
4.6 Permeability Averaging 58

	4.7	Transmissibility	61
	4.8	Measures of Permeability Heterogeneity	63
	4.9	Darcy's Law with Directional Permeability	65
	CS.4	Valley Fill Case Study: Permeability	68
		Exercises	69
5	**Geophysics**		**71**
	5.1	Reservoir Scales	71
	5.2	Physics of Waves	74
	5.3	Propagation of Seismic Waves	76
	5.4	Acoustic Impedance and Reflection Coefficients	78
	5.5	Seismic Data Acquisition, Processing, and Interpretation	80
	5.6	Seismic Resolution	82
	5.7	Stratigraphy	85
	CS.5	Valley Fill Case Study: V_P/V_S Model	87
		Exercises	88
6	**Petrophysics**		**89**
	6.1	Elastic Constants	89
	6.2	Elasticity Theory	93
	6.3	The Petroelastic Model	96
	6.4	The Geomechanical Model	101
	6.5	Time-Lapse (4-D) Seismology	104
	CS.6	Valley Fill Case Study: Bulk Moduli	107
		Exercises	108
7	**Well Logging**		**109**
	7.1	Drilling and Well Logging	109
	7.2	Direct Measurement Logs	113
	7.3	Lithology Logs	114
	7.4	Porosity Logs	115
	7.5	Resistivity Logs	118
	7.6	Other Types of Logs	120
	7.7	Reservoir Characterization Issues	121
	CS.7	Valley Fill Case Study: Well logs	123
		Exercises	124
8	**Well Testing**		**125**
	8.1	Pressure Transient Testing	125
	8.2	Oil Well Pressure Transient Testing	126
	8.3	Gas Well Pressure Transient Testing	136
	8.4	Well Test Capabilities	141
	CS.8	Valley Fill Case Study: Well Pressures	143
		Exercises	144
9	**Production Evaluation Techniques**		**145**
	9.1	Decline Curve Analysis	145

    9.2   Gas Well Deliverability   147
    9.3   Material Balance   149
    9.4   Production Performance Ratios and Drive Mechanisms   153
    9.5   Production Stages   155
    9.6   Tracer Tests   157
    9.7   Tracer Test Design   160
    CS.9 Valley Fill Case Study: Production   164
    Exercises   165

10 **Rock–Fluid Interactions**   **167**
    10.1 Interfacial Tension   167
    10.2 Wettability   168
    10.3 Capillary Pressure   169
    10.4 Correlation of Capillary Pressure to Rock Properties   173
    10.5 Equivalent Height and Transition Zone   174
    10.6 Effective Permeability and Relative Permeability   176
    10.7 Mobility, Relative Mobility, and Flow Capacity   181
    CS.10 Valley Fill Case Study: Rock–Fluid Interaction Data   183
    Exercises   184

11 **Reservoir Characterization**   **187**
    11.1 Flow Units   187
    11.2 Traditional Mapping   190
    11.3 Computer-Generated Mapping   192
    11.4 Geostatistics and Kriging   194
    11.5 Geostatistical Modeling   198
    11.6 Visualization Technology   202
    CS.11 Valley Fill Case Study: Reservoir Structure   203
    Exercises   204

12 **Fluid Displacement**   **205**
    12.1 Fractional Flow   205
    12.2 The Buckley-Leverett Theory   209
    12.3 Welge's Method   211
    12.4 Frontal Advance   213
    12.5 Linear Stability Analysis   216
    12.6 Well Patterns   218
    CS.12 Valley Fill Case Study: Conceptual Model   221
    Exercises   221

13 **Reservoir Simulation**   **223**
    13.1 Continuity Equation   223
    13.2 The Convection–Dispersion Equation   226
    13.3 The Navier-Stokes Equation   228
    13.4 Black Oil Model Equations   229

13.5   Integrated Flow Model Equations   232
13.6   The Well Model   236
CS.13 Valley Fill Case Study: Layering of Reservoir
       Flow Model   240
Exercises   241

14   Data Management   **243**
14.1   Sources of Rock Data   243
14.2   Sources of Fluid Data   244
14.3   Sources of Field Performance Data   249
14.4   Data Management   250
14.5   Data Preparation   253
CS.14 Valley Fill Case Study: Input Data Uncertainty   255
Exercises   256

15   Reservoir Flow Modeling   **257**
15.1   Green Field Modeling   257
15.2   Brown Field Modeling   263
15.3   Deterministic Reservoir Forecasting   264
15.4   Probabilistic Reservoir Forecasting   267
15.5   Guidelines for Modern Flow Modeling   275
CS.15 Valley Fill Case Study: Deterministic History Match
       and Base Case Prediction   276
Exercises   277

16   Modern Reservoir Management Applications   **279**
16.1   Improved Oil Recovery   279
16.2   Unconventional Fossil Fuels   282
16.3   Geothermal Reservoir Management   285
16.4   Sequestration   288
16.5   Compressed Air Energy Storage   289
CS.16 Valley Fill Case Study: Waterflood Prediction   291
Exercises   292

Appendix A Unit Conversion Factors   **295**
Appendix B IFLO User's Manual   **299**
B.1   Introduction to IFLO   301
B.2   Initialization Data   303
B.3   Recurrent Data   332
B.4   Program Output   341
References   **343**
Index   **355**

Preface

The primary objective of this book, *Integrated Reservoir Asset Management: Principles and Best Practices*, is to introduce the topic of reservoir management to those with diverse technical backgrounds. Modern reservoir management relies on asset management teams composed of people from a variety of scientific and engineering disciplines. In addition to geologists, geophysicists, and reservoir engineers, asset management teams can include chemists, physicists, biologists, production engineers, flow assurance engineers, drilling engineers, facilities engineers, mechanical engineers, electrical engineers, and environmental engineers. This book is designed to present concepts and terminology for topics that are often encountered by members of reservoir asset management teams and professionals. This book can be used as an introduction to reservoir management for science and engineering students, practicing scientists and engineers, continuing education classes, industry short courses, or self-study.

Included in the book is an update of the material in *Shared Earth Modeling* (2002), which was a compilation of material that I taught in reservoir characterization courses for geoscientists and petroleum engineers at the Colorado School of Mines. The change in title from *Shared Earth Modeling* to *Integrated Reservoir Asset Management* recognizes the technical diversity now found in modern asset management teams, and it changes the focus from the shared earth model to reservoir management. Exercises have been added that allow the reader apply a flow simulator (IFLO) as part of a case study that is used to illustrate and integrate the material in the book. The flow simulator was originally provided with the book *Principles of Applied Reservoir Simulation, Third Edition* (Elsevier, 2006).

Chapter 1 presents an overview of reservoir management. Chapter 2 discusses fluids that may be contained in reservoirs. Chapter 3 reviews geological principles used to characterize the subsurface environment, and Chapter 4 introduces two key reservoir parameters (porosity and permeability). Chapters 5 through 9 describe methods used to acquire information about the subsurface environment. Chapter 10 reviews rock–fluid interaction relationships that are needed for a realistic formulation of multi-phase fluid flow equations. Chapter 11 discusses how to distribute properties throughout the reservoir, and Chapter 12 presents fluid displacement concepts. An introduction to fluid flow equations used in reservoir simulation is presented in Chapter 13. Data management is discussed in Chapter 14. Chapter 15 introduces modern reservoir flow modeling workflows, and Chapter 16 describes a variety of reservoir management applications, including some that are relevant to sustainable energy systems. A Valley Fill Case Study is used to show the reader how the information in each chapter can be applied as part of an integrated reservoir management study. Exercises are provided at the end of each chapter.

Two types of units are commonly found in petroleum literature: oil field units and metric (SI) units. The units used in this book are typically oil field units. In Appendix A, the process of converting from one set of units to another is simplified by providing frequently used factors for conversion between oil field units and metric units. A flow simulator (IFLO) is used in several exercises; see *www.elsevierdirect.com/9780123820884*. The user's manual for the flow simulator is provided in Appendix B.

My colleagues in industry and academia, as well as the students in my multidisciplinary classes, helped me identify important and relevant topics that cross disciplinary lines. I am, of course, responsible for the final selection of topics. I would especially like to thank Kathy Fanchi and Chris Fanchi for their efforts in the preparation of this manuscript.

John R. Fanchi, Ph.D.

About the Author

John R. Fanchi is a professor in the Department of Engineering and Energy Institute at Texas Christian University in Fort Worth, Texas. He holds the Ross B. Matthews Chair of Petroleum Engineering and teaches courses in energy and engineering. Before this appointment, he taught petroleum and energy engineering courses at the Colorado School of Mines and has worked in the technology centers of four energy companies.

He co-edited the General Engineering volume of the *Petroleum Engineering Handbook* published by the Society of Petroleum Engineers, and he is the author of several books, including *Principles of Applied Reservoir Simulation, Third Edition* (Elsevier, 2006), *Math Refresher for Scientists and Engineers, Third Edition* (Wiley, 2006), *Energy in the 21st Century, Second Edition* (World Scientific, 2010), *Energy: Technology and Directions for the Future* (Elsevier–Academic Press, 2004), *Shared Earth Modeling* (Elsevier, 2002), *Integrated Flow Modeling* (Elsevier, 2000), and *Parametrized Relativistic Quantum Theory* (Kluwer, 1993).

1 Introduction

Oil and gas are essential sources of energy in the modern world. They are found in subsurface reservoirs in many challenging environments. Modern reservoir management relies on asset management teams composed of people from a variety of scientific and engineering backgrounds to produce oil and gas. The purpose of this book is to introduce people with diverse technical backgrounds to reservoir management. The book is a reference to topics that are often encountered by members of multidisciplinary reservoir asset management teams and professionals with an interest in managing subsurface resources. These topics are encountered in many applications, including oil and gas production, coalbed methane production, unconventional hydrocarbon production, geothermal energy production, and greenhouse gas sequestration. This chapter presents an overview of reservoir management.

1.1 Life Cycle of a Reservoir

The analysis of the costs associated with the development of an energy source should take into account the initial capital expenditures and annual operating expenses for the life of the system. This analysis is life cycle analysis, and the costs are life cycle costs. Life cycle costing requires the analysis of all direct and indirect costs associated with the system for the entire expected life of the system. In the case of a reservoir, the life cycle begins when the field becomes an exploration prospect, and it does not end until the field is properly abandoned.

The first well in the field is the discovery well. Reservoir boundaries are established by seismic surveys and delineation wells. Delineation wells are originally drilled to define the size of the reservoir, but they can also be used for production or injection later in the life of the reservoir. The production life of the reservoir begins when fluid is withdrawn from the reservoir. Production can begin immediately after the discovery well is drilled or years later after several delineation wells have been drilled. The number of wells used to develop the field, the location of the wells, and their flow characteristics are among the many issues that must be addressed by reservoir management.

1.1.1 History of Drilling Methods

The first method of drilling for oil in the modern era was introduced by Edwin Drake in the 1850s and is known as cable-tool drilling. In this method, a rope connected to a wood beam had a drill bit attached to the end. The beam was raised

Integrated Reservoir Asset Management. DOI: 10.1016/B978-0-12-382088-4.00001-3

and lowered, which lifted and dropped the bit and dug a hole into the ground. Cable-tool drilling does not work in soft-rock formations, where the sides of the hole might collapse. Cable-tool drilling has been largely replaced by rotary drilling.

Developed in France in the 1860s, rotary drilling was first used in the United States in the 1880s because it could drill into the soft-rock formations of the Corsicana oil field in Texas. Rotary drilling uses a rotating drill bit with nozzles for shooting out drilling mud to penetrate into the earth. Drilling mud is designed to carry rock cuttings away from the bit and lift them up the wellbore to the surface.

Rotary drilling gained great popularity after Captain Anthony F. Lucas drilled the Lucas 1 well at Spindletop, near Beaumont, Texas. The Lucas 1 well was a discovery well and a "gusher." Gas and oil flowed up the well and engulfed the drilling derrick. Instead of flowing at the expected 50 barrels of oil per day, the well produced up to 75,000 barrels per day. The Lucas gusher began the Texas oil boom (Yergin, 1992, pp. 83–85). Since then, rotary drilling has become the primary means of drilling.

Once a hole has been drilled, it is necessary to "complete" the well. A well is completed when it is prepared for production. The first well of the modern era was completed in 1808 when two American brothers, David and Joseph Ruffner, used wooden casings to prevent low-concentration saltwater from diluting the high-concentration saltwater they were extracting from deeper in their saltwater well (Van Dyke, 1997).

It is sometimes necessary to provide energy to extract oil from reservoirs. Oil can be lifted using pumps or by injecting gas into the wellstream to increase the buoyancy of the gas-oil mixture. The earliest pumps used the same wooden beams that were used for cable-tool drilling. Oil companies developed central pumping power in the 1880s. Central pumping power used a prime mover—a power source—to pump several wells. In the 1920s, demand for the replacement of on-site rigs led to the use of a beam pumping system for pumping wells. A beam pumping system is a self-contained unit that is mounted at the surface of each well and operates a pump in the hole. More modern techniques include gas-lift and electric submersible pumps.

1.1.2 Modern Drilling Methods

Advances in drilling technology are extending the options available for prudently managing subsurface reservoirs and producing fossil fuels, especially oil and gas. Modern drilling methods include horizontal wells, multilateral wells, and infill drilling.

A well is a string of connected, concentric pipes. The path followed by the string of pipes is called the trajectory of the well. Historically, wells were drilled vertically into the ground, and the well trajectory was essentially a straight, vertical line. Today, wells can be drilled so that the well trajectory is curved. A curved wellbore trajectory is possible because the length of each straight pipe that makes up the well is small compared to the total well length. The length of a typical section of pipe in

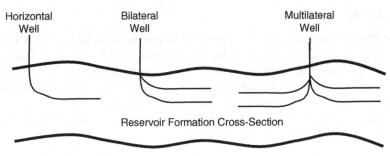

Figure 1.1 Multilateral wells.

a well is 30 to 40 feet. Wells with one or more horizontal trajectories are shown in Figure 1.1.

A well can begin as a vertical well and then later be modified to a horizontal or multilateral well. The vertical section of the well is called the main (mother) bore or trunk. The point where the main bore and a lateral meet is called a junction. When the vertical segment of the well reaches a specified depth called the kick-off point (KOP), mechanical wedges (whipstocks) or other downhole tools are used to change the direction of the drill bit and alter the well path. The beginning of the horizontal segment is the heel, and the end of the horizontal segment is the toe. The distance, or reach, of a well from the drilling rig to final bottomhole location can exceed six miles. Wells with unusually long reach are called extended reach wells.

Wells with more than one hole can be drilled. Each hole is called a lateral or branch, and the well itself is called a multilateral well. For example, a bilateral well is a well with two branches. Figure 1.1 shows examples of modern multilateral well trajectories.

Multilateral wells make it possible to connect multiple well paths to a common wellbore, and they have many applications. For example, multilateral wells are used in offshore environments where the number of well slots is limited by the amount of space available on a platform. They are also used to produce fluids from reservoirs that have many compartments. A compartment in a reservoir is a volume that is isolated from other parts of the reservoir by barriers to fluid flow such as sealing faults.

Horizontal, extended reach, and multilateral wellbores that follow subsurface formations provide access to more parts of the reservoir from fewer well locations. This provides a means of minimizing the environmental impact associated with drilling and production facilities, either on land or at sea. Extended reach wells make it possible to extract petroleum from beneath environmentally or commercially sensitive areas by drilling from locations outside of the environmentally sensitive areas. Extended reach wells make it possible to produce offshore fields from onshore drilling locations and reduce the environmental impact of drilling by reducing the number of surface drilling locations.

Infill Drilling

Infill drilling is the process of increasing the number of wells in an area by drilling wells in spaces between existing wells. The increase in well density, or number of wells per unit area, can improve recovery efficiency by providing fluid extraction points in parts of the reservoir that have not been produced. Changes to well patterns and the increase in well density can alter flow patterns in displacement processes and enable the displacement of *in situ* fluids by injected fluids. Infill drilling is especially useful in heterogeneous reservoirs.

Geosteering

Geosteering is the technology that makes it possible to accurately steer the well to its targeted location and is a prerequisite for successful extended reach drilling. Microelectronics is used in the drilling assembly to provide information to drill rig operators at the surface about the location of the drill bit as it bores a hole into the earth. Operators can modify the trajectory of the well while it is being drilled based on information from these measurement-while-drilling (MWD) systems. Geosteering and extended reach drilling can reduce costs associated with the construction of expensive, new offshore platforms by expanding the volume of the reservoir that is directly accessible from a given drilling location. In some cases, wells drilled from onshore drilling rigs can be used to produce coastal offshore fields that are within the range of extended reach drilling.

1.1.3   Production Systems

A production system can be thought of as the collection of subsystems illustrated in Figure 1.2. Fluids are taken from the reservoir using wells, which must be drilled and completed. The performance of the well depends on the properties of the reservoir rock, the interaction between the rock and the fluids in the reservoir, and the properties of the fluids in the reservoir. Reservoir fluids include the fluids originally

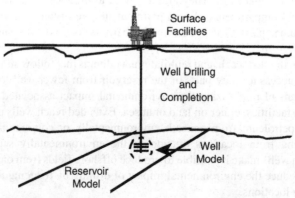

Figure 1.2 A production system.

contained in the reservoir, as well as fluids that may be introduced as part of the reservoir management program. Well performance also depends on the properties of the well itself, such as its cross-section, length, trajectory, and type of completion. The completion of the well establishes the connection between the well and the reservoir. A completion can be as simple as an open-hole completion where fluids are allowed to drain into the wellbore from consolidated reservoir rock, to completions that require the use of tubing with holes punched through the walls of the tubing using perforating guns.

Surface facilities are needed to drill, complete, and operate wells. Drilling rigs may be moved from one location to another on trucks, ships, or offshore platforms; or drilling rigs may be permanently installed at specified locations. The facilities may be located in desert climates in the Middle East, stormy offshore environments in the North Sea, arctic climates in Alaska and Siberia, and deepwater environments in the Gulf of Mexico and off the coast of West Africa.

Produced fluids must be recovered, processed, and transported to storage facilities and eventually to the consumer. Processing can begin at the well site where the produced wellstream is separated into oil, water, and gas phases. Further processing at refineries separates the hydrocarbon fluid into marketable products, such as gasoline and diesel fuel. Transportation of oil and gas may be by a variety of means, including pipelines, tanker trucks, double-hulled tankers, and liquefied natural gas transport ships.

1.2 Reservoir Management

Modern reservoir management is generally defined as a continuous process that optimizes the interaction between data and decision making during the life cycle of a field (Saleri, 2002). This definition covers the management of hydrocarbon reservoirs and other reservoir systems, including geothermal reservoirs and reservoirs used for geological sequestration. Geological sequestration is the long-term storage of greenhouse gases, such as carbon dioxide, in geological formations. The reservoir management plan should be flexible enough to accommodate technological advances, changes in economic and environmental factors, and new information obtained during the life of the reservoir, and it should be able to address all relevant operating issues, including governmental regulations.

Many disciplines contribute to the reservoir management process. In the case of a hydrocarbon reservoir, successful reservoir management requires understanding the structure of the reservoir, the distribution of fluids within the reservoir, drilling and maintaining wells that can produce fluids from the reservoir, transport and processing of produced fluids, refining and marketing the fluids, safely abandoning the reservoir when it can no longer produce, and mitigating the environmental impact of operations throughout the life cycle of the reservoir. Properly constituted asset management teams include personnel with the expertise needed to accomplish all of these tasks. These people are often specialists in their disciplines. They must be able to communicate with one another and work together toward a common objective.

Reservoir management studies are important when significant choices must be made. The choices can range from "business as usual" to major changes in investment strategy. For example, decision makers may have to choose between investing in a new project or investing in an existing project that requires changes in operations to maximize return on investment. By studying a range of scenarios, decision makers will have information that can help them decide how to commit limited resources to activities that can achieve management objectives.

Reservoir flow modeling is the most sophisticated methodology available for generating production profiles. A production profile presents fluid production as a function of time. Fluid production can be expressed as flow rates or cumulative production. By combining production profiles with hydrocarbon price forecasts, it is possible to create cash flow projections. The combination of production profile from flow modeling and price forecast from economic modeling yields economic forecasts that can be used to compare the economic value of competing reservoir management concepts. This is essential information for the management of a reservoir, and it can be used to determine reservoir reserves. The definition of reserves is summarized in Table 1.1 (SPE-PRMS, 2007).

The probability distribution associated with the SPE-PRMS reserves definitions can be illustrated using a normal distribution. We assume that several statistically independent models of the reservoir have been developed and used to estimate reserves. In the absence of data to the contrary, a reasonable first approximation is that each model has been sampled from a normal distribution of reserves. Given this assumption, an average μ and standard derivation σ may be calculated to

Table 1.1 SPE/WPC Reserves Definitions

Proved Reserves	• Those quantities of petroleum, which by analysis of geoscience and engineering data, can be estimated with reasonable certainty to be commercially recoverable, from a given date forward, from known reservoirs and under defined economic conditions, operating methods, and government regulations. • There should be at least a 90 percent probability (P_{90}) that the quantities actually recovered will equal or exceed the estimate.
Probable Reserves	• Those additional reserves that analysis of geoscience and engineering data indicate are less likely to be recovered than proved reserves but more certain to be recovered than possible reserves. • There should be at least a 50 percent probability (P_{50}) that the quantities actually recovered will equal or exceed the estimate.
Possible Reserves	• Those additional reserves that analysis of geoscience and engineering data suggest are less likely to be recoverable than probable reserves. • There should be at least a 10 percent probability (P_{10}) that the quantities actually recovered will equal or exceed the estimate.

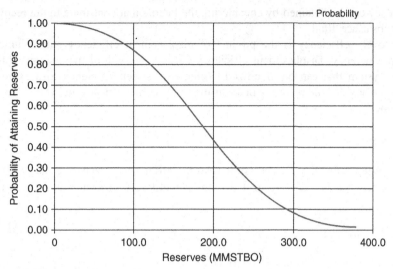

Figure 1.3 The production system.

prepare a normal distribution of reserves. For a normal distribution with mean μ and standard deviation σ, the SPE-PRMS reserves definitions are

$$
\begin{aligned}
\text{Proved reserves} &= P_{90} = \mu - 1.28\sigma \\
\text{Probable reserves} &= P_{50} = \mu \\
\text{Possible reserves} &= P_{10} = \mu + 1.28\sigma
\end{aligned}
\qquad (1.2.1)
$$

Figure 1.3 shows a normal distribution for a mean of 189 MMSTBO and a standard deviation of 78 MMSTBO. The SPE-PRMS reserves from this distribution are

$$
\begin{aligned}
\text{Proved reserves} &= P_{90} = 88\,\text{MMSTBO} \\
\text{Probable reserves} &= P_{50} = 189\,\text{MMSTBO} \\
\text{Possible reserves} &= P_{10} = 289\,\text{MMSTBO}
\end{aligned}
\qquad (1.2.2)
$$

In this case, the normal distribution is used to associate an estimate of the likelihood of occurrence of any particular prediction case with its corresponding economic forecast. For example, we use Figure 1.3 to see that a reserves estimate of 200 MMSTBO corresponds to a probability of approximately P_{43}.

1.3 Recovery Efficiency

One of the objectives of reservoir management is to develop a plan for maximizing recovery efficiency. Recovery efficiency is a measure of the amount of resource recovered relative to the amount of resource originally in place. It is defined by comparing initial and final *in situ* fluid volumes. An estimate of expected recovery efficiency can be obtained by considering the factors that contribute to the recovery of a subsurface fluid.

Recovery efficiency is the product of displacement efficiency and volumetric sweep efficiency. Displacement efficiency E_D is a measure of the amount of fluid in the system that can be mobilized. Volumetric sweep efficiency E_{Vol} expresses the efficiency of fluid recovery in terms of areal sweep efficiency and vertical sweep efficiency:

$$E_{Vol} = E_A \times E_V \tag{1.3.1}$$

Areal sweep efficiency E_A and vertical sweep efficiency E_V measure the degree of contact between *in situ* and injected fluids. Areal sweep efficiency is defined as

$$E_A = \frac{\text{swept area}}{\text{total area}} \tag{1.3.2}$$

and vertical sweep efficiency is defined as

$$E_V = \frac{\text{swept net thickness}}{\text{total net thickness}} \tag{1.3.3}$$

Recovery efficiency RE is the product of these efficiencies:

$$RE = E_D \times E_{Vol} = E_D \times E_A \times E_V \tag{1.3.4}$$

Each of the recovery efficiencies is a fraction that varies from 0 to 1. If one or more of the factors that enter into the calculation of recovery efficiency is small, recovery efficiency will be small. On the other hand, each of the factors can be relatively large, and the recovery efficiency will still be small because it is a product of factors that are less than one. In many cases, technology is available for improving recovery efficiency, but it may not be implemented because it is not economic. The application of technology and the ultimate recovery of fossil fuels depend on the economic value of the resource.

1.4   Reservoir Management and Economics

The definition of reservoir management presented previously recognizes the need to consider the economics of resource development. The economic value of a project is influenced by many factors, some of which can be measured. An economic measure that is typically used to evaluate cash flow associated with reservoir management options is net present value (NPV). The cash flow of an option is the net cash generated or expended on the option as a function of time. The time value of money is included in economic analyses by applying a discount rate to adjust the value of money to the value during a base year. Discount rate is the adjustment factor, and

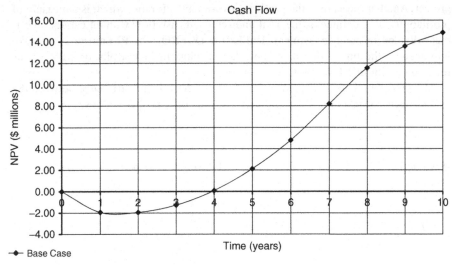

Figure 1.4 Typical cash flow.

the resulting cash flow is called the discounted cash flow. The NPV of the cash flow is the value of the cash flow at a specified discount rate. The discount rate at which NPV is zero is called the discounted cash flow return on investment (DCFROI) or internal rate of return (IRR).

Figure 1.4 shows a typical plot of NPV as a function of time. The early time part of the figure shows a negative NPV and indicates that the project is operating at a loss. The loss is usually associated with initial capital investments and operating expenses that are incurred before the project begins to generate revenue. The reduction in loss and eventual growth in positive NPV is due to the generation of revenue in excess of expenses. The point in time on the graph where the NPV is zero after the project has begun is the discounted payout time. Discounted payout time in Figure 1.4 is approximately four years.

Table 1.2 presents the definitions of several commonly used economic measures. DCFROI and discounted payout time are measures of the economic viability of a

Table 1.2 Definitions of Selected Economic Measures

Economic Measure	Definition
Discount Rate	Factor to adjust the value of money to a base year
Net Present Value (NPV)	Value of cash flow at a specified discount rate
DCFROI or IRR	Discount rate at which NPV = 0
Discounted Payout Time	Time when NPV = 0
Profit-to-Investment (PI) Ratio	Undiscounted cash flow without capital investment divided by total investment

project. Another measure is the profit-to-investment (PI) ratio, which is a measure of profitability. It is defined as the total undiscounted cash flow without capital investment divided by total investment. Unlike the DCFROI, the PI ratio does not take into account the time value of money. Useful plots include a plot of NPV versus time and a plot of NPV versus discount rate.

The preceding ideas are quantified as follows. *NPV* is the difference between the present value of revenue R and the present value of expenses E; thus,

$$NPV = R - E \tag{1.4.1}$$

If we define $\Delta E(k)$ as the expenses incurred during a time period k, then E may be written as

$$E = \sum_{k=0}^{N \times Q} \frac{\Delta E(k)}{\left(1 + \frac{i'}{Q}\right)^k} \tag{1.4.2}$$

where i' is the annual inflation rate, N is the number of years of the expenditure schedule, and Q is the number of times interest is compounded each year. A similar expression is written for revenue R:

$$R = \sum_{k=0}^{N \times Q} \frac{\Delta R(k)}{\left(1 + \frac{i}{Q}\right)^k} \tag{1.4.3}$$

where $\Delta R(k)$ is the revenue obtained during time period k, and i is the annual interest or discount rate. Equations (1.4.2) and (1.4.3) include the assumptions that i and i' are constants over the life of the project, but i and i' are not necessarily equal. These assumptions let us compute the present value of money expended relative to a given inflation rate i' and compare the result to the present value of revenue associated with a specified interest or discount rate i.

Net present value takes into account the time value of money. NPV for an oil and/or gas reservoir may be calculated for a specific discount rate using the equation

$$NPV = \sum_{n=1}^{N} \frac{P_{on}Q_{on} + P_{gn}Q_{gn} - CAPEX_n - OPEX_n - TAX_n}{(1 + r)^n} \tag{1.4.4}$$

where

  N = Number of years
  P_{on} = Oil price during year n
  Q_{on} = Oil production during year n
  P_{gn} = Gas price during year n
  Q_{gn} = Gas production during year n

$CAPEX_n$ = Capital expenses during year n
$OPEX_n$ = Operating expenses during year n
TAX_n = Taxes during year n
r = Discount rate

In many cases, resource managers have little influence on taxes and prices. On the other hand, most resource managers can exert considerable influence on production performance and expenses. Several strategies may be used to affect NPV. Some strategies include accelerating production, increasing recovery, and lowering operating costs. One reservoir management challenge is to optimize economic measures like NPV.

Revenue stream forecasts are used to prepare both short- and long-term budgets. They provide the production volumes needed in the NPV calculation. For this reason, the asset management team may be expected to generate flow predictions using a combination of reservoir parameters that yield a range of recoveries. Uncertainty analysis is a useful process for determining the likelihood that any one set of parameters will be realized and estimating the probability distribution of reserves.

Reservoir management must consider how much money will be available to pay for wells, compressors, pipelines, platforms, processing facilities, and any other items that are needed to implement the plan represented by the model. The revenue stream is used to pay taxes, capital expenses, and operating expenses. The economic performance of the project depends on the relationship between revenue and expenses. Several economic criteria may be considered in the evaluation of a project, such as NPV, internal rate of return, and profit-to-investment ratio. The selection of economic criteria is typically a management function. Once the criteria are defined, they can be applied to a range of possible operating strategies. The strategies should include assessment of both tangible and intangible factors. A comparative analysis of different operating strategies gives decision-making bodies valuable information for making informed decisions.

1.5   Reservoir Management and the Environment

The impact of a project on the environment must be considered when developing a reservoir management strategy. Environmental studies should consider such topics as pollution evaluation and prevention, and habitat preservation in both onshore and offshore environments. An environmental impact analysis provides a baseline on existing environmental conditions and provides an estimate of the impact of future operations on the environment. Forecasts of environmental impact typically require risk assessment, with the goal of identifying an acceptable risk for implementing a project (Wilson and Frederick, 1999). Computer flow models are often used to prepare forecasts as well as guide remedial work to reclaim the environment.

A well-managed field should be compatible with both the surface and subsurface environment. Failure to adequately consider environmental issues can lead to tangible and intangible losses. Tangible losses have more readily quantifiable economic consequences. For example, if potable water is contaminated, the cost to remediate can adversely affect project economics. Intangible losses are more difficult to quantify, but they can include loss of public support for an economically attractive project. For example, the poor public image of the oil industry in the United States has contributed to political opposition to oil industry development of land regulated by the federal government. In some cases, the intangible loss can take the form of active opposition to an otherwise economically viable project. In many parts of the world, it is necessary to provide an environmental impact statement as part of the reservoir management plan.

Environmental issues must always be considered in the development of a reservoir management strategy. For example, the Louisiana Offshore Oil Production (LOOP) facility is designed to keep the transfer of hydrocarbons between pipelines and tankers away from sensitive coastal areas. Periodic water sampling of surface and produced waters can ensure that freshwater sources are not contaminated. In addition, periodic testing for the excavation or production of naturally occurring radioactive materials helps ensure environmental compliance.

The advantages of operating a field with prudent consideration of environmental issues can pay economic dividends. In addition to improved public relations, sensitivity to environmental issues can minimize adverse environmental effects that may require costly remediation and financial penalties. Remediation often takes the form of cleanup, such as the cleanup required after the oil spill from the *Exxon Valdez* oil tanker in Alaska. Technologies are being developed to improve our ability to clean up environmental pollutants. For example, bioremediation uses living microorganisms or their enzymes to accelerate the rate of degradation of environmental pollutants (Westlake, 1999).

It becomes a question of business ethics whether a practice that is legal but can lead to an adverse environmental consequence should nonetheless be pursued because a cost-benefit analysis showed that economic benefits exceeded economic liabilities. Typically, arguments to pursue an environmentally undesirable practice based on cost-benefit analyses do not adequately account for intangible costs. For example, the decision by Shell to dispose of the Brent Spar platform by sinking it in the Atlantic Ocean led to public outrage in Europe in 1995. Reversing the decision and disassembling the platform for use as a quay in Norway resolved the resulting public relations problem, but the damage had been done. The failure to anticipate the public's reaction reinforced a lack of public confidence in the oil and gas industry, and it helped motivate government action to regulate the decommissioning of offshore platforms in northwestern Europe (Wilkinson, 1997; Offshore Staff, 1998).

1.5.1 *Sustainable Development*

The concept of sustainable development was introduced in 1987 in a report prepared by the United Nations' World Commission on Environment and Development (Brundtland, 1987). The commission, known as the Brundtland Commission, after chairwoman Gro Harlem Brundtland of Norway, said that societies should adopt a policy of sustainable development that allows them to meet their present needs while preserving the ability of future generations to meet their own needs. The three components of sustainable development are economic prosperity, social equity, and environmental protection.

Sustainable development is intended to preserve the rights of future generations. It is possible to argue that future generations have no legal rights to current natural resources and are not entitled to any. From this perspective, each generation must do the best it can with available resources. On the other hand, many societies are choosing to adopt the value of preserving natural resources for future generations. National parks are examples of natural resources that are being preserved.

1.5.2 Global Climate Change

One environmental concern that is facing society currently is global climate change. Measurements of ambient air temperature show a global warming effect that corresponds to an increase in the average temperature of the earth's atmosphere. The increase in atmospheric temperature has been linked to the combustion of fossil fuels (Wigley et al., 1996).

When a carbon-based fuel burns, carbon can react with oxygen and nitrogen in the atmosphere to produce carbon dioxide (CO_2), carbon monoxide, and nitrogen oxides (often abbreviated as NOx). The combustion by-products, including water vapor, are emitted into the atmosphere in gaseous form. Some of the gaseous byproducts are called greenhouse gases because they contribute to the greenhouse effect, illustrated in Figure 1.5 (Fanchi, 2004). Some of the incident solar radiation from the Sun is absorbed by the earth, some is reflected into space, and some is captured by greenhouse gases in the atmosphere and reradiated as infrared radiation (heat). The reradiated energy would escape the earth as reflected sunlight if

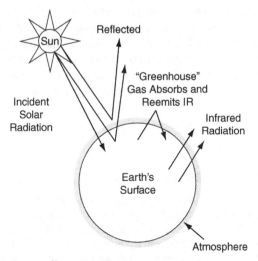

Figure 1.5 The greenhouse effect.

greenhouse gases were not present in the atmosphere. Greenhouse gases include carbon dioxide, methane, and nitrous oxide, as well as other gases such as volatile organic compounds and hydrofluorocarbons.

Carbon dioxide (CO_2) is approximately 83 percent of the greenhouse gases emitted by the United States as a percent of the mass of carbon or carbon equivalent. Wigley and colleagues (1996) projected ambient CO_2 concentration through the twenty-first century. Pre-industrial atmospheric CO_2 concentration was approximately 288 parts per million, and the current atmospheric CO_2 concentration is 340 parts per million. The concentration of CO_2 that would establish an acceptable energy balance is considered to be 550 parts per million. To achieve the acceptable concentration of CO_2 through the next century, societies would have to reduce the volume of greenhouse gases entering the atmosphere.

Many scientists attribute global climate change to the greenhouse effect. The Kyoto Protocol is an international treaty that was negotiated in Kyoto, Japan, in 1997 to establish limits on the amount of greenhouse gases a country can emit into the atmosphere. The Kyoto Protocol has not been accepted worldwide. Some countries believe the greenhouse gas emission limits are too low and would adversely impact national and world economies without solving the problem of global warming. Another criticism of the Kyoto Protocol is that it does not apply to all nations. For example, China is exempt from greenhouse gas emission limitations in the Kyoto Protocol even though it has one of the world's fastest-growing economies and the world's largest population.

Concern about global climate change has motivated a change in the definition of pollution. For example, it used to be an acceptable practice to release natural gas into the atmosphere by flaring the gas. This practice is now prohibited in many parts of the world as an undesirable practice because natural gas is a greenhouse gas. One proposed method for reducing the climatic greenhouse effect is to collect and store carbon dioxide in geologic formations as part of a process known as CO_2 sequestration. The sequestration of CO_2 in subsurface formations is a gas storage process that must satisfy the three primary objectives in designing and operating natural gas storage reservoirs: verification of injected gas volume, monitoring of injected gas migration, and determination of gas injectivity. The goal of geologic carbon sequestration and similar programs is to provide economically competitive and environmentally safe options to offset all of the projected growth in baseline emissions of greenhouse gases.

CS.1 Valley Fill Case Study: Introduction

The primary purpose of the Valley Fill case study from a pedagogical perspective is to show how to apply reservoir management concepts using a realistic example. The incised valley model is useful for describing reservoirs in both mature and frontier basins around the world (e.g., Bowen et al., 1993; Peijs-van Hilten et al., 1998). Each chapter presents information that is integrated into the reservoir management example. The reservoir of interest is an oil reservoir that has been producing for a year. Wells in the field are shown in Figure CS.1A.

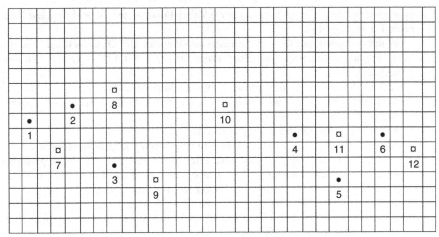

Figure CS.1A Well locations in an area that is 6,000 feet long by 3,000 feet wide.

Exercises

1-1. List several questions you would want to have answered if you were trying to decide how to manage the Valley Fill reservoir.

1-2. Suppose displacement efficiency is 27 percent, areal sweep efficiency is 60 percent, and vertical sweep efficiency is 75 percent. Estimate recovery efficiency.

1-3. We want to drill a 5,000-foot-deep vertical well. We know from previous experience in the area that the drill bit will be effective for 36 hours before it has to be replaced. The average drill bit will penetrate 20 feet of rock in the area for each hour of drilling. Again, based on previous experience, we expect the average trip to replace the drill bit to take about 8 hours. A "trip" is the act of withdrawing the drill pipe, replacing the drill bit, and then returning the new drill bit to the bottom of the hole. Given this information, estimate how long it will take to drill the 5,000-foot-deep vertical well. *Hint:* Prepare a table like the following.

Incremental Time (hrs)	Incremental Depth (ft)	Cumulative Time (hrs)	Cumulative Depth (ft)

1-4. Complete the following table and estimate the proved, the probable, and the possible reserves. Assume the reserves are normally distributed. *Hint:* Reserves = OOIP × Recovery Factor.

Model	OOIP (MMSTB)	Recovery Factor	Reserves (MMSTB)
1	700	0.42	
2	650	0.39	
3	900	0.45	
4	450	0.25	
5	725	0.43	

1-5. Any reader interested in participating in the Valley Fill case study should complete Exercises 1-5 through 1-8.

A three-dimensional, three-phase reservoir simulator (IFLO) is included with this book. Prepare a folder on your hard drive for running IFLO using the following procedure.
- Make a directory on your computer called RMSE/VALLEY.
- Go to the website *http://www.bh.com/companions/0750675225* and copy the zip file to RMSE/VALLEY.
- Extract all of the files to RMSE/VALLEY.
- Some of the files may be labeled "Read Only" when you copy the files to RMSE/VALLEY. To remove this restriction, select the file(s) and change the properties of the file(s) by removing the check symbol adjacent to the "Read Only" attribute.

What is the size of the executable file IFLO.EXE in megabytes (MB)?

1-6. Several example data files are provided with IFLO. Make a list of the data files (files with the extension "DAT"). Unless stated otherwise, all exercises assume that IFLO and its data files reside in the RMSE/VALLEY directory.

1-7. The program IFLO runs the file called "ITEMP.DAT". To run a new data file, such as NEWDATA.DAT, copy NEWDATA.DAT to ITEMP.DAT. In this exercise, copy VFILL1_HM.DAT to ITEMP.DAT, and run IFLO by double clicking on the IFLO.EXE file on your hard drive. Select option "Y" to write the run output to files. When the program ends, it will print "STOP". Close the IFLO window. You do not need to save any changes. Open run output file ITEMP.ROF, and find the line reading "MAX # OF AUTHORIZED GRID BLOCKS". How many grid blocks are you authorized to use with the simulator provided with this book?

1-8. The program 3DVIEW may be used to view the reservoir structure associated with IFLO data files. 3DVIEW is a visualization program that reads IFLO output files with the extension "ARR". To view a reservoir structure, proceed as follows:
- Use your file manager to open your folder containing the IFLO files. Unless stated otherwise, all mouse clicks use the left mouse button.
  - **a.** Start 3DVIEW (double click on the application entitled 3DVIEW.EXE).
  - **b.** Click on the button "File".
  - **c.** Click on "Open Array File".
  - **d.** Click on "ITEMP.ARR" in the file list.
  - **e.** Click on "OK".
- At this point you should see a structure in the middle of the screen. The structure is an oil-filled channel sand. To see the channel, use the left mouse button to select Model/Select Active Attribute/SO. This displays oil saturation in the channel.
- To view different perspectives of the structure, hold the left mouse button down and move the mouse. With practice, you can learn to control the orientation of the structure on the screen.
- The grid block display may be able to be smoothed by selecting Project/Smooth Model Display.

To exit 3DVIEW, click on the "File" button and then click "Exit".

2 Fluid Properties

Reservoir pore spaces are occupied by fluids that range from *in situ* water to hydro-carbon liquid and gas. In addition, reservoirs are being used as permanent storage locations for gases that people do not want to emit into the atmosphere, such as carbon dioxide and hydrogen sulfide. Properties of fluids are needed to determine the amount of fluid that can be contained in a reservoir and to determine fluid flow characteristics. This chapter describes the leading theory of the origin of fossil fuels, defines the concepts needed to understand and classify fluid properties, shows how to identify sources of fluid data, and explains how to prepare a quantitative representation of fluid properties.

2.1 The Origin of Fossil Fuels

The mainstream theory of the origin of fossil fuels is the biogenic theory. It says that oil and gas formed as a result of the death, burial, decay, and compaction of organisms that lived on or near the surface of the earth. When vegetation dies and decays in aqueous environments such as swamps, it can form a carbon-rich organic material called peat. If peat is buried by subsequent geological activity, the buried peat is subjected to increasing temperature and pressure. Peat can eventually be transformed into coal by the process of diagenesis. A similar diagenetic process is thought to be the origin of petroleum and is illustrated in Figure 2.1.

A petroleum fluid is a mixture of hydrocarbon molecules and inorganic impurities, such as nitrogen, carbon dioxide, and hydrogen sulfide. According to the biogenic theory, petroleum is formed by the death of organisms such as algae and bacteria. The remains of the organisms settle into the sediments at the base of an aqueous environment as organic debris. Lakebeds and seabeds are examples of favorable sedimentary environments. Subsequent sedimentation buries the organic debris. As burial continues, the organic material is subjected to increasing temperature and pressure and is transformed by bacterial action into oil and gas. Petroleum fluids are usually less dense than water and will migrate upward until they encounter impermeable barriers and are collected in traps. The accumulation of hydrocarbon in a geologic trap forms a petroleum reservoir.

The composition of a petroleum fluid depends on such factors as the composition of its source, reservoir temperature, and reservoir pressure. Petroleum can exist as a solid, liquid, or gas. The phase depends on fluid composition and the temperature and pressure of its surroundings. Oil is a liquid mixture of hydrocarbons. Natural gas is typically methane with lesser amounts of heavier hydrocarbon molecules like

Integrated Reservoir Asset Management. DOI: 10.1016/B978-0-12-382088-4.00002-5

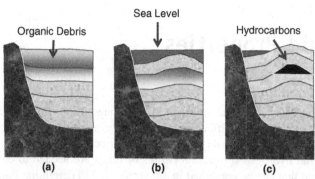

Figure 2.1 The biogenic origin of oil and gas: (a) death and burial, (b) bacterial and geologic action, and (c) migration and trapping.

ethane and propane. The elemental mass content of petroleum fluids ranges from approximately 84 to 87 percent carbon and 11 to 14 percent hydrogen, which is comparable to the carbon and hydrogen content of life. This is one piece of evidence for the origin of petroleum from biological sources.

2.2 Description of Fluid Properties

Several concepts are needed to provide a description of fluid properties that can be used for geoscience and engineering calculations. Some of the most fundamental concepts of fluid properties are defined in the following subsections for ease of reference. For more details, see the extensive literature on fluid properties, including Amyx et al. (1960), Pederson et al. (1989), McCain (1990, 1991), Dake (2001), Ahmed (2000), Whitson and Brulé (2000), Walsh and Lake (2003), and the collection of articles in Fanchi (2006b).

2.2.1 Temperature

Temperature is a measure of the average kinetic energy of a system. The most commonly used temperature scales are the Fahrenheit and Celsius scales. The relationship between these scales is

$$T_C = \frac{5}{9}(T_F - 32) \tag{2.2.1}$$

where T_C and T_F are temperatures in degrees Celsius and degrees Fahrenheit, respectively.

Applications of the equations of state described in the following require the use of absolute temperature scales. Absolute temperature may be expressed in terms of degrees Kelvin or degrees Rankine. The absolute temperature scale in degrees Kelvin is related to the Celsius scale by

$$T_K = T_C + 273 \tag{2.2.2}$$

where T_K is temperature in degrees Kelvin. The absolute temperature scale in degrees Rankine is related to the Fahrenheit scale by

$$T_R = T_F + 460 \qquad (2.2.3)$$

where T_R is temperature in degrees Rankine.

2.2.2 Density

Density ρ is the mass of a substance M divided by the volume V it occupies:

$$\rho = M/V \qquad (2.2.4)$$

Fluid density depends on pressure, temperature, and composition.

Mass and volume are examples of extensive properties, which are properties that depend on the amount of material. Density, temperature, and pressure are examples of intensive properties. An intensive property is a fluid property that is independent of the amount of material. For example, suppose we subdivide a cell of gas with volume V into two halves by a vertical partition. If the gas was initially in an equilibrium state, the gas in each half of the cell after inserting the partition should have the same pressure and temperature as it did before the partition was inserted. The mass and volume in each half of the cell will be one-half the original mass and volume, but their ratio—the density—will remain unchanged.

2.2.3 Composition

The composition of a fluid is determined by the types of molecules that comprise the fluid. A pure fluid consists of a single type of molecule, such as water or methane, whereas a fluid mixture contains several types of molecules. Petroleum is a mixture of hydrocarbon compounds, and *in situ* water usually contains dissolved solids and may contain dissolved gases. The composition of a fluid is specified by listing the molecular components contained in the fluid and their relative amounts. The relative amount of each component in a mixture is defined as the concentration of the component. Concentration may be expressed in such units as volume fraction, weight fraction, or molar fraction. The unit of concentration should be clearly expressed to avoid errors. Concentration is often expressed as mole fraction, but it is wise to confirm the unit if a unit is not explicitly specified.

The symbols $\{x_i, y_i, z_i\}$ are often used to denote the mole fraction of component i in the liquid phase, gas phase, and wellstream, respectively. The mole fraction of component i in a gas mixture is the number of moles n_{Vi} of the component in the gas (vapor phase) divided by the total number of moles in the gas:

$$y_i = \frac{n_{Vi}}{\displaystyle\sum_{j=1}^{N_c} n_{Vj}} \qquad (2.2.5)$$

where N_c is the number of components in the mixture. The liquid phase mole fraction x_i is the number of moles in the liquid n_{Li} divided by the total number of moles in the liquid:

$$x_i = \frac{n_{Li}}{\sum_{j=1}^{N_c} n_{Lj}}$$

(2.2.6)

The amount of component i in the gas phase relative to the oil phase is expressed as the equilibrium K value, which is the ratio

$$K_i = y_i/x_i$$

(2.2.7)

The possible range of the equilibrium K value may be determined by considering two special cases. If component i exists entirely in the oil phase, then the gas phase mole fraction y_i is 0 and K_i is 0. Conversely, if component i exists entirely in the gas phase, then the liquid phase mole fraction x_i is 0 and K_i approaches infinity. Thus, the equilibrium K value for component i has the range $0 \leq K_i \leq \infty$. An equilibrium K value can be calculated for each distinct molecular species in a fluid, including both inorganic and organic molecules.

2.2.4 Pressure

Suppose an external force is applied to a surface. The component of the force that is acting perpendicularly to the surface is the normal force. The total normal force applied to the surface divided by the area of the surface is the average pressure on the surface. Pascal's law says that pressure applied to an enclosed fluid will be transmitted without a change in magnitude to every point of the fluid and to the walls of the container. The pressure at any point in the fluid is equal in all directions. If the fluid is at rest in the pore space of a rock, the pressure is equal at all points in the fluid at the same depth. The pressure in the pore space is often referred to as the pore pressure. If fluid is injected or withdrawn from the pore space, the rate of transmission of the change in pore pressure throughout the enclosed system can be used to obtain information about the system.

2.2.5 Compressibility

Compressibility is a measure of the change in volume resulting from the external pressure applied to the system. The fractional volume change of a system is the ratio of the change in volume ΔV to the initial volume V. The fractional volume change $\Delta V/V$ may be estimated from

$$\frac{\Delta V}{V} \approx -c\Delta P$$

(2.2.8)

where

c = the compressibility of the system
ΔP = the applied external pressure

The minus sign is applied so an increase in the applied external pressure ($\Delta P > 0$) will result in a decrease in the volume of the system. Similarly, a decrease in the applied external pressure ($\Delta P < 0$) will result in an increase in the volume of the system.

2.2.6 Formation Volume Factor

Formation volume factor is the ratio of the volume occupied by a fluid phase at reservoir conditions divided by the volume occupied by the fluid phase at surface conditions. Surface conditions are typically stock tank or standard conditions. Traditional oil field standard conditions are 14.7 psia and 60°F.

The volume of a fluid phase can have a sensitive dependence on changes in pressure and temperature. For example, gas formation volume factor is often determined with reasonable accuracy using the real gas equation of state

$$PV = ZnRT \tag{2.2.9}$$

where

R = the gas constant
n = the number of moles of gas in volume V at pressure P and temperature T

The gas is an ideal gas if gas compressibility factor $Z = 1$. If $Z \neq 1$, the gas is considered a real gas.

The volume of a petroleum mixture depends on changes in composition as well as changes in temperature and pressure. For example, a barrel of oil at reservoir conditions (relatively high pressure and temperature) will shrink as the barrel is brought to the surface (relatively low pressure and temperature). The shrinkage, or reduction in volume of the barrel of oil as it moves from reservoir to surface conditions, is due to the release of solution gas as the pressure and temperature of the oil decline. Shrinkage and expansion of fluids can take place within the reservoir as a result of changes in reservoir pressure, temperature, or composition during the life of the reservoir.

Formation volume factors for oil and water usually range from 1 to 2 units of reservoir volume for each unit of volume at surface conditions. Gas formation volume factor varies over a wider range because gas volume is more sensitive to changes in pressure.

2.2.7 Specific Gravity

Specific gravity is the ratio of the density of a fluid divided by a reference density. Gas specific gravity is calculated at standard conditions using air density as the reference density. The specific gravity of gas is as follows:

$$\gamma_g = \frac{M_a(gas)}{M_a(air)} \approx \frac{M_a(gas)}{29} \tag{2.2.10}$$

where M_a is apparent molecular weight. Apparent molecular weight is calculated as

$$M_a = \sum_{i=1}^{N_c} y_i M_i \tag{2.2.11}$$

where

  N_c = the number of components
  y_i = the mole fraction of component i
  M_i = the molecular weight of component i

Oil specific gravity is calculated at standard conditions using the density of fresh water as the reference density. The American Petroleum Institute characterizes oil in terms of API gravity, which is calculated from oil specific gravity γ_o at standard temperature and pressure by the equation

$$API = \frac{141.5}{\gamma_o} - 131.5 \tag{2.2.12}$$

If specific gravity $\gamma_o > 1$, the oil is denser than water and API < 10. If specific gravity $\gamma_o < 1$, the oil is less dense than water and API > 10. Heavy oils with API < 20 do not contain much gas in solution and have a relatively large molecular weight and specific gravity γ_o. By contrast, light oils with API > 30 typically contain a large amount of gas in solution and have a relatively small molecular weight and specific gravity γ_o. The equation for API gravity shows that heavy oil has a relatively low API gravity because it has a large γ_o, while light oils have a relatively high API gravity.

2.2.8 Heating Value

The heating value of a gas can be estimated from the composition of the gas and heating values associated with each component of the gas. The heating value of the mixture H_m is defined as

$$H_m = \sum_{i=1}^{N_c} y_i H_i \tag{2.2.13}$$

where

  N_c = the number of components
  y_i = the mole fraction of component i
  H_i = the heating value of component i

The heating value of a typical natural gas is often between 1,000 BTU/SCF to 1,200 BTU/SCF. Heating values of molecular components in a mixture are tabulated in reference handbooks.

2.2.9 Gas–Liquid Ratio

The gas–liquid ratio is the ratio of a volume of gas divided by a volume of liquid at the same temperature and pressure. The gas–liquid ratio, or GLR, is useful for characterizing the behavior of a reservoir. The choice of GLR depends on the fluids in the reservoir. Two commonly used gas–liquid ratios are the gas–oil ratio (GOR) and the gas–water ratio (GWR).

The ratio of gas volume to water volume at the same temperature and pressure, or gas–water ratio, is a sensitive indicator of the behavior of a gas reservoir connected to a water source, such as an aquifer or injected water. The gas–oil ratio, or the ratio of gas volume to oil volume at the same temperature and pressure, provides information about the behavior of an oil reservoir.

2.2.10 Viscosity

The coefficient of viscosity is a measure of resistance to flow of the fluid. In general, gas viscosity is less than liquid viscosity. The inverse of viscosity is called fluidity (McCain, 1990). Thus, a fluid with a large viscosity has a low fluidity. Two types of viscosity are commonly used: dynamic viscosity μ and kinematic viscosity v. Dynamic viscosity is related to kinematic viscosity by the equation $\mu = \rho v$, where ρ is the density of the fluid. The unit of dynamic viscosity μ is centipoise. If fluid density ρ has the unit of g/cc, then kinematic viscosity v has the unit of centistoke. Thus, 1 centistoke equals 1 centipoise divided by 1 g/cc.

Dynamic viscosity μ is used in Darcy's law to calculate the rate of fluid flow in porous media. The relationship between viscosity and flow rate defines the rheology of the fluid. A fluid is considered a non-Newtonian fluid if the viscosity of the fluid depends on flow rate. If the viscosity does not depend on flow rate, the fluid is considered a Newtonian fluid.

2.3 Classification of Petroleum Fluids

Petroleum fluids are typically mixtures of organic, and often inorganic, molecules. The elemental composition of petroleum is primarily carbon (84–87 percent by mass) and hydrogen (11–14 percent by mass). Petroleum contains other elements as well, including sulphur, nitrogen, oxygen, and various metals. A petroleum fluid may be called "sweet" if it contains only negligible amounts of sulphur compounds such as hydrogen sulfide (H_2S) or mercaptans. A mercaptan is also known as a thiol. It is an organic molecule containing a sulfur–hydrogen functional group. If the petroleum fluid contains a sulphur compound such as H_2S or a mercaptan, it is called "sour."

Petroleum fluids are predominantly hydrocarbons. The most common hydrocarbon molecules are paraffins, napthenes, and aromatics. These molecules are relatively stable at pressures and temperatures commonly found in reservoirs. Paraffin molecules such as methane and ethane have a single bond between carbon atoms and are considered saturated hydrocarbons. Paraffins have the general chemical formula C_nH_{2n+2}. Napthenes have the general chemical formula C_nH_{2n} and are saturated hydrocarbons with a ringed structure, as in cyclopentane. Aromatics are unsaturated hydrocarbons with a ringed structure. A well-known example of an aromatic is benzene. Aromatics have multiple bonds between the carbon atoms, and their unique ring structure makes aromatics relatively stable and unreactive.

2.3.1 The P–T Diagram

The phase behavior of a fluid is generally presented as a function of the three variables pressure, volume, and temperature. The resulting PVT diagram is often simplified for petroleum fluids by preparing a pressure–temperature (P–T) projection, and example of which is shown in Figure 2.2.

The P–T diagram in the figure displays both single-phase and two-phase regions. The curve separating the single-phase region from the two-phase region is called the phase envelope. The pressures associated with the phase envelope are called saturation pressures. A petroleum fluid at a temperature below the critical point temperature T_c and at pressures above the saturation pressure exists as a single-phase liquid. Saturation pressures at temperatures below T_c are called bubble point pressures. If the pressure drops below the bubble point pressure, the single-phase liquid will make a transition to two phases: gas and liquid. At temperatures below T_c and pressures above the bubble point pressure, the single-phase liquid is referred to as black oil (point A in Figure 2.2). If we consider pressures in the single-phase region and move to the right of the diagram by letting temperature increase toward the critical point, we encounter volatile oils.

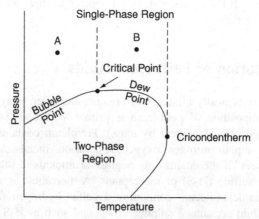

Figure 2.2 A pressure-temperature diagram: (A) black oil and (B) condensate.

Table 2.1 Rules of Thumb for Classifying Fluid Types

Fluid Type	Separator GOR (MSCF/STB)	Pressure Depletion Behavior in Reservoir
Dry gas	No surface liquids	Remains gas
Wet gas	>100	Remains gas
Condensate	3–100	Gas with liquid drops out
Volatile oil	1.5–3	Liquid with significant gas
Black oil	0.1–1.5	Liquid with some gas
Heavy oil	≈0	Negligible gas formation

The behavior of the petroleum fluid at temperatures above the critical point depends on the location of the cricondentherm. The cricondentherm is the maximum temperature at which a fluid can exist in both the gas and liquid phases. If the temperature is less than the cricondentherm but greater than T_c, reservoir fluids are condensates (point B in Figure 2.2). When the temperature exceeds the cricondentherm, we encounter gas reservoirs. A summary of these fluid types is given in Table 2.1. Separator gas–oil ratio (GOR) is a useful indicator of fluid type. The unit MSCF/STB equals 1,000 standard cubic feet of gas per stock tank barrel of oil.

Changes in phase behavior as a result of changes in pressure can be anticipated using the P–T diagram. Suppose a reservoir contains hydrocarbons at a pressure and temperature corresponding to the single-phase black oil region. If the reservoir pressure declines at a constant temperature, the reservoir pressure will eventually cross the bubble point pressure curve and enter the two-phase gas–oil region. A free gas phase will form in the two-phase region. Similarly, if we start with a single-phase condensate and allow the reservoir pressure to decline at a constant temperature, the reservoir pressure will eventually cross the dew point pressure curve to enter the two-phase region. In this case, a free-phase liquid drops out of the condensate gas, and once liquid drops out, it is very difficult to recover the liquid. If the pressure declines further, some hydrocarbon compositions will undergo retrograde condensation.

The P–T diagram may also be applied to temperature and pressure changes in a wellbore. Reservoir fluid moves from relatively high temperature and pressure at reservoir conditions to relatively low temperature and pressure at surface conditions. As a result, it is common to see single-phase reservoir fluids become two-phase fluids by the time they reach the surface. If the change from single-phase to two-phase occurs quickly in the wellbore, which is common, then the fluid is said to have undergone a flash from one to two phases.

The P–T diagram in Figure 2.3 shows two-phase envelopes for four types of fluids (Pederson et al., 1989). Reservoir fluids can change from one fluid type to another during the life of a reservoir. As an example, suppose we inject dry gas into a black oil reservoir. Dry gas injection increases the relative amount of low-molecular-weight components in the black oil and causes the two-phase envelope to rotate in a counterclockwise direction in the P–T diagram as the relative amount of lower-molecular-

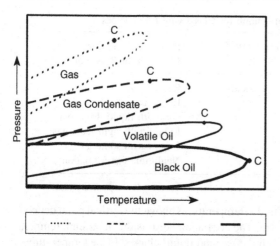

Figure 2.3 Typical two-phase P–T envelopes for different fluid types.

weight components increases. The decision to inject or to not inject a fluid into the reservoir is a reservoir management decision. The relationship between fluid behavior and reservoir management decisions is an example of a topic that should be considered by an asset management team.

Different compositions for typical fluid types are shown in Table 2.2. The notation C_N is used to indicate that there are N carbon atoms in the molecule, and N is the carbon number of the molecule. For example, methane C_1 has one carbon atom and carbon number 1, ethane C_2 has two carbon atoms and carbon number 2, and so on. The notation C_{6+} refers to the set of molecules with six or more carbon atoms. The molar composition in each column should add up to 100 percent.

Table 2.2 illustrates the general observation that methane content (C_1) is a good indicator of fluid type. Methane content tends to decrease as fluids change from dry gas to black oil. As we move from left to right in the table, we also see an increase in higher-molecular-weight components. Dry gas usually contains lower-molecular-weight components, especially methane, while the addition of higher-molecular-weight components such as organic molecules with three or more carbon atoms will eventually yield black oil. Volatile oil and black oil have significant amounts of intermediate- and high-molecular-weight components.

2.3.2 Miscibility

The concept of miscibility is important when considering the phase behavior of two fluids that are brought into contact with each other. For example, suppose carbon dioxide (CO_2) is injected into a cell containing oil. At low pressure and temperature, the CO_2 is a gas, the oil is a liquid, and an interface exists between the gas and the liquid. Two phases separated by an interface are considered immiscible when they do not mix.

Changes in pressure, temperature, and composition can weaken the interfacial forces separating the phases. If the interface between the two phases vanishes, a

Table 2.2 Typical Molar Compositions of Petroleum Fluid Types

Component	Gas	Gas Condensate	Volatile Oil	Black Oil
N_2	0.3	0.71	1.67	0.67
CO_2	1.1	8.65	2.18	2.11
C_1	90.0	70.86	60.51	34.93
C_2	4.9	8.53	7.52	7.00
C_3	1.9	4.95	4.74	7.82
iC_4+nC_4	1.1	2.00	4.12	5.48
iC_5+nC_5	0.4	0.81	2.97	3.80
iC_6+nC_6	C_{6+}: 0.3	0.46	1.99	3.04
C_7		0.61	2.45	4.39
C_8		0.71	2.41	4.71
C_9		0.39	1.69	3.21
C_{10}		0.28	1.42	1.79
C_{11}		0.20	1.02	1.72
C_{12}		0.15	C_{12+}: 5.31	1.74
C_{13}		0.11		1.74
C_{14}		0.10		1.35
C_{15}		0.07		1.34
C_{16}		0.05		1.06
C_{17}		C_{17+}: 0.37		1.02
C_{18}				1.00
C_{19}				0.90
C_{20}				C_{20+}: 9.18

Source: Adapted from Pedersen et al. (1989).

single phase forms and the fluids that comprised the two phases are considered miscible. Returning to the CO_2–oil example, if we increase pressure at suitably high temperatures, the interface between CO_2 and oil weakens. At a high enough pressure, called the minimum miscibility pressure (MMP), the interface disappears. When cell pressure is at or above the MMP, the gas and oil mix to form a single phase. The properties of the single phase depend on the pressure, temperature, and composition of the mixture. See Section 10.1 for more discussion of interfacial tension.

2.4 Representation of Fluid Properties

Black oil simulators and compositional simulators are the two most common types of computer programs that model fluid flow in reservoirs. From a physical perspective, they differ primarily in the way they model fluid properties. The two different types of fluid models are described next.

2.4.1   The Black Oil Model

Typical gas and oil properties for a standard black oil model are shown in
Figures 2.4 and 2.5. Gas and oil properties in a black oil model depend on pressure
only. Gas phase properties are gas formation volume factor (B_g), gas viscosity (μ_g),
and liquid yield (r_s). Oil phase properties are oil formation volume factor (B_o), oil
viscosity (μ_o), and solution GOR (R_{so}). Phase changes occur at the saturation
pressures. Single-phase oil becomes two-phase gas–oil when pressure drops below
the bubble point pressure (P_b), and single-phase gas becomes two-phase gas and
gas condensate when pressure drops below the dew point pressure (P_d). Both
saturated and undersaturated curves are modeled as functions of pressure only.

Flow simulators are most efficient when fluid property data are smooth curves.
Numerical difficulties can arise if a discontinuity is present in a fluid property curve.
Realistic fluid properties are ordinarily smooth functions of pressure except at points
where phase transitions occur. Slope discontinuities between saturated and undersaturated conditions are shown at the bubble point pressure P_b in Figure 2.5. As a practical matter, it is usually wise to plot input PVT data to verify the smoothness of the
data. Most simulators reduce the nonlinearity of the gas formation volume factor B_g
by using the inverse $b_g = 1/B_g$ to interpolate gas properties.

Oil properties from a laboratory experiment must usually be corrected for use in
a black oil simulator (Moses, 1986). The justification for the correction is based on
the argument that flow in the reservoir is a relatively slow, or differential, process.
In a differential process, pressure changes in small increments over a period of time.
Reservoir fluid flow corresponds to a differential process in the laboratory. When oil
is produced through the wellbore, however, it is subjected to a rapid change in pressure. Flow up a wellbore is considered a flash process. Corrections to fluid properties used in the model are designed to more adequately represent fluids as they
flow differentially in the reservoir prior to being flashed to surface conditions as
the fluid moves up the wellbore. The corrections alter the solution gas–oil ratio
and the oil formation volume factor. Given the separator oil formation volume factor
B_o and the solution gas–oil ratio R_{so}, the corrections are

$$B_o(P) = B_{od}(P)\frac{B_{ofbp}}{B_{odbp}} \tag{2.4.1}$$

and

$$R_{so}(P) = R_{sofbp} - \left(R_{sodbp} - R_{sod}(P)\right)\frac{B_{ofbp}}{B_{odbp}} \tag{2.4.2}$$

where P is pressure. Subscript d refers to differential liberation data, f refers to flash
data, and bp refers to the bubble point. Walsh and Lake (2003) discuss other methods for adjusting fluid property data.

Water properties are almost always needed in a flow simulator. Ideally, water
properties will be obtained from laboratory analyses of produced water samples.

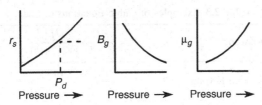

Figure 2.4 Gas phase properties.

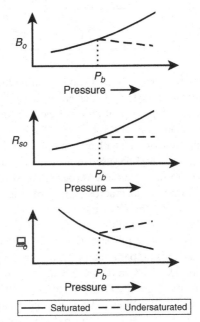

Figure 2.5 Oil phase properties.

Correlations are often sufficiently accurate for describing the behavior of water if water samples are not readily available. If reliable fluid data are missing for one or more of the reservoir fluids, fluid properties from analogous fields or from correlations can be used. For a review of the use of correlations to describe fluid properties, see McCain (1991), Towler (2006), and Sutton (2006).

2.4.2 The Compositional Model

Compositional models use equations of state and represent hydrocarbon phases in terms of components or pseudocomponents. Fluid composition is usually expressed in terms of wellstream composition z_i of component i, or in terms of the mole fractions of component i in the liquid (x_i) and gas (y_i) phases. The ratio y_i/x_i is the equilibrium K value (K_i) for component i and is a measure of how component i is distributed between gas and liquid phases at equilibrium conditions.

Table 2.3 Examples of Cubic Equations of State

Redlich-Kwong	$P = \dfrac{RT}{V-b} - \dfrac{a/T^{1/2}}{V(V+b)}$
Soave-Redlich-Kwong	$P = \dfrac{RT}{V-b} - \dfrac{a(T)}{V(V+b)}$
Peng-Robinson	$P = \dfrac{RT}{V-b} - \dfrac{a(T)}{V(V+b)+b(V-b)}$
Zudkevitch-Joffe	$P = \dfrac{RT}{V-b(T)} - \dfrac{a(T)/T^{1/2}}{V[V+b(T)]}$

Fluids are usually described in one of two ways: by a multicomponent equation of state, or as a pseudocomponent system (Pedersen et al., 1989; Whitson and Brulé, 2000). A pseudocomponent is a mixture of pure components that is treated as a single component to expedite computing. In general, fluid behavior is most often represented using a multicomponent equation of state with one or more pseudocomponents. Table 2.3 shows some cubic equations of state (EoS) used in commercial compositional simulators.

Equations of state used to quantitatively represent fluid properties are based on the thermodynamic postulate that all macroscopic properties of a fluid system can be expressed in terms of pressure (P), temperature (T), and composition only. The EoS shown in Table 2.3 are examples of EoS that can be used to model the behavior of oil and gas. They relate pressure (P), volume (V), gas constant R, temperature (T), and a set of adjustable parameters $\{a, b\}$ that may be functions of temperature and composition. The EoS in Table 2.3 are called "cubic" because they yield a cubic equation for the compressibility factor $Z = PV/RT$. In the case of an ideal gas, $Z = 1$. Following van der Waals, the parameter b adjusts for the finite size of an atom or molecule in the first term on the right side of each EoS. The second term accounts for interactions.

Equations of state are used to calculate equilibrium relations in a compositional model. This entails tuning parameters such as EoS parameters $\{a, b\}$ in Table 2.3. For a mixture with N_c components, the parameters $\{a, b\}$ have the form

$$a = \sum_{i=1}^{N_c} \sum_{j=1}^{N_c} a_i a_j x_i x_j \left(1 - \delta_{ij}\right) \tag{2.4.3}$$

and

$$b = \sum_{i=1}^{N_c} b_i x_i \tag{2.4.4}$$

where $\{a_i, b_i\}$ refer to equation of state values for the ith component with mole fraction x_i, and Δ_{ij} is a symmetric array of numbers called the binary interaction parameters. The binary interaction parameter Δ_{ij} represents the interaction between component i and component j. In principle, binary interaction parameters are determined by fitting EoS parameters to fluid property measurements on a mixture of two components. In practice, they are determined by a regression technique.

Several regression techniques exist for tuning an EoS. They usually differ in the choice of EoS parameters that are to be varied in an attempt to match lab data with the EoS. The modification of EoS parameters is called tuning the EoS. The justification for tuning an EoS is that the EoS parameters are determined for systems with one or two components only, while petroleum is generally a mixture with many components. The EoS parameter adjustments attempt to match the multicomponent behavior of the fluid system.

A black oil model can be thought of as a compositional model with two pseudocomponents. The number of pseudocomponents in a compositional model typically ranges from six to ten. For comparison, EoS models in process engineering require on the order of 20 components or more to model fluid behavior in surface facilities. Compositional model costs increase dramatically with increases in the number of specified components, but the additional components make it possible to calculate more accurate fluid properties.

CS.2 Valley Fill Case Study: Fluid Properties

Oil, water, and gas fluid property data are shown in Table CS.2A. The oil contains 480 SCF/STB dissolved gas at the bubble point pressure. Saturated oil property data in Table CS.2A have been corrected from differential to flash form using the separator test information provided in Table CS.2B (Moses, 1986). Undersaturated oil properties are shown in Table CS.2C.

Table CS.2A Valley Fill Fluid Properties

| Pressure | Oil* | | | Gas | | Water | |
| | Vis | FVF | R_{so} | Vis | FVF | Vis | FVF |
Psia	cp	RB/STB	SCF/STB	cp	RCF/SCF	cp	RB/STB
14.7	1.04	1.06	1	0	0.9358	0.5	1.019
514.7	0.910	1.111	89	0.0112	0.0352	0.5005	1.0175
1014.7	0.830	1.192	208	0.0140	0.0180	0.5010	1.0160
1514.7	0.765	1.256	309	0.0165	0.0120	0.5015	1.0145
2014.7	0.695	1.320	392	0.0189	0.0091	0.5020	1.0130
2514.7	0.641	1.380	457	0.0208	0.0074	0.5025	1.0115
3014.7	0.594	1.426	521	0.0228	0.0063	0.5030	1.0100
4014.7	0.510	1.472	586	0.0260	0.0049	0.5040	1.0070
5014.7	0.450	1.490	622	0.0285	0.0040	0.5050	1.0040
6014.7	0.410	1.500	650	0.0300	0.0034	0.5060	1.0010

*Corrected from differential to flash form.
Selected Units: RB = Reservoir Barrels; STB = Stock Tank Barrel; SCF = Standard Cubic Feet; RCF = Reservoir Cubic Feet

Table CS.2B Valley Fill Separator Test

Variable	Value
B_{ofbp}	1.5 RB/STB
R_{sofbp}	650 SCF/STB
B_{odbp}	1.63 RB/STB
R_{sodbp}	760 SCF/STB

Note: Flash from a saturation pressure of 6,000 psig to 0 psig.

Table CS.2C Valley Fill Undersaturated Oil Properties

Pressure (psia)	Corrected B_{obp} (RB/STB)	μ_o (cp)	Remarks
2015	1.3200	0.695	Bubble point
4015	1.2740	0.787	Undersaturated values

Exercises

2-1. Calculate the specific gravity of a gas with the following composition.

Component	Mole Fraction	Molecular Weight
Methane (C_1)	0.85	16
Ethane (C_2)	0.09	30
Propane (C_3)	0.04	44
n-Butane ($n - C_4$)	0.02	58
Total	1.00	

2-2. Typical reservoir values for formation, oil, water, and gas compressibilities are $c_f = 3 \times 10^{-6}$/psia, $c_o = 10 \times 10^{-6}$/psia, $c_w = 3 \times 10^{-6}$/psia, and $c_g = 500 \times 10^{-6}$/psia. Estimate the fractional volume change of each fluid for a pressure difference $\Delta P = P_{final} - P_{initial} = -100$ psia. The minus sign indicates a reduction in pressure.

2-3. Calculate the laboratory differential oil formation volume factor $B_{od}(P)$ as a function of pressure using the data in Tables CS-2A and CS-2B. The correction is discussed in Section 2.4.

2-4. Plot the oil and gas PVT data in Table CS-2A as a function of pressure. Make a semilog plot of gas formation volume factor versus pressure to see more detail at higher pressures.

2-5. Plot API gravity as a function of oil specific gravity for the range $0.7 \leq$ oil specific gravity ≤ 1.0

2-6. As a rule of thumb, the temperature of the earth's crust increases by about $1°F$ for every 100 feet of depth. Use this information to estimate the temperature of the earth at two miles. Assume the temperature at the earth's surface is $60°F$.

2-7. Open VFILL1_HM.dat and find the PVT data for the model. What is the oil formation volume factor at the bubble point pressure?

3 Geology

The ability to manage subsurface resources depends in part on knowledge of the geologic history of the earth and theories of the earth's formation and maturation. We outline theories of the earth's formation and plate tectonics in this chapter and then present concepts from development geology. For more details, refer to Montgomery (1990), Levin (1991), Selley (1998), Press and Siever (2001), Gluyas and Swarbrick (2004), Holstein (2007), and Reynolds et al. (2008).

3.1 The Geologic History of the Earth

Many scientists believe that the earth first began to form 4 to 5 billion years ago. It is believed that the earth was a large body of hot, gaseous matter that whirled through space for several hundred million years. This body of matter slowly condensed and cooled. A solid crust gradually formed around the molten interior. A cross-section that bisects the earth shows that the earth consists of an iron core wrapped inside a mantle of rock with a thin crust at the surface. The density of rock in the mantle is higher than in the crust because of an abundance of iron and magnesium. Rock density in the mantle increases with depth as you move from the shallower upper mantle to the deeper lower mantle. The thickness of the crust, which consists of oceanic crust and continental crust, is small compared to the diameter of the earth. The mobile part of the upper mantle and crust is called the lithosphere. Lithospheric plates drift on a denser, partially molten material called the asthenosphere.

As the earth cooled from its hot, gaseous state, the surface was subjected to forces that caused great changes in its topography, including the formation of continents and the uplift of mountain ranges. Pressure from the earth's interior could crack the sea floor and allow less dense molten material to flow onto the sea floor to form subsea ridges.

Magma from the mantle can flow through sea floor cracks and emerge as lava. The lava extruded from a sea floor crack spreads laterally on each side of the crack to form the subsea ridge. The symmetry of the material spread on each side of the sea floor crack supports the contention that the material was in a molten state as it gradually moved outward from the crack. As the material cooled, magnetic constituents within the molten material aligned themselves in conjunction with the polarity of the earth's magnetic field at the time that the material solidified. Several periods of polarity have been identified and dated.

Integrated Reservoir Asset Management. DOI: 10.1016/B978-0-12-382088-4.00003-7

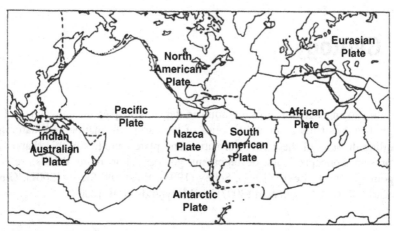

Figure 3.1 Tectonic plates.

Satellite measurements of the earth's gravitational field have identified sea floor ridges and boundaries between continents. The shapes of the boundaries between continents are suggestive of vast plates, as depicted in Figure 3.1. Only the largest of the known plates, called tectonic plates, are shown in the figure. Sea floor ridges are long subsea mountain ranges that can serve as plate boundaries.

In the theory of plate tectonics, the entire crust of the earth is considered a giant, ever-shifting jigsaw puzzle. Tectonic plates move in relation to one another at the rate of up to 4 inches per year. Tectonic plates are often associated with continental land masses. Many of the plate boundaries can be directly observed by gravimetric surveys of the earth's surface from satellites.

The historical movement of tectonic plates is responsible for much of the geologic heterogeneity that is found in hydrocarbon-bearing reservoirs. Figure 3.2 shows the hypothesized movement of tectonic plates during the past 225 million years. It began at the time that all surface land masses were thought to be coalesced into a single land mass known as Pangaea. Pangaea was not the initial state of surface land masses. Geoscientists believe that Pangaea was formed by the movement of tectonic plates, and the continued movement of plates led to the breakup of the single land mass into the surface features we see today. Tectonic plates are driven by forces that originate in the earth's interior. As the plates pull apart or collide, they can cause such geologic activities as volcanic eruptions, earthquakes, and mountain range formation.

Along with the occasional impact of a meteor or asteroid, it is theorized that the movement and position of tectonic plates caused extensive environmental changes. These environmental changes are not just the local volcano destroying a city or creating an island where there was none, but also include global sea level and atmospheric changes. Plate movement can lower the sea level, creating a period of vast erosion and deposition. The biosphere is also affected by these changes. Plants and animals may thrive in one set of conditions and readily become extinct when

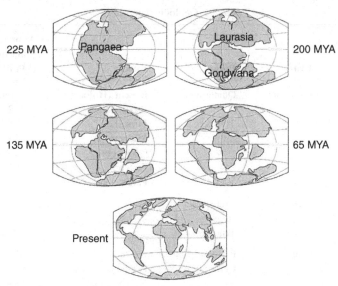

Figure 3.2 Tectonic plate movement.
Source: USGS website, 2001.

the conditions change. Plate movement also provides a mechanism to help explain the geographic distribution of organisms around the world. Based on these changes, geologists have found that the geologic history of the earth can be broken into convenient periods.

The most encompassing segment of time is the eon. Eons are subdivided into eras, which are further subdivided into periods. On a finer scale, periods are separated into epochs. Table 3.1 shows an abridged version of the geologic time scale since the earth's formation. The acronym MYBP stands for "millions of years before the present." Levin (1991), Ridley (1996), and Press and Siever (2001) all write about the range of starting times of selected intervals. A comparison of the geologic time scales reported in the literature indicates that the actual chronology of the earth is approximate.

The oldest terrestrial rocks formed during the Precambrian Eon. Microscopic organisms are believed to have originated during this eon. Organisms with cellular nuclei (eukaryotes) appeared during the Proterozoic Era of the Precambrian. The Paleozoic is the era when life began to blossom. The demise of trilobites during the Permo-Triassic extinction marked the end of the Permian Period at the end of the Paleozoic Era. The Mesozoic Era includes the age of dinosaurs, and its end was marked by the Cretaceous–Tertiary (K–T) extinction. Cenozoic rocks are relatively new, being less than 70 million years old. Mammals began to flourish during the Cenozoic. The variations in life forms give rise to variations in the fossil record that can help geoscientists characterize the stratigraphic column, which shows the vertical location of rock units in the region of interest.

Table 3.1 Geologic Time Scale

Eon	Era	Period	Epoch	Approximate Start of Interval (MYBP)
Phanerozoic	Cenozoic	Quaternary	Recent (Holocene)	0.01
			Pleistocene	2
		Tertiary	Pliocene	5
			Miocene	24
			Oligocene	35–37
			Eocene	54–58
			Paleocene	65–66
	Mesozoic	Cretaceous		144
		Jurassic		206–213
		Triassic		245–251
	Paleozoic	Permian		286–300
		Carboniferous	Pennsylvanian	320
			Mississippian	354–360
		Devonian		408–409
		Silurian		438–439
		Ordovician		505–510
		Cambrian		543–590
Precambrian	Proterozoic			2,500
	Archean			3,800
	Hadean			

3.2 Rock Formations and Facies

The movement of tectonic plates across the surface of the globe generated forces powerful enough to cause rocks to form and change shape. We can think of the process of rock creation as a cycle. The beginning of the cycle occurs with the cooling of molten magma and subsequent hardening into rock. Typically, the formation of new rock occurs at plate boundaries, but it can also occur over "hot spots" within the earth's mantle. As plates collide, pressure and heat may cause part of the plate to melt and result in molten rock being thrust to the surface. After cooling, surface rock is subjected to atmospheric phenomena.

Chemical and physical processes cause exposed rock to break into smaller and smaller particles. Wind and water transport these particles from their source location in a process called erosion. The particles become finer as they collide with other objects during the transport process. A particle will be deposited along with other particles when the energy of the wind or water dissipates to the point where there is not enough energy to transport the particle. The accumulation of particles thickens as particles are deposited in a specific location.

Slowly—over millions of years—tectonic plates move up and down relative to sea level, alternately causing erosion and deposition. Deposition can range from thousands of feet of sediment in an area to none at all. Erosion can carve canyons, level once jagged mountains, or remove all traces of a formation that was once hundreds of feet thick. High pressure and temperature can cause rocks to change character in a process called metamorphism. Particles may become fused together to form considerably larger objects. Given enough time, pressure, and heat, rocks will melt and start the cycle again. Based on this rock cycle, geologists recognize three primary types of rocks: igneous, sedimentary, and metamorphic. Igneous rocks are formed by the cooling of material that has been molten. Sedimentary rocks form when materials at the surface of the earth are weathered, transported, deposited, and cemented together. If rocks are subjected to heat and pressure, metamorphic rocks may form.

Sedimentary rocks are important because they are often the porous medium associated with commercially important reservoirs. The key attributes used to classify sedimentary rock are mineral composition, grain size, color, and structure. When a rock is described, it is desirable to convey exactly what the rock looks like. Well-sorted, well-rounded, coarse-grained quartz sandstone would probably make a better reservoir than a poorly sorted, angular, fine-grained arkose, which is sandstone with both quartz and feldspar. Each of these descriptive terms tells us something about the rock.

Sorting refers to the uniformity of grain size. Fluids will typically flow better through a well-sorted rock than through a poorly sorted rock. The ability to flow is related to a rock property known as permeability, which is discussed later. If the grains have sharp edges, the grains making up the rock probably did not get transported very far. Rounded grains indicate a longer period of transport. Rocks made up of rounded grains may have better permeability than rocks composed of grains that are flat or have sharp edges. Coarse-grained sandstone is made up of particles that are approximately 0.5 to 1.0 mm in diameter, while fine-grained sandstone particles have diameters between 0.125 and 0.25 mm. As a rule, larger particles allow easier passage of fluids through interstitial pore spaces than smaller particles.

Mineral content is another very important characteristic of a rock. A mineral is a naturally occurring, inorganic solid with a specific chemical and crystalline structure. For example, quartz is much less reactive than the feldspars of an arkose, which is why quartz withstands weathering so well. A quartz grain may be able to withstand multiple cycles of erosion and deposition. As another example, the presence of clays in the pore space can cause reduced productivity if a change in the salinity of the formation causes clay swelling. Drilling mud and other fluids introduced into the reservoir through the wellbore can react with clays to swell and plug clay-bearing formations. The mineralogy of the rock is the collection of minerals within the rock.

The grains that form sedimentary rocks are created by weathering processes at the surface of the earth. Weathering creates particles that can be practically any size, shape, or composition. A glacier may create and transport a particle the size of a house, while a desert wind might create a uniform bed of very fine sand. The particles, also known as sediments, are transported to the site of deposition, usually by aqueous processes. Sometimes the particles are transported very far. In these cases, only the most durable particles survive the transport. The grains of sand roll and bump along the transport pathway. Grains that started out as angular chunks of rock slowly become smaller and more rounded. A grain of quartz, being fairly hard, may even be able to withstand multiple cycles of deposition and erosion. This leaves a grain that is very rounded. The minerals that make up a sedimentary rock will depend on many factors. The source of the minerals, the rate of mineral breakdown, and the environment of deposition are important factors to consider in characterizing the geologic environment.

3.2.1 Formations

The environment under which a rock forms is called the environment of deposition. As the environment, such as a shoreline, moves from one location to another, it leaves a laterally continuous progression of rock that is distinctive in character. These progressions of rocks can extend for hundreds of miles. If the progression is large enough to be mapped, it can be called a formation.

Formations are the basic descriptive unit for a sequence of sediments. The formation represents a recognizable, mappable rock unit that was deposited under a uniform set of conditions at one time. Formations should represent a dominant set of depositional conditions, even though a formation may consist of more than one rock type. If the different rock types within the formation are mappable, they are referred to as members. Formations can be a few feet thick or hundreds of feet thick. The thickness is related to the length of time an environment was in a particular location and how much subsidence was occurring during that period.

Each environment of deposition causes a different type of rock sequence to be deposited. A fluvial (river) environment of deposition may cause sandstone to be deposited. A deltaic environment will also cause sandstone to be deposited. However, the sandstone that each environment produces has a very different character. For example, a fluvial system can produce rocks that meander along, while a delta tends to stay in one place to enable a large volume of sediment to be deposited. In a fluvial system, the sandstones have certain characteristics. The grain size of sand deposited by a fluvial system becomes finer at shallower depths in a process called fining upwards. In addition, there is a coarser deposit at the base of the formation that distinguishes the fluvial system from a delta deposit. The surrounding rock types for a particular environment are also predictable. Lying above the sandstones of a meandering river are typically mudstones that may have dessication cracks and traces of roots. The sandstones are interpreted to be formed in the river channel, while the mudstones are considered floodplain deposits.

3.2.2 Facies

The term *facies* refers to those characteristics of rocks that can be used to identify their depositional environment. In the fluvial environment of deposition example, the sandstone would be considered one facies, while the mudstone would be another. The characteristics of sandstone that indicate its environment of deposition also define it as a facies. A facies is distinguished by characteristics that can be observed in the field. If pebbly sandstone is seen independently of surrounding rocks, it could easily be classified into a fluvial environment. Adding information that it is surrounded by hemipelagic (deep-sea) mud, a geologist may revise the interpretation to be a turbidite depositional environment in which sediments are suspended in a fast-flowing current. The turbidite depositional environment is associated with high-energy turbidity currents. Integration of all of the available information will lead to better characterization of the rock.

3.3 Structures and Traps

After hydrocarbons have been generated, they migrate along pathways. Typically hydrocarbons are lighter than water and will tend to migrate to the surface. Petroleum will accumulate in traps, which are locations where oil and gas can no longer migrate to the surface. The two primary types of traps are structural and stratigraphic. A structural trap is present when the geometry of the reservoir prevents fluid movement. Structural traps occur where the reservoir beds are folded and perhaps faulted into shapes that can contain commercially valuable fluids like oil and gas. Anticlines are a common type of structural trap. Folding and faulting can be caused by tectonic or regional activity. Tectonic activity is a consequence of moving plates, while regional activity is exemplified by the growth of a salt dome. Formation of a structural trap by regional activity is also known as a diapiric trap.

Stratigraphic traps occur where the fluid flow path is blocked by changes in the formation's character. The formation changes in such a manner that hydrocarbons can no longer move upward. Types of stratigraphic traps include sand thinning out or porosity reduction because of diagenetic changes. Diagenesis refers to processes in which the lithology of a formation is altered at relatively low pressures and temperatures when compared with the metamorphic formation of rock. Diagenesis includes processes such as compaction, cementation, and dolomitization. Dolomitization is the process of replacing a calcium atom in calcite (calcium carbonate) with a magnesium atom to form dolomite. The resulting dolomite occupies less volume than the original calcite and results in the formation of additional porosity in carbonate reservoirs. This porosity is called secondary porosity.

In addition to structural and stratigraphic traps, many traps are formed by a combination of structural and stratigraphic features and are called combination traps. An example of a combination trap is the Prudhoe Bay Field on the North Slope of Alaska (Selley, 1998). It is an anticlinal trap that has been truncated and bounded by impermeable shale.

3.4 Petroleum Occurrence

Petroleum reservoirs are normally found in sedimentary rocks. Rarely, petroleum is found in fractured igneous or metamorphic rocks. Igneous and metamorphic rocks originate in high-pressure and high-temperature conditions that do not favor the formation of petroleum reservoirs. They usually do not have the interconnected pore space, or permeability, needed to form a conduit for petroleum to flow to a wellbore. While a metamorphic rock may have originated as a piece of sandstone, it has been subjected to heat and pressure. Any petroleum fluid that might once have occupied the pores is cooked away. Therefore, the most likely type of rock for containing commercial volumes of petroleum is sedimentary rock.

Several key ingredients must be present for a hydrocarbon reservoir to develop. First, a source rock for the hydrocarbon must be present. It is commonly thought that hydrocarbons form from the remains of aquatic life. The remains accumulate in a sedimentary environment such as shale that becomes a source rock. Second, the pressure and temperature of the source rock should be suitable for the generation of oil or gas from the organic mixture. If the conditions are not right, the generation of oil or gas may not be optimal. For example, if the temperature is too high, decaying organic material can be overcooked. The result could be the generation of gas and carbon residue.

A third factor is the presence of a reservoir rock and a conduit from the source rock to the reservoir rock. Rock is considered reservoir rock if fluids can be confined in a volume of the rock and the rock can support economic flow rates. Typically, hydrocarbons are formed in rocks that are not very amenable to modern production techniques. For hydrocarbons to be producible, they must be able to flow into wells. The flow rate must be large enough to make the wells economically viable.

Two important characteristics control the economical viability of the reservoir: porosity and permeability. Porosity is the ratio of the volume of void space to the total volume. Porosity is a factor that determines the volume of reservoir that can be used to store fluid. Permeability is a measure of the connectivity of pore spaces that allows fluids to flow. Both porosity and permeability are discussed in greater detail later.

If hydrocarbon fluid is not stopped from migrating, buoyancy and other forces will cause it to move toward the surface. Two factors are needed to confine the fluid: vertical confinement of fluid by the presence of an impermeable cap rock and prevention of lateral fluid movement by a trapping mechanism. The major categories of traps are structural traps, stratigraphic traps, and combination traps.

Overriding all these factors is timing. Without a correct timing sequence, nothing can happen. A source rock can provide billions of barrels of oil to a reservoir, but if the trap does not form until a million years after the oil has passed through the reservoir, not much will be found except perhaps an oil-stained rock.

3.5 Geochemistry

Geochemistry is the application of chemistry to geologic systems. It can be used to understand the composition of rocks and the chemical interaction that can occur between rocks and other materials, such as organic fluids and water. A rock is an aggregate of one or more minerals; and a mineral, according to Press and Siever, is a "naturally occurring, solid crystalline substance, generally inorganic, with a specific chemical composition" (2001, p. 26). For example, sandstone often contains the mineral quartz, which is composed of silicon and oxygen in the ratio of one atom of silicon to two atoms of oxygen, so quartz is silicon dioxide (SiO_2). Limestone is a sedimentary rock composed primarily of the mineral calcite, which is calcium carbonate ($CaCO_3$). Rock fragments are often held together by cementation. Quartz, carbonates, and clays can fill the interstices between rock fragments and form new rocks by cementing the fragments together. The cementation can also be undone in some cases and create smaller rock fragments that can move into a wellbore or increase the flow capacity of fractures or faults that were formerly filled with a cementation material.

The constituent molecules in a mineral may be chemically active with materials contained in the pore space of a rock. Naturally occurring water contains dissolved solids, such as salt (NaCl), in ionic form (Na^+ and Cl^-), as well as other elements, including potassium and magnesium. The resulting brine can dissolve some types of rock. Halite, for example, is rock salt (NaCl). Water can be used to leach out or create a cavern in rock salt by dissolving the rock. Fluids in the pore space can chemically alter the stability of rock and change associated rock properties such as porosity and permeability.

We noted earlier that petroleum is thought to be formed by the decay of biological organisms. The resulting organic material is a mixture of hydrocarbons that are usually less chemically reactive with rock than brine. Changes in pressure, temperature, or composition can result in a chemical transformation from a petroleum liquid to a highly viscous tar or asphaltene. This process is a geochemical process because it depends on the organic material's chemistry, the effect of organisms that may be catalyzing the decay process, and the sedimentary environment.

Geochemistry plays a major role in fluid injection processes. For example, the injection of water into an oil reservoir is a waterflood process. The injected water must be chemically compatible with reservoir lithology—notably shale—to prevent undesirable effects such as clay swelling and the subsequent reduction in pore space and flow capacity. Another example is the enhanced oil recovery process known as chemical flooding. In this process, chemicals are injected with water into an oil reservoir to reduce the forces that are preventing the oil from flowing. The success of a chemical flood process depends on the chemical interaction among the injected fluids, the ionic environment associated with the *in situ* brine, and the chemical composition of petroleum in the reservoir. If a polymer such as a large chain sugar (polysaccharide) is injected, there is a chance that microorganisms will consume the polymer unless a biocide is injected with the polymer. This example illustrates the need to understand biological factors as well as geochemical factors.

Another application of geochemistry is the disposal of greenhouse gases such as carbon dioxide (CO_2) and acid gases that can be composed of CO_2 and hydrogen sulfide (H_2S). These gases can dissolve in reservoir water to form an *in situ* acid that can corrode reservoir rock and oil field equipment. The asset management team may need to consider using corrosion-resistant alloys or corrosion inhibitors to mitigate the impact of acid gas injection.

The preceding examples demonstrate a range of geochemical effects. Geochemistry can have an impact on the petroleum system in many ways, including chemical reactions, dissolution, precipitation, and solubility. For more information, the reader should consult Wilhite (1986), Warner (2007), Lake (1989), and Green and Wilhite (1998).

3.6 Basin Modeling

The occurrence of petroleum requires several geologic factors, including the presence of source rock, reservoir rock, cap rock, a trap, and appropriate pressure and temperature. The generation of petroleum from source rocks is one aspect of a petroleum system. Another aspect of a petroleum system is the sedimentary basin. Selley defines a sedimentary basin as "an area on the earth's surface where sediments have accumulated to a greater thickness than they have in adjacent areas" (1998, p. 363). Sedimentary basins are formed by the deposition of sediments in large crustal regions that have subsided or are subsiding. The areal extent of a sedimentary basin is on the order of 10,000 km^2 or larger. The combination of sediment deposition and crustal subsidence leads to the formation of thick accumulations of sedimentary rock (Press and Siever, 2001). Increasing pressure and temperature associated with the accumulation of sediment as the basin subsides can lead to the formation of sedimentary rock containing organic material. Source rock for petroleum generation and reservoir rock are found in sedimentary basins. The rock can exhibit significant heterogeneity, including faulting and fracturing, because of variations in pressure, density, and composition of deposited material, and application of tectonic and regional forces.

Selley (1998) reported that the volume of petroleum generated in a sedimentary basin depends on the area of the sedimentary basin, the average total thickness of source rock, and a factor called the transformation ratio. The transformation ratio is the ratio of the amount of petroleum formed in the basin to the amount of material that was available for generating petroleum. One method for estimating the transformation ratio is pyrolysis, or heating, source rock such as shale. Studies have shown that the transformation ratio should exceed 0.1 to generate significant oil, and major petroleum provinces have a transformation ratio in the range 0.3 to 0.7.

Computer programs have been used to study the formation and distribution of petroleum in sedimentary basins (Gluyas and Swarbrick, 2004). These programs couple geochemistry and fluid flow algorithms to model such processes as sedimentation, compaction, heat transfer, the maturation of source rock, subsequent expulsion of hydrocarbons formed from organic material, and migration of hydrocarbons from source rock to reservoir rock. Basin modeling can be used to study the distribution of source rock, the compaction and heating of source rock, the timing of the generation of petroleum in source rock, the subsequent expulsion of petroleum from source rock and its migration to reservoir rock, and the composition of oil and gas in the reservoir rock. This information is valuable for guiding exploration activities. It can help assess drilling prospects and provide predrilling information about the pressure and temperature that may be encountered by an exploratory well.

3.7 Porous Media

Subsurface reservoirs in the Earth are examples of porous media, which is a medium that contains solid material and pore space. Figure 3.3 shows a block of rock with grains of sand filling the block. Bulk volume is the volume of the block and includes both grain volume and the volume of space, or pore volume. Bulk volume V_B of the porous medium shown in Figure 3.3 is the product of area A in the horizontal plane and gross thickness H:

$$V_B = AH \qquad (3.7.1)$$

The volume that is not occupied by grains of sand is the pore space available for occupation by fluids such as oil, gas, and water. Porosity ϕ is defined as the ratio of pore volume to bulk volume. Pore volume V_P is the volume remaining when the volume of grains V_G is subtracted from the bulk volume—thus,

$$\phi = \frac{V_P}{V_B} = \frac{V_B - V_G}{V_B} \qquad (3.7.2)$$

If the void spaces in a porous medium are connected and form a flow path to a wellbore, the connected void space is referred to as effective porosity. Void space that cannot communicate with the wellbore is considered ineffective porosity. The original porosity resulting from sediment deposition is primary porosity. Secondary porosity is an increase in primary porosity due to the chemical dissolution of reservoir rocks, especially carbonates. Primary and secondary porosity can be both effective and ineffective. Total porosity is a combination of ineffective porosity and effective (interconnected) porosity.

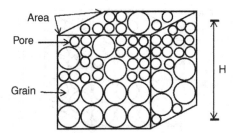

Figure 3.3 Porous medium.

Table 3.2 The Dependence of Porosity on Rock Type

Rock Type	Porosity Range (%)	Typical Porosity (%)
Sandstone	15–35	25
Unconsolidated sandstone	20–35	30
Carbonate		
Intercrystalline limestone	5–20	15
Oolitic limestone	20–35	25
Dolomite	10–25	20

Porosity values depend on rock type, as illustrated in Table 3.2. The two basic techniques for directly measuring porosity are core analysis in the laboratory and well logging. Laboratory measurements tend to be more accurate, but they sample only a small fraction of the reservoir. Also, changes in rock properties can occur when the core is brought from the reservoir to the surface. Well log measurements sample a much larger portion of the reservoir than core analysis. Ideally, a correlation can be established between *in situ* measurements such as well logging and surface measurements such as core analysis.

3.7.1 Tortuosity

When fluids are produced from a reservoir, they travel through reservoir rock from a place of higher energy to a place of lower energy. The route that the fluid takes can be a straight path or very circuitous. The property of the rock that measures the length of the path from one point A to another point B relative to a straight line is called tortuosity. If the path is a straight line, which can occur in a linear fracture, the tortuosity is 1. In most cases, the flow path between points A and B will be longer than a straight line, so tortuosity is greater than 1. Figure 3.4 illustrates the concept of tortuosity.

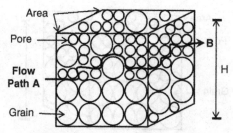

Figure 3.4 Tortuosity.

3.7.2 Saturation

We can rearrange Eq. (3.7.2) to show that pore volume is the product of bulk volume and porosity:

$$V_P = \phi V_B \tag{3.7.3}$$

When sedimentary rocks are being deposited, the pore space is usually filled with water. The pores of a rock can be filled at a later time with commercially important fluids ranging from potable water to oil and gas. Bulk volume and pore volume are measures of the gross volume in a system. To determine the volume of the system that is commercially significant, the gross volume must be adjusted by introducing the concept of net thickness.

Gross thickness H is the thickness of the formation, and net thickness h is the thickness of the formation that is commercially significant. The net-to-gross ratio η_{NTG} is the ratio of net thickness h to gross thickness H:

$$\eta_{NTG} = h/H, 0 \leq \eta_{NTG} \leq 1 \tag{3.7.4}$$

The inequality highlights the fact that net thickness is always less than or equal to gross thickness. The volume of net pay is the product of pore volume and net to gross ratio:

$$V_{pay} = \eta_{NTG}V_P = \eta_{NTG}HA\phi = hA\phi \tag{3.7.5}$$

The saturation S_ℓ of phase ℓ is the fraction of the volume of pore space occupied by phase ℓ. Oil, water, and gas are the most common fluid phases. The volume V_ℓ of phase ℓ in the pay interval is the product of net pay volume V_{pay} and phase saturation:

$$V_\ell = S_\ell V_{pay} = S_\ell hA\phi \tag{3.7.6}$$

The sum of the saturations in the pay interval must equal 1. If the system has N_ℓ phases, the saturation constraint is

$$1 = \sum_{\ell=1}^{N_P} S_\ell \tag{3.7.7}$$

For an oil-water-gas system, the saturation constraint is $S_o + S_w + S_g = 1$, where the subscripts $\{o, w, g\}$ refer to oil, water, and gas, respectively.

3.7.3 Permeability

We are interested in void spaces that are connected with other void spaces. Connected pore spaces form a conduit for fluid flow. Permeability is a measure of the connectivity of pore spaces. A rock is considered impermeable if no connectivity between pore spaces is present. Permeability is discussed in Chapter 4.

3.8 Volumetric Analysis

Estimates of fluid volume in a reservoir are values that can be obtained from a variety of calculation procedures and data sources. Consequently, the estimates of fluid volume serve as a quality control point at the interface between disciplines. Volumetric analysis is used to determine volume from static information (see, for example, Craft et al., 1991; Dake, 2001; Tearpock and Bischke, 1991; Towler, 2002; and Walsh and Lake, 2003). Static information is information that is relatively constant with respect to time, such as reservoir volume and original saturation and pressure distributions. By contrast, dynamic information such as pressure changes and fluid production is information that changes with respect to time. Material balance and reservoir flow modeling techniques use dynamic data to obtain original fluid volumes. An accurate characterization of the reservoir should yield consistent estimates of the fluid volumes that are originally in place in the reservoir regardless of the method chosen to determine the fluid volumes. In this section, we present the equations for volumetric estimates of original oil and gas in place.

Original oil in place (OOIP) in an oil reservoir is given by

$$N = \frac{7758 \phi A h_o S_{oi}}{B_{oi}} \qquad\qquad (3.8.1)$$

where

N = original oil in place (STB)
ϕ = reservoir porosity (fraction)
A = reservoir area (acres)
h_o = net thickness of oil zone (feet)
S_{oi} = initial reservoir oil saturation (fraction)
B_{oi} = initial oil formation volume factor (RB/STB)

Associated gas, or gas in solution, is the product of solution gas–oil ratio R_{so} and original oil in place N.

Original free gas in place for a gas reservoir is given by

$$G = \frac{7758 \phi A h_g S_{gi}}{B_{gi}} \qquad\qquad (3.8.2)$$

where

G = original free gas in place (SCF)
h_g = net thickness of gas zone (feet)
S_{gi} = initial reservoir gas saturation (fraction)
B_{gi} = initial gas formation volume factor (RB/SCF)

Equation (3.8.2) is often expressed in terms of initial water saturation S_{wi} by writing $S_{gi} = 1 - S_{wi}$. Initial water saturation can be determined from a well log or core analysis.

CS.3 Valley Fill Case Study: Geologic Model

The Valley Fill model presented here illustrates the application of an integrated flow model to a Valley Fill system. An incised valley is formed by the incision and fluvial erosion of an existing facies (Peijs-van Hilten et al., 1998; Weimer and Sonnenberg, 1989). The incised valley forms during a fall in relative sea level. The receding sea level exposes older deposits to incisement by drainage. The base of the incised valley is a sequence boundary that is referred to as the LSE, or Lowstand Surface of Erosion.

If the sea level starts to rise again, the initial deposition into the incised valley is typical of flooded systems. During this period of transgression, the incised valley is filled by a variety of fluvial, estuarine, and marine environments. When the period of transgression ends, the surface of the filled valley is covered by a new depositional layer associated with flooding. The top of the Valley Fill is a second sequence boundary that is referred to as the TSE, or Transgressive Surface of Erosion. A typical incised valley is characterized by a set of fluvial system tracts bounded below by an LSE and above by a TSE. The LSE and TSE are key surfaces in the description of the geologic system.

The Valley Fill reservoir in our case study is water-wet sandstone. It is interpreted as a meandering channel through an incised valley that has a regional dip. Six wells are productive in the channel. Six of the other wells and their seismic data help delineate the channel boundaries. Locations of wells in the area are shown in Figure CS.1A in Chapter 1. Wells 1 through 6 are productive. The top and base of the reservoir in each productive well are listed in Table CS.3A. Details supporting the geologic interpretation are presented in subsequent sections.

Table CS.3A Top and Base of the Valley Fill Reservoir

Well Number	Depth to Top of Reservoir (ft)	Depth to Base of Reservoir (ft)
1	8,435	8,555
2	8,430	8,550
3	8,450	8570
4	8,440	8,560
5	8,455	8,575
6	8,440	8,560

Exercises

3-1. Tectonic plates move relative to one another at the rate of up to 4 inches per year. How far apart would two plates move in 135 million years if moving at the maximum rate? Express your answer in miles.

3-2. South America and Africa are about 4,500 miles apart. If they began to separate from Pangaea about 135 million years ago, what is their rate of separation?

3-3. Based on the data in Table CS.3A, estimate the gross thickness of the Valley Fill reservoir.

3-4. Prepare a structure contour map of the Valley Fill reservoir using Figure CS.1A and the data in Table CS.3A.

3-5. A sandstone core sample is cleanly cut and carefully measured in a laboratory. The cylindrical core has a length of 3 inches and a diameter of 0.75 inch. The core is dried and weighed. The dried core weighs 125.00 grams. The core is then saturated with fresh water. The water-saturated core weighs 127.95 grams. What is the porosity of the sandstone core? Ignore the weight of air in the dried core and assume the density of water is 1 g/cc at laboratory temperature and pressure.

3-6. Suppose a formation consists of two 20-foot sections of impermeable shale and 60 feet of permeable sandstone. What are the gross thickness and net-to-gross ratio of the formation?

3-7. Displacement efficiency for oil depends on the difference between the initial volume of oil and the final volume of oil. Consider a reservoir with pore volume V_P, initial oil saturation S_{oi} at initial formation volume factor B_{oi}, and oil saturation at abandonment S_{oa} with formation volume factor B_{oa}. An estimate of displacement efficiency is

$$E_D = \frac{\dfrac{V_p S_{oi}}{B_{oi}} - \dfrac{V_p S_{or}}{B_{oa}}}{\dfrac{V_p S_{oi}}{B_{oi}}} = \frac{\dfrac{S_{oi}}{B_{oi}} - \dfrac{S_{or}}{B_{oa}}}{\dfrac{S_{oi}}{B_{oi}}} = \frac{S_{oi} - S_{or}\dfrac{B_{oi}}{B_{oa}}}{S_{oi}} = 1 - \frac{S_{or}}{S_{oi}}\frac{B_{oi}}{B_{oa}}$$

What is the displacement efficiency if $S_{oi} = 0.7$ at $B_{oi} = 1.43$, and $S_{oa} = 0.3$ at $B_{oa} = 1.39$? *Note*: Set $S_{or} = S_{oa}$ for this calculation.

3-8. An oil reservoir has average porosity $= 0.2$ in an area of 640 acres with a net thickness of 100 feet, initial oil saturation of 75 percent, and an initial oil formation volume factor of 1.4 RB/STB. Use Eq. (3.8.1) to estimate OOIP.

3-9. Open VFILL5_HM.dat and find GRID BLOCK LENGTHS. The construct $n*y$ means that there are n entries with the value y. For example, 6*0 is the same as six entries of 0, so 0 0 0 0 0 0. Create a table with 30 columns and 15 rows. Enter the nonzero net thickness values into the table for the first layer. Compare with the contour map drawn in Exercise 3-4.

3-10. Open VFILL5_HM.dat and find DEPTH TO TOP OF UPPER SAND. Notice that the depth is increasing by 5 feet for each 200-foot-wide row. Estimate the north-to-south dip angle from this information.

4 Porosity and Permeability

Porosity and permeability are measures of the storage capacity and flow capacity, respectively, in a porous medium. Many fundamental concepts are needed to understand storage and flow capacity. They include bulk volume, pore volume, net volume, porosity, porosity compressibility, transmissibility, and permeability. Averaging techniques for estimating a scalar value of permeability are presented in this chapter, but permeability is in general a tensor and can be approximated as a scalar in many cases. The tensor concept of directional permeability is discussed.

4.1 Bulk Volume and Net Volume

The bulk volume of a cell that is shaped like a parallelepiped, shown in Figure 4.1, is given by the triple scalar product

$$V_B = \left| \vec{U} \cdot (\vec{V} \times \vec{W}) \right| \tag{4.1.1}$$

The vectors $\{\vec{U}, \vec{V}, \vec{W}\}$ are aligned along the axes of the parallelepiped and have magnitudes that are equal to the lengths of the sides of the parallelepiped. In Cartesian coordinates, we have the result

$$V_B = \left| L_x \hat{i} \cdot \left(L_y \hat{j} \times L_z \hat{k} \right) \right| = L_x L_y L_z \tag{4.1.2}$$

in terms of the lengths $\{L_x, L_y, L_z\}$ of the sides of the Cartesian cell. The vectors $\hat{i}, \hat{j}, \hat{k}$ are unit vectors in Cartesian coordinates.

Bulk volume V_B of a Cartesian cell is the product of cross-sectional area A times gross thickness H:

$$V_B = AH \tag{4.1.3}$$

The area of a cell is the product of the x-direction cell length Δx and the y-direction cell length Δy; thus $A = \Delta x \, \Delta y$. Gross thickness H is the cell length in the z-direction.

Bulk volume is a measure of the gross volume in the system. It includes both rock volume and pore volume. To determine the volume of the system that is commercially significant, the gross volume must be adjusted by introducing the concept

Integrated Reservoir Asset Management. DOI: 10.1016/B978-0-12-382088-4.00004-9

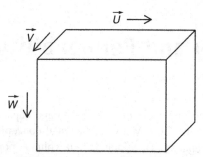

Figure 4.1 Volume of a parallelepiped.

of net thickness. Net thickness h is the thickness of the commercially significant formation. The net-to-gross ratio η is the ratio of net thickness h to gross thickness H:

$$\eta = h/H,\ 0 \leq \eta \leq 1 \tag{4.1.4}$$

Net thickness is always less than or equal to gross thickness.

4.2 Porosity and Grain Volume

Porosity is the fraction of a porous medium that is void space. The bulk volume V_B of a porous medium is the sum of pore volume V_P and grain volume V_G; thus

$$V_B = V_P + V_G \tag{4.2.1}$$

Porosity is the ratio of pore volume to bulk volume:

$$\phi = V_P/V_B \tag{4.2.2}$$

Dividing Eq. (4.2.1) by V_B and using the definition of porosity expresses the grain volume in terms of porosity as

$$\frac{V_G}{V_B} = 1 - \phi \tag{4.2.3}$$

4.3 Effective Pore Volume

Most porous media contain a fraction of pores that are not in communication with the flow path. This pore volume is ineffective. Effective pore volume $(V_P)_{eff}$ is the interconnected pore volume that communicates with a well. Effective porosity is defined as the ratio of effective pore volume to bulk volume; thus

$$\phi_{eff} \equiv \frac{(V_P)_{eff}}{V_B} \tag{4.3.1}$$

Unless stated otherwise, any further discussion using porosity will assume that the porosity of interest is effective porosity, and the subscript "*eff*" is not written.

4.4   Porosity Compressibility

Porosity compressibility is a measure of the change in porosity ϕ as a function of fluid pressure P. It is defined as

$$c_\phi = \frac{1}{\phi}\frac{d\phi}{dP} \tag{4.4.1}$$

If ϕ_0 is porosity at pressure P_0 and ϕ is porosity at pressure P, the integral of Eq. (4.4.1) yields the relationship

$$\phi = \phi_0 \exp\left[\int_{P_0}^{P} c_\phi dP\right] \tag{4.4.2}$$

If porosity compressibility is constant with respect to pressure, the integral in Eq. (4.4.2) can be evaluated and gives

$$\phi = \phi_0 \exp\left[c_\phi \Delta P\right] \tag{4.4.3}$$

where $\Delta P = P - P_0$. The first-order approximation to Eq. (4.4.3) is

$$\phi \approx \phi_0\left[1 + c_\phi \Delta P\right] = \phi_0\left[1 + c_\phi(P - P_0)\right] \tag{4.4.4}$$

Equation (4.4.4) is used in many reservoir flow simulators to calculate the change in porosity with respect to changes in fluid pressure. A first-order approximation is reasonable in many cases because reservoir rock compressibility is typically on the order of $c_\phi = 3 \times 10^{-6}$/psia and a typical pressure change is on the order of $P - P_0 \approx 1000$ psia, so the bracketed term in Eq. (4.4.4) is on the order of 1.003. There are reservoirs where the approximation is not as valid because the product of rock compressibility and pressure change can be an order of magnitude larger.

4.5   Darcy's Law and Permeability

The basic equation that describes fluid flow in porous media is Darcy's law. Darcy found that flow rate was proportional to pressure gradient. Darcy's equation for calculating volumetric flow rate q for linear, horizontal, single-phase flow is

$$q = -0.001127\frac{KA}{\mu}\frac{\Delta P}{\Delta x} \tag{4.5.1}$$

The units of the physical variables determine the value of the constant (0.001127) in Eq. (4.5.1). The constant 0.001127 corresponds to variables expressed in the following oil field units:

q = volumetric flow rate (bbl/day)
K = permeability (md)
A = cross-sectional area (ft^2)
P = pressure (psi)
μ = fluid viscosity (cp)
Δx = length (ft)

Figure 4.2 illustrates the terms in Darcy's law for a cylindrical core of rock. The movement of a single-phase fluid through a porous medium depends on cross-sectional area A that is normal to the direction of fluid flow, pressure difference ΔP across the length Δx of the flow path, and viscosity μ of the flowing fluid. The minus sign indicates that the direction of fluid flow is opposite to the direction of increasing pressure: the fluid flows from high pressure to low pressure in a horizontal (gravity-free) system. The proportionality constant in Eq. (4.5.1) is permeability.

If we rearrange Eq. (4.5.1) and perform a dimensional analysis, we see that permeability has dimensions of area (L^2), where L is a unit of length:

$$K = \frac{\text{rate} \times \text{viscosity} \times \text{length}}{\text{area} \times \text{pressure}} = \frac{\left(\dfrac{L^3}{\text{time}}\right)\left(\dfrac{\text{force} \times \text{time}}{L^2}\right)L}{L^2\left(\dfrac{\text{force}}{L^2}\right)} = L^2 \qquad (4.5.2)$$

The areal unit (L^2) is physically related to the cross-sectional area of pore throats in rock. A pore throat is the opening that connects two pores. The size of a pore throat depends on grain size and distribution. For a given grain distribution, the cross-sectional area of a pore throat will increase as grain size increases. Relatively large pore throats imply relatively large values of L^2 and correspond to relatively large values of permeability.

Permeability typically ranges from 1 md (1.0×10^{-15} m^2) to 1 Darcy (1,000 md or 1.0×10^{-12} m^2) for commercially successful oil and gas fields. Permeability can be much less than 1 md in unconventional reservoirs such as tight gas and shale gas reservoirs. Advances in well stimulation technology and increases in oil and gas prices have improved the economics of low-permeability reservoirs.

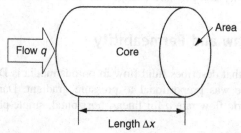

Figure 4.2 Darcy's law.

Darcy's law shows that flow rate and pressure difference are linearly related. The pressure gradient from the point of fluid injection to the point of fluid withdrawal is found by rearranging Eq. (4.5.1):

$$\frac{\Delta P}{\Delta x} = -\left(\frac{q}{0.001127A}\right)\frac{\mu}{K} \qquad (4.5.3)$$

4.5.1 Superficial Velocity and Interstitial Velocity

Superficial velocity is the volumetric flow rate q in Darcy's law divided by the cross-sectional area $A_\perp$ normal to flow (Bear, 1972; Lake, 1989); thus $u = q/A_\perp$ in appropriate units. The interstitial, or "front," velocity v of the fluid through the porous rock is the actual velocity of a fluid element as the fluid moves through the tortuous pore space. Interstitial velocity v is the superficial velocity u divided by porosity ϕ, or

$$v = \frac{u}{\phi} = \frac{q}{\phi A_\perp} \qquad (4.5.4)$$

Interstitial velocity is larger than superficial velocity since porosity is a fraction between 0 and 1.

4.5.2 The Validity of Darcy's Law

Darcy's law is valid when fluid flow is laminar. Laminar fluid flow represents one type of flow regime. Three types of flow regimes may be defined: laminar flow regime with low flow rate; inertial flow regime with moderate rate; and turbulent flow regime with high flow rate. Flow regimes are classified in terms of the dimensionless Reynolds number (Fancher and Lewis, 1933). The Reynolds number is the ratio of inertial (fluid momentum) forces to viscous forces. It has the form

$$N_{Re} = 1488\frac{\rho v_D d_g}{\mu} \qquad (4.5.5)$$

where

ρ = fluid density (lbm/ft^3)
v_D = superficial velocity (ft/sec)
d_g = average grain diameter (ft)
μ = absolute viscosity (cp)

The flow regime is determined by calculating the Reynolds number. A Reynolds number that is low corresponds to laminar flow, and a high Reynolds number corresponds to turbulent flow. Govier (1978) presented the classification in Table 4.1 that expresses flow regime in terms of the Reynolds number.

Table 4.1 Classification of Flow Regimes

Flow Regime	Description
Laminar	Low flow rates ($N_{Re} < 1$)
Inertial	Moderate flow rates ($1 < N_{Re} < 600$)
Turbulent	High flow rates ($N_{Re} > 600$)

The linear relationship between pressure gradient and rate in Darcy's law is valid for many flow systems, but not all. Fluid flow in a porous medium can have a nonlinear effect that is represented by the Forchheimer equation (Govier, 1978). Forchheimer observed that turbulent flow in high-flow-rate gas wells has the quadratic dependence

$$\frac{\Delta P}{\Delta x} = -\left(\frac{q}{0.001127A}\right)\frac{\mu}{K} + \beta\rho\left(\frac{q}{A}\right)^2 \qquad (4.5.6)$$

for fluid with density ρ and turbulence factor β. A minus sign and conversion unit is inserted in the first-order rate term on the right-hand side of Eq. (4.5.6) to be consistent with the rate convention used in Eq. (4.5.1). Equation (4.5.6) is called the Forchheimer equation. The nonlinear effect becomes more important in high-flow-rate environments such as some gas wells and in hydraulic fracturing (Barree and Conway, 2005). Darcy's law correctly describes laminar flow and may be used as an approximation of turbulent flow. Permeability calculated from Darcy's law is less than true rock permeability at turbulent flow rates.

4.5.3 Radial Flow of Liquids

Darcy's law for steady-state, radial, horizontal, single-phase liquid flow in a porous medium is

$$Q = -\frac{0.00708Kh(P_w - P_e)}{\mu B \ln(r_e/r_w)} \qquad (4.5.7)$$

where

Q = liquid flow rate (STB/D)
r_w = wellbore or inner radius (ft)
r_e = outer radius (ft)
K = permeability (md)
h = formation thickness (ft)
P_w = pressure at inner radius (psi)
P_e = pressure at outer radius (psi)
μ = viscosity (cp)
B = formation volume factor (RB/STB)

The formation volume factor that is in the denominator on the right-hand side of Eq. (4.5.7) converts volumetric flow rate from reservoir to surface conditions. The rate Q is positive for a production well $\{P_w < P_e\}$ and negative for an injection well $\{P_w > P_e\}$.

Different procedures may be used to estimate the outer radius r_e. The outer radius r_e is equated to the drainage radius of the well when analyzing the pressure at a well. Alternatively, the value of r_e in a reservoir flow model depends on the size of the grid block containing the flow rate term (Peaceman, 1978; Fanchi, 2006a). The flow rate is less sensitive to an error in the estimate of r_e than a similar error in a parameter like permeability because the radial flow calculation depends on the logarithm of r_e. It is therefore possible to tolerate larger errors in r_e than other flow parameters and still obtain a reasonable value for radial flow rate.

4.5.4 Radial Flow of Gases

Consider Darcy's law in radial coordinates for a single phase:

$$q_r = -0.006328 \frac{2\pi rhK}{\mu} \frac{dP_r}{dr} \tag{4.5.8}$$

where the radial distance r increases as we move away from the well, and

q_r = gas rate (rcf/d)
r = radial distance (ft)
h = zone thickness (ft)
μ = gas viscosity (cp)
K = permeability (md)
P_r = reservoir pressure (psia)

The cross-sectional area in Eq. (4.5.8) is the cross-sectional area $2\pi rh$ of a cylinder enclosing the wellbore. Let subscripts r and s denote reservoir and surface conditions, respectively. To convert from reservoir to surface conditions, we divide the gas rate at reservoir conditions by the gas formation volume factor; thus

$$q_s = \frac{q_r}{B_g} \tag{4.5.9}$$

where

q_s = gas rate (scf/d)
B_g = gas formation volume factor (rcf/scf)

Gas formation volume factor B_g is a function of pressure P, temperature T, and gas compressibility factor Z from the real gas equation of state:

$$B_b = \frac{P_s T_r Z_r}{P_r T_s Z_s} \tag{4.5.10}$$

The rate at surface conditions is found by substituting Eqs. (4.5.8) and (4.5.10) into (4.5.9) to get

$$q_s = -0.03976 \frac{rhK}{\mu} \frac{P_r}{P_s} \frac{T_s}{T_r} \frac{Z_s}{Z_r} \frac{dP_r}{dr} \tag{4.5.11}$$

and q_s has units of scf/d.

If we assume a constant rate, we can rearrange Eq. (4.5.11) and integrate from the inner radius to the outer radius to get

$$q_s \int_{r_w}^{r_e} \frac{dr}{r} = q_s \ln \frac{r_e}{r_w} = -0.03976 \frac{KhT_sZ_s}{P_sT_r} \int_{P_w}^{P_e} \frac{P_r}{\mu Z_r} dP_r \tag{4.5.12}$$

Subscripts w and e denote values at the wellbore radius and external radius, respectively. Equation (4.5.12) can be written in a simpler form by introducing the real gas pseudopressure $m(P)$:

$$m(P) = 2 \int_{P_{ref}}^{P} \frac{P'}{\mu_g Z} dP' \tag{4.5.13}$$

where

P_{ref} = a reference pressure
P' = a dummy variable of integration

The integrand of Eq. (4.5.13) has a nonlinear dependence on pressure. It is often necessary to solve the integral numerically because of the nonlinear dependence of gas viscosity and gas compressibility on pressure. Given $m(P)$, the radial form of Darcy's law becomes

$$q_s = -0.01988 \frac{KhT_sZ_s}{P_sT_r \left(\ln \frac{r_e}{r_w} \right)} [m(P_e) - m(P_w)] \tag{4.5.14}$$

Specifying the standard conditions

$Z_s = 1$
$P_s = 14.7 \text{ psia}$
$T_s = 60°\text{F} = 520°\text{R}$

in Eq. (4.5.14) gives Darcy's law for the radial flow of gas:

$$q_s = -0.703 \frac{Kh}{T_r \left(\ln \frac{r_e}{r_w} \right)} [m(P_e) - m(P_w)] \tag{4.5.15}$$

Solving Eq. (4.5.15) for real gas pseudopressure at the external radius gives

$$m(P_e) = m(P_w) - \frac{1.422T_r \left(\ln \dfrac{r_e}{r_w} \right)}{Kh} q_s \qquad (4.5.16)$$

Equation (4.5.16) shows that $m(P_e)$ is proportional to q_s and inversely proportional to permeability.

4.5.5 Klinkenberg's Effect

Klinkenberg found that the permeability for gas flow in a porous medium depends on pressure according to the relationship

$$k_g = k_{abs} \left(1 + \frac{b}{\bar{P}} \right) \qquad (4.5.17)$$

where

k_g = apparent permeability calculated from gas flow tests
k_{abs} = true absolute permeability of rock
$\bar{P}$ = mean flowing pressure of gas in the flow system
b = Klinkenberg's effect

The factor b is a positive constant for gas in a specific porous medium. Equation (4.5.17) shows that when the factor $(1 + b/\bar{P}) \geq 1$, we have the inequality $k_g \geq k_{abs}$. As pressure increases, the factor $(1 + b/\bar{P})$ approaches 1 and the apparent permeability to gas k_g approaches the true absolute permeability of the rock k_{abs}.

The dependence of k_g on pressure is called Klinkenberg's effect and is attributed to the "slippage" of gas molecules along pore walls. The interaction between gas molecules and pore walls is greater at low pressures than at high pressures. Conversely, slippage along pore walls is greater at high pressures than at low pressures. At low pressures, the calculated permeability for gas flow k_g may be greater than true rock permeability. Measurements of k_g are often conducted with air and are not corrected for Klinkenberg's effect. This should be considered when comparing k_g with permeability estimates from other sources such as well tests.

4.5.6 Properties of Permeability

A micro scale measurement of grain size distribution shows that different grain sizes and shapes affect permeability. Permeability may be viewed as a mathematical convenience for describing the statistical behavior of a given flow experiment. In this context, transient testing gives the best measure of permeability over a large volume. Despite their importance to the calculation of flow, permeability and its distribution will not be known accurately. Seismic data can help define the distribution of permeability between wells if a good correlation exists between seismic amplitude and a rock quality measurement that includes permeability.

Permeability depends on rock type. The two most common reservoir rock types are clastic reservoirs and carbonate reservoirs. The permeability in a clastic reservoir depends on pore size, which is seldom controlled by secondary solution vugs. Compacted and cemented sandstone rocks tend to have lower permeabilities than clean, unconsolidated sands. Productive sandstone reservoirs usually have permeabilities in the range of 10 to 1,000 md. The permeability in tight gas and coalbed methane reservoirs is less than 1 md.

Carbonate reservoirs are generally less homogeneous than clastic reservoirs and have a wider range of grain size distribution. The typical matrix permeability in a carbonate reservoir tends to be relatively low. Significant permeability in a carbonate reservoir may be associated with secondary porosity features such as vugs and oolites. The presence of clay can adversely affect permeability. Clay material may swell on contact with fresh water, and the resulting swelling can reduce rock permeability by several orders of magnitude.

Natural or manmade fractures can significantly increase flow capacity in both carbonate and clastic reservoirs. An extensive natural fracture system can provide high-flow-capacity conduits for channeling flow from the reservoir matrix to a wellbore. Naturally fractured reservoirs are usually characterized by relatively high-permeability, low-porosity fractures, and by a relatively low-permeability, high-porosity matrix. Most of the fluid is stored in the matrix, while flow from the reservoir to the wellbore is controlled by the permeability in the fracture system.

4.6   Permeability Averaging

Permeability averaging poses a problem in the estimation of a representative average permeability for use in Darcy's equation. Permeability can be obtained from core plugs and well tests. Core plug permeability and well test permeability are measurements of permeability at different scales. Several techniques exist for estimating an average value of permeability. Some practical averaging techniques are presented below.

4.6.1   Parallel Beds—Linear Flow

Linear flow through parallel beds of differing permeability is shown in Figure 4.3. Pressure is constant at each end of the flow system, and total flow rate is the sum of the rates q_i in each layer i:

$$q = \sum_i q_i \qquad (4.6.1)$$

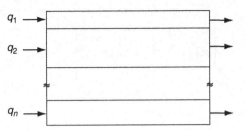

Figure 4.3 Beds in parallel.

Suppose layer i has length L, width w, net thickness h_i, and permeability k_i. Applying Darcy's law for linear flow of a fluid with viscosity μ gives

$$\frac{k_{ave}h_t w(p_1 - p_2)}{\mu L} = \sum_i \frac{k_i h_i w(p_1 - p_2)}{\mu L} \qquad (4.6.2)$$

where the sum is over all layers. After canceling common terms, we obtain the expression

$$k_{ave}h_t = \sum_i k_i h_i \qquad (4.6.3)$$

where

h_t = total thickness
k_{ave} = an average permeability

Solving for k_{ave} gives

$$k_{ave} = \frac{\sum_i k_i h_i}{\sum_i h_i} \qquad (4.6.4)$$

The average permeability for parallel flow through beds of differing permeabilities equals the thickness weighted average permeability. If each bed has the same thickness, k_{ave} is the arithmetic average.

4.6.2 Parallel Beds—Radial Flow

The average permeability for radial flow in parallel beds is the same relationship as linear flow—namely, the thickness weighted average

$$k_{ave} = \frac{\sum_i k_i h_i}{\sum_i h_i} \qquad (4.6.5)$$

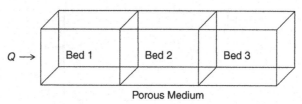

Figure 4.4 Flow in beds in series.

4.6.3 Beds in Series—Linear Flow

Figure 4.4 illustrates beds in series. The average permeability for beds in series is the harmonic average:

$$k_{ave} = \frac{\sum_i L_i}{\sum_i L_i/k_i} \qquad (4.6.6)$$

where bed i has length L_i and permeability k_i.

4.6.4 Beds in Series—Radial Flow

Radial flow in beds in series treats beds as concentric rings around the wellbore. For a system with three beds, the average permeability for radial flow in beds in series is the harmonic average:

$$k_{ave} = \frac{\ln(r_e/r_w)}{\dfrac{\ln(r_e/r_2)}{k_3} + \dfrac{\ln(r_2/r_1)}{k_2} + \dfrac{\ln(r_1/r_w)}{k_1}} \qquad (4.6.7)$$

where

r_w = the radius of the well
r_e = the radius to the outer ring

The radius r_e corresponds to the drainage radius of the system.

4.6.5 Random Flow

For permeability values distributed randomly, the average permeability is a geometric average:

$$k_{ave} = \left(k_1^{h_1} \cdot k_2^{h_2} \cdot k_3^{h_3} \cdot \ldots k_n^{h_n} \right)^{\left(1/\sum_{i=1}^{n} h_i \right)} \qquad (4.6.8)$$

where

> h_i = the thickness of interval i with permeability k_i
> n = the number of intervals

4.6.6 Permeability Averaging in a Layered Reservoir

The average permeability for a layered reservoir can be estimated using the following procedure:

- Determine the geometric average for each layer.
- Determine the arithmetic average of the geometric averages, weighted by the thickness of each layer.

Several other procedures exist for determining the average permeability of a layered reservoir.

One method that can be applied with relative ease is to perform a flow model study using two models. One model is a cross-section model with all of the geologic layers treated as model layers. The other model is a single-layer model with all the geologic layers combined into a single layer. Flow performance from the cross-section model is compared with the flow performance of the single-layer model. The permeability in the single-layer model is adjusted until the performance of the single-layer model is approximately equal to the performance of the cross-section model. The resulting permeability is an "upscaled," or average, permeability for the cross-section model.

4.7   Transmissibility

Flow between neighboring blocks is treated as a series application of Darcy's law. We are concerned with the movement of fluids between two blocks such as those shown in Figure 4.5. If we assume that the conditions needed for Darcy flow are satisfied and simplify the problem by considering single-phase flow for a phase with formation volume factor B_ℓ, we have

$$\bar{Q}_\ell = \overline{KA}_c \frac{1}{\mu_\ell B_\ell} \frac{P_{i-1} - P_i}{\dfrac{\Delta x_{i-1} + \Delta x_i}{2}} \tag{4.7.1}$$

where

> $\bar{Q}_\ell$ = the average volumetric flow rate of phase ℓ
> $\bar{K}$ = the absolute permeability associated with a pressure drop from x_{i-1} to x_i
> $\bar{A}_c$ = the cross-sectional area between x_{i-1} and x_i

To use Eq. (4.7.1), it is necessary to express the product $\overline{KA}_c$ in terms of known variables—namely, x_j, K_j, and A_{cj}, where subscript j refers to blocks $i-1$ and i.

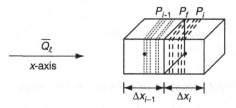

Figure 4.5 Flow between blocks.

We begin by using Darcy's law to write the flow rates through each shaded volume element. The flow rate of phase ℓ is

$$\bar{Q}_\ell = K_{i-1} A_{c,i-1} \frac{1}{\mu_\ell B_\ell} \frac{P_{i-1} - P_f}{\dfrac{\Delta x_{i-1}}{2}} \qquad (4.7.2)$$

for block $i-1$ and

$$\bar{Q}_\ell = K_i A_{c,i} \frac{1}{\mu_\ell B_\ell} \frac{P_f - P_i}{\dfrac{\Delta x_i}{2}} \qquad (4.7.3)$$

for block i. The pressure P_f refers to the pressure at the interface between block $i-1$ and block i. Solving for $(P_{i-1} - P_f)$ and $(P_f - P_i)$ in Eqs. (4.7.2) and (4.7.3), respectively, and then adding the results yields

$$P_{i-1} - P_i = \bar{Q}_\ell \left[\frac{\dfrac{\Delta x_{i-1}}{2} \dfrac{1}{\mu_\ell B_\ell}}{K_{i-1} A_{c,i-1}} + \frac{\dfrac{\Delta x_i}{2} \dfrac{1}{\mu_\ell B_\ell}}{K_i A_{c,i}} \right] \qquad (4.7.4)$$

Inserting this expression into Eq. (4.7.1) and solving for $\overline{KA}_c$ gives

$$\overline{KA}_c = \frac{\Delta x_{i-1} + \Delta x_i}{\left(\dfrac{\Delta x_{i-1}}{K_{i-1} A_{c,i-1}} \right) + \left(\dfrac{\Delta x_i}{K_i A_{c,i}} \right)} \qquad (4.7.5)$$

Substituting this expression back into Eq. (4.7.1) gives, after some simplification, the flow rate equation

$$\bar{Q}_\ell = \frac{1}{\mu_\ell B_\ell} \left[\frac{2(KA_c)_{i-1} \cdot (KA_c)_i}{\Delta x_{i-1}(KA_c)_i + \Delta x_i(KA_c)_{i-1}} \right] (P_{i-1} - P_i) \qquad (4.7.6)$$

Equation (4.7.6) has the form

$$\bar{Q}_\ell = A'_{\ell,\,i-1/2} \cdot (P_{i-1} - P_i) \tag{4.7.7}$$

where $A'_{\ell,\,i-1/2}$ is the Darcy phase transmissibility between blocks $i-1$ and i.

Notice that transmissibility depends on properties in both blocks $i-1$ and i. A similar procedure is used to obtain transmissibility values for the y and z directions. Transmissibilities may be used to define sealing or partially sealing faults and to define high-permeability channels.

4.8 Measures of Permeability Heterogeneity

It is often useful to represent permeability heterogeneity with a single number. This number is referred to here as a measure of permeability heterogeneity. Several such measures exist (Lake and Jensen, 1989). The Dykstra-Parsons coefficient and the Lorenz coefficient are described in this section as illustrations. The procedure outlined in the following makes some simplifying assumptions that are not too restrictive in practice but make it possible to calculate permeability heterogeneity measures with relative ease. These measures may be used to verify that the permeability distribution used in a model has comparable heterogeneity to the permeability distribution observed in an analysis of field data.

4.8.1 The Dykstra-Parsons Coefficient

The Dykstra-Parsons coefficient can be estimated for a log normal permeability distribution as

$$V_{DP} = 1 - \exp\left[-\sqrt{\ln(k_A/k_H)}\right] \tag{4.8.1}$$

where k_A is the arithmetic average

$$k_A = \frac{1}{n}\sum_{i=1}^{n} k_i \tag{4.8.2}$$

and k_H is the harmonic average

$$\frac{1}{k_H} = \frac{1}{n}\sum_{i=1}^{n}\frac{1}{k_i} \tag{4.8.3}$$

The Dykstra-Parsons coefficient should be in the range $0 \le V_{DP} \le 1$. For a homogeneous reservoir, $V_{DP} = 0$ because $k_A = k_H$. An increase in heterogeneity increases V_{DP}. Typical values of the Dykstra-Parsons coefficient are in the range $0.4 \le V_{DP} \le 0.9$.

As an example of a Dykstra-Parsons coefficient calculation, suppose we have a three-layer case with the following permeabilities:

Layer 1: $k = 35$ md
Layer 2: $k = 48$ md
Layer 3: $k = 126$ md

The arithmetic average is

$$k_A = \frac{1}{n} \sum_{i=1}^{n} k_i = \frac{1}{3}(126 + 48 + 35) = 69.7$$

and the harmonic average is

$$\frac{1}{k_H} = \frac{1}{n} \sum_{i=1}^{n} \frac{1}{k_i} = \frac{1}{3}\left(\frac{1}{126} + \frac{1}{48} + \frac{1}{35}\right)$$

or $k_H = 52.3$ md. Using these average values, we estimate the Dykstra-Parsons coefficient for a log normal permeability distribution to be

$$V_{DP} = 1 - \exp\left[-\sqrt{\ln(69.7/52.3)}\right] = 0.415$$

4.8.2 The Lorenz Coefficient

The Lorenz coefficient requires a bit more work than the Dykstra-Parsons coefficient. We begin by calculating the following quantities:

$$F_m = \text{cum flow capacity} = \sum_{i=1}^{m} k_i h_i \bigg/ \sum_{i=1}^{n} k_i h_i; \ m = 1, \ldots, n \qquad (4.8.4)$$

and

$$H_m = \text{cum thickness} = \sum_{i=1}^{m} h_i \bigg/ \sum_{i=1}^{n} h_i; \ m = 1, \ldots, n \qquad (4.8.5)$$

for $n = $ number of reservoir layers. Layers should be arranged in order of decreasing permeability; thus $i = 1$ has thickness h_1 and the largest permeability k_1, while $i = n$ has thickness h_n and the smallest permeability k_n. By definition, cumulative flow capacity should be in the range $0 \le F_m \le 1$, and cumulative thickness should be in the range $0 \le H_m \le 1$ for $0 < m < n$.

The Lorenz coefficient is defined in terms of a plot of F_m versus H_m, shown in Figure 4.6. The Lorenz coefficient L_c is two times the area enclosed between the Lorenz curve ABC in the figure and the diagonal AC. The range of the Lorenz coefficient is $0 \le L_c \le 1$. For a homogeneous reservoir, the Lorenz coefficient satisfies the equality $L_c = 0$. An increase in heterogeneity increases the value of the Lorenz coefficient L_c. Typical values of the Lorenz coefficient are in the range $0.2 \le L_c \le 0.6$.

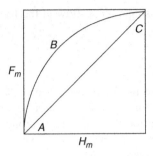

Figure 4.6 The Lorenz plot.

An estimate of the Lorenz coefficient is obtained by assuming that all of the permeabilities have equal probability so that the trapezoidal rule can be used to estimate area. The result is

$$L_c = \frac{1}{2n} \left[\left(\sum_{i=1}^{n} \sum_{j=1}^{n} |k_i - k_j| \right) / \sum_{i=1}^{n} k_i \right] \tag{4.8.6}$$

It is not necessary to order permeabilities using this estimate. Notice that in the homogeneous case, all of the permeabilities are equal so that we have the relationship $k_i = k_j$. Substituting this equality into Eq. (4.8.6) gives the Lorenz coefficient $L_c = 0$. In the ideal homogeneous case, the Lorenz coefficient is zero, as just indicated.

4.9 Darcy's Law with Directional Permeability

Permeability can be a complex function of spatial location and orientation. Spatial and directional variations of a function are described in terms of homogeneity, heterogeneity, isotropy, and anisotropy. If the value of a function does not depend on spatial location, it is called homogeneous. The function is heterogeneous if its value changes from one spatial location to another. If the value of a function depends on directional orientation—that is, the value is larger in one direction than another—then the function is anisotropic. The function is isotropic if its value does not depend on directional orientation. Permeability is a function that can be both heterogeneous and anisotropic. To account for heterogeneity and anisotropy, the simple 1-D form of Darcy's law must be generalized.

In general, flow occurs in dipping beds. To account for the effect of gravity, we define a variable called the potential of phase ℓ as

$$\Phi_\ell = P_\ell - \gamma_\ell(\Delta z) \tag{4.9.1}$$

where

Δz = depth from a datum
P_ℓ = pressure of phase ℓ
γ_ℓ = pressure gradient associated with the gravity term

If we write Darcy's law for single-phase flow in the form

$$q = -\frac{0.001127KA}{\mu}\frac{d\Phi}{dz} \tag{4.9.2}$$

we find that no vertical movement can occur when $d\Phi/dz = 0$. Thus, Eq. (4.9.2) expresses the movement of fluids in a form that accounts for gravity equilibrium.

Darcy's law in one dimension says that rate is proportional to pressure gradient. This can be extended to three dimensions using vector notation. Darcy's law for single-phase flow in three dimensions is

$$q_x = -0.001127K\frac{A}{\mu}\frac{\partial\Phi}{\partial x}$$

$$q_y = -0.001127K\frac{A}{\mu}\frac{\partial\Phi}{\partial y} \tag{4.9.3}$$

$$q_z = -0.001127K\frac{A}{\mu}\frac{\partial\Phi}{\partial z}$$

where the gradient of potential accounts for gravity effects. In vector notation we have

$$\vec{q} = -0.001127K\frac{A}{\mu}\nabla\Phi \tag{4.9.4}$$

Equation (4.9.3) can be written in matrix notation as

$$\begin{bmatrix} q_x \\ q_y \\ q_z \end{bmatrix} = -0.001127K\frac{A}{\mu}\begin{bmatrix} \partial\Phi/\partial x \\ \partial\Phi/\partial y \\ \partial\Phi/\partial z \end{bmatrix} \tag{4.9.5}$$

where permeability K and cross-sectional area A are treated as constants with respect to direction. A more general extension of Eq. (4.9.5) is

$$\begin{bmatrix} q_x \\ q_y \\ q_z \end{bmatrix} = -0.001127\frac{A}{\mu}\begin{bmatrix} K_{xx} & K_{xy} & K_{xz} \\ K_{yx} & K_{yy} & K_{yz} \\ K_{zx} & K_{zy} & K_{zz} \end{bmatrix}\begin{bmatrix} \partial\Phi/\partial x \\ \partial\Phi/\partial y \\ \partial\Phi/\partial z \end{bmatrix} \tag{4.9.6}$$

where permeability is now treated either as a 3×3 matrix with nine elements or as a tensor of rank two (Fanchi, 2006c). The diagonal permeability elements $\{K_{xx}, K_{yy}, K_{zz}\}$ represent the usual dependence of rate in one direction on pressure differences in the same direction. The off-diagonal permeability elements $\{K_{xy}, K_{xz}, K_{yx}, K_{yz}, K_{zx}, K_{zy}\}$ account for the dependence of rate in one direction on pressure differences in orthogonal directions. Expanding Eq. (4.9.6) gives the corresponding set of equations (Eq. (4.9.7)) that demonstrates this dependence:

$$q_x = -0.001127\frac{A}{\mu}\left[K_{xx}\frac{\partial\Phi}{\partial x} + K_{xy}\frac{\partial\Phi}{\partial y} + K_{xz}\frac{\partial\Phi}{\partial z}\right]$$

$$q_y = -0.001127\frac{A}{\mu}\left[K_{yx}\frac{\partial\Phi}{\partial x} + K_{yy}\frac{\partial\Phi}{\partial y} + K_{yz}\frac{\partial\Phi}{\partial z}\right] \qquad (4.9.7)$$

$$q_z = -0.001127\frac{A}{\mu}\left[K_{zx}\frac{\partial\Phi}{\partial x} + K_{zy}\frac{\partial\Phi}{\partial y} + K_{zz}\frac{\partial\Phi}{\partial z}\right]$$

In many practical situations it is mathematically possible to find a coordinate system $\{x', y', z'\}$ in which the permeability tensor has the diagonal form

$$\begin{bmatrix} K_{x'x'} & 0 & 0 \\ 0 & K_{y'y'} & 0 \\ 0 & 0 & K_{z'z'} \end{bmatrix}$$

The coordinate axes $\{x', y', z'\}$ are called the principal axes of the tensor, and the diagonal form of the permeability tensor is obtained by a principal axis transformation. The flow equations along the principal axes are

$$q_{x'} = -0.001127\frac{A}{\mu}\left[K_{x'x'}\frac{\partial\Phi}{\partial x'}\right]$$

$$q_{y'} = -0.001127\frac{A}{\mu}\left[K_{y'y'}\frac{\partial\Phi}{\partial y'}\right] \qquad (4.9.8)$$

$$q_{z'} = -0.001127\frac{A}{\mu}\left[K_{z'z'}\frac{\partial\Phi}{\partial z'}\right]$$

The principal axes in a field can vary from one point of the field to another because of permeability heterogeneity.

The form of the permeability tensor depends on the properties of the porous medium. The medium is said to be anisotropic if two or more elements of the diagonalized permeability tensor are different. The permeability of the medium is isotropic if the elements of the diagonalized permeability tensor are equal—that is,

$$K_{x'x'} = K_{y'y'} = K_{z'z'} = K \qquad (4.9.9)$$

If the medium is isotropic, permeability does not depend on direction. If the isotropic permeability does not change from one position in the medium to another, the

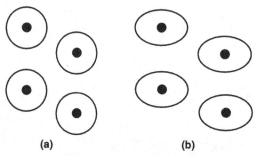

Figure 4.7 The effect of permeability anisotropy on a drainage area: (a) isotropic ($K_x = K_y$) and (b) anisotropic ($K_x \neq K_y$).

medium is said to be homogeneous in permeability. On the other hand, if the values of the elements of the permeability tensor vary from one point in the medium to another, both the permeability tensor and the medium are considered heterogeneous. Virtually all reservoirs exhibit some degree of anisotropy and heterogeneity, but the flow behavior in many reservoirs can be approximated as homogeneous and isotropic. In Figure 4.7, (a) is the drainage area of four production wells with isotropic permeability, and (b) is the drainage area of four production wells with anisotropic permeability.

4.9.1 Vertical Permeability

Permeability for flow in a direction that is perpendicular to gravity is horizontal permeability. By contrast, vertical permeability is the permeability for flow in the direction aligned with the direction of the gravitational field. Vertical permeability can be measured in the laboratory or in pressure transient tests conducted in the field. In many cases vertical permeability is not measured and must be assumed. A rule of thumb is to assume vertical permeability to be approximately one-tenth of horizontal permeability. These are reasonable assumptions when there is no data to the contrary. It is preferable from a technical point of view to make direct measurements of all relevant reservoir data. Sometimes it to be difficult to justify the cost or logistics of obtaining direct measurements. If it is necessary to use a rule of thumb or data from an analogous formation to estimate a particular variable, the sensitivity of the shared earth model to changes in the estimated variable should be considered.

CS.4 Valley Fill Case Study: Permeability

The permeability distribution in the Valley Fill reservoir is assumed to be isotropic and homogeneous because the spatial dependence of permeability is not known. Horizontal permeability is 150 md based on a pressure transient test. Vertical permeability was not measured and is therefore assumed to be one-tenth of horizontal permeability.

Exercises

4-1. Suppose gross thickness includes 4 feet of impermeable shale and 16 feet of permeable sandstone. What are the gross thickness, net thickness, and net-to-gross ratio?

4-2. Consider a three-layer reservoir with the following permeabilities:

Layer 1: k = 35 md
Layer 2: k = 48 md
Layer 3: k = 126 md

Calculate the arithmetic average, the harmonic average, and the Dykstra-Parsons coefficient for a log normal permeability distribution.

4-3. Consider a three-layer reservoir with the following permeabilities:

Layer 1: k = 35 md
Layer 2: k = 48 md
Layer 3: k = 126 md

Estimate the Lorenz coefficient using Eq. (4.8.6).

4-4A. Consider a linear flow system with area = 25 ft². End point A is 5 feet higher than end point B, and the distance between end points is 15 feet. Suppose the system contains oil with viscosity = 0.8 cp, gravity = 35° API (γ_o = 0.85), and FVF = 1.0 RB/STB. If the end point pressures are $P_A = P_B = 20$ psia, is there flow? If so, how much flow is there and in what direction? Use Darcy's law with the gravity term and dip angle α:

$$q = -0.001127 \frac{kA}{\mu B} \left[\frac{P_A - P_B}{L} + \rho g \sin\alpha \right]$$

4-4B. Using the data in Exercise 4-4A, calculate the pressure P_B that would prevent fluid flow.

4-5A. Suppose the pressure P_1 of a water-bearing formation at depth z_1 = 10,000 feet is 4,000 psia. If the pressure gradient for water is 0.433 psia/ft, calculate the pressure P_2 at depth z_2 = 11,000 feet. Calculate the phase potentials Φ_1 and Φ_2 at depths z_1 and z_2, respectively.

4-5B. Use the pressures and potentials in Exercise 4-5A to estimate the derivatives dP/dz and $d\Phi/dz$. Will there be vertical flow?

4-6. Estimate the volumetric flow rate using Darcy's law that is given by Eq. (4.5.1). Assume the permeability in a cylindrical core is 100 md, the length of the core is 6 inches, the diameter of the core is 3 inches, the pressure drop across the core is 10 psi, and the viscosity of liquid passing through the core is 2 cp. Express volumetric flow rate in bbl/day or cc/s.

4-7. A barrier island is a large sand body. Consider a barrier island that averages 3 miles wide, 10 miles long, and 30 feet thick. The porosity of the sand averages almost 25 percent. What is the pore volume of the barrier island? Express your answer in bbl and m³.

5 Geophysics

Different measurements provide information on different scales. Geophysical seismic measurements provide a picture of the large-scale structure of the reservoir by initiating a disturbance that propagates through the earth's crust. Part of the disturbance is reflected at boundaries between formations with different acoustic properties. The reflected signal is acquired, processed, and interpreted. This chapter defines reservoir scales, discusses the physics of waves, describes the propagation of a seismic wave through the subsurface medium, defines a set of geophysical attributes, reviews the process for acquiring and analyzing seismic data, discusses the resolution of seismic information, and describes how seismic data are used to understand geologic layering.

5.1 Reservoir Scales

Rock properties that are used to characterize a reservoir are measured on samples of a specific size and scale. The scale associated with data measurements depends on the type of measurement, the sampling technique, and the size of the sample. For example, core and well log information samples a very small part of the reservoir. A seismic section expands the volume of the subsurface environment that is sampled, but the interpretation of seismic data is less precise. Surface seismic data can cover the entire region of interest, but surface seismic data are often viewed as "soft data" because surface seismic measurements are indirect measurements. The reliability of surface seismic data can be improved when correlated with "hard data" such as core and well log measurements. Figure 5.1 illustrates the ranges.

Many experts have stressed the importance of sample size in characterizing rock properties. As early as 1961, Collins showed that variance in porosity distribution depends on sample size. Bear (1972) stressed the need to identify an appropriate averaging volume for assigning a macroscopic property to a porous medium. Haldorsen and Lake (1989) defined four conceptual scales to account for variations in the range of data applicability. Their scheme is schematically illustrated in Figure 5.2. Although scale definitions are not universally accepted, they are useful because they illustrate the relative scale associated with reservoir property measurements. Scale sizes range from the very big to the microscopic. Figure 5.2 shows the system of reservoir scales that is adopted for use here.

The measured values of rock properties are usually averages over smaller elements of the rocks that make up the sampled region. For example, core plug measurements are macro scale averages over micro scale elements, whereas pressure transient testing

Integrated Reservoir Asset Management. DOI: 10.1016/B978-0-12-382088-4.00005-0

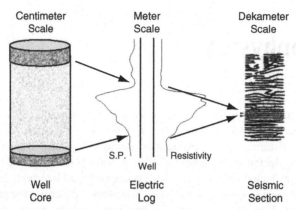

Figure 5.1 Range of data sampling techniques.
Source: Adapted from Richardson et al., 1987a.

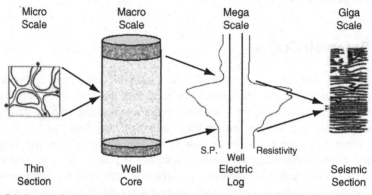

Figure 5.2 Reservoir scales.
Source: Adapted from Haldorsen and Lake, 1989.

provides a mega scale average over both macro scale elements and, on a finer scale, micro scale elements. These four scales are related to averaging volume in Table 5.1.

The giga scale includes information associated with geophysical measurement techniques. An example of a giga scale measurement is the imaging of reservoir architecture. Theories of regional characterization, such as plate tectonics, provide

Table 5.1 Reservoir Characterization Scales

Micro Scale	The size of a few pores only
Macro Scale	The size of conventional core plugs
Mega Scale	The size of grid blocks in full field flow models
Giga Scale	The total formation or regional scale

an intellectual framework within which giga scale measurement techniques, like seismic and satellite data, can be interpreted. The mega scale is comparable in size to the reservoir being characterized. Mega scale measurement techniques include well logging, well testing, and 3-D seismic analysis. Data sampling at the level of core analysis and fluid property analysis provides macro scale information. Thin section analysis and measurements of grain size distribution provide pore size information at the micro scale.

5.1.1 Reservoir Heterogeneity

The description of reservoir heterogeneity depends on how the values of the property of interest vary spatially in the reservoir and the size of the samples used to measure the rock property. Every measurement of any rock property is made on a sample of some specific size. For example, porosity may be measured using core plugs of various sizes. The value of the rock property depends on the number of measurements made, which depends on the number of samples. Thus, the accuracy of a characterization of the reservoir depends on the number of samples. In addition, the value of a rock property is a function of the averaging volume. This observation is illustrated in Figure 5.3, which shows that the value of a rock property can vary within a scale and from one scale to another. An important task of reservoir characterization and associated reservoir flow modeling is to find a suitable scale for developing flow models that provide reasonable performance predictions for use in a reservoir management study.

Rock property values have meaning only for samples of rock of some macroscopic size that contain large numbers of pores and grains. These values will not be unique because of random events in the depositional process and subsequent diagenetic history. We can expect to find a statistical distribution of values associated with measurements of any rock property within a given depositional unit. A depositional unit can be defined as homogeneous in a statistical sense if the same frequency distribution of a rock property is found throughout the unit. Given this definition of homogeneity, we define heterogeneity as the absence of homogeneity.

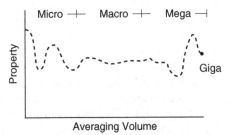

Figure 5.3 Rock property as a function of averaging volume.

5.1.2   The Geologic Model, the Shared Earth Model, the Static Model, and the Dynamic Model

Geologic models are renderings of the subsurface based on interpretations of available data. One example of a geologic model is a shared earth model. Tearpock and Brenneke define a shared earth model as "a single model of a portion of the earth that seamlessly incorporates the observations, interpretations, and data of each specialist involved in its development" (2001, p. 84). The shared earth model is developed from static geological information such as reservoir structure and dynamic information such as fluid production history. A shared earth model may change as static and dynamic data are integrated into the model.

Another term for the geologic model is the static model, which implies the dependence on information gathered at one point in time. In this terminology, the representation of fluid flow is modeled in a reservoir flow simulator, which is called the dynamic model. The dynamic model combines the static model with time-varying data such as fluid production and reservoir pressure changes. The integration of dynamic data and the static model can justify alterations of the static model in an iterative process. Modern modeling techniques recognize that multiple static and dynamic models may be needed to quantify uncertainty in reservoir management. This is discussed in more detail in Chapter 13.

5.2   Physics of Waves

Seismic waves are vibrations, or oscillating displacements from an undisturbed position, that propagate from a source, such as an explosion or mechanical vibrator, through the earth. The energy released by the disturbance propagates away from the source of the disturbance as seismic waves. The seismic waves that propagate through the earth are called body waves. Most of the energy propagates as P waves or S waves. P waves are longitudinal waves, and S waves are transverse waves. Longitudinal waves are a class of waves in which the particles of the disturbed medium are displaced in a direction that is parallel to the direction of propagation of the wave. Transverse waves are a class of waves in which the particles of the disturbed medium are displaced in a direction that is perpendicular to the direction of propagation of the wave. The velocity of P waves depends on the elastic properties and density of the medium. The velocity of S waves depends on the shear modulus and density of the medium. S waves do not travel as fast as P waves. Other names exist for P waves and S waves. P waves are sometimes called compressional, primary, or pressure waves, while S waves are sometimes called shear or secondary waves.

Seismic waves can be detected by a seismometer, which is often referred to as a geophone on land or a hydrophone in a marine environment. The output from the seismometer is transmitted to a recording station, where the signals are recorded by a seismograph. The graph representing the motion of a single seismometer, such as a geophone or hydrophone, is called the trace of the seismometer. The display of the output is called the seismogram or seismic section.

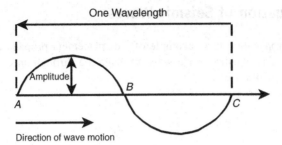

Figure 5.4 Wave motion.

A prerequisite to understanding seismic waves is a familiarity with the physics of waves. Figure 5.4 shows a single wave, where the length of the wave from one point on the wave to an equivalent point is called the wavelength. For example, the distance from point *A* to point *C* is one wavelength. The number of waves that pass a particular point—say, point *B* in Figure 5.4—in a specified time interval is the frequency of the wave.

The concept of a wave is easily illustrated. Suppose we tie a rope to the doorknob of a closed door. Hold the rope tight and move it up and then back to its original position. A pulse like the one shown in Figure 5.5a should travel toward the door. The pulse is reflected back when the pulse strikes the door as shown in Figure 5.5b. This pulse is one-half of a wave, and its length is one-half a wavelength. To make a whole wave, move the rope up, back to its original position, down, and then up to its original position in a continuous, smooth motion. The resulting pulse should look something like Figure 5.5c, which shows a complete wave. We can make many of these waves by moving the rope up and down rhythmically. The ensuing series of waves is called a wavetrain. The mathematical equation for describing the motion of the rope has the same form as the equation for describing the motion of a vibration propagating through the earth. The wave equation for seismic vibrations is discussed in the next section.

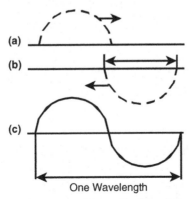

Figure 5.5 Creating a wave: (a) a pulse, (b) a reflection of the pulse, and (c) a complete wave.

5.3   Propagation of Seismic Waves

Seismic wave propagation is an example of a displacement propagating through an elastic medium. The equation for a wave propagating through an elastic, homogeneous, isotropic medium is

$$\rho \frac{\partial^2 u}{\partial t^2} = (\lambda + 2\mu)\nabla(\nabla \cdot u) - \mu\nabla \times (\nabla \times u) \tag{5.3.1}$$

where

ρ = the mass density of the medium
λ and μ = Lamé's constants
u = the displacement measurement of the medium from its undisturbed state (Tatham and McCormack, 1991)

The following derivation shows how a displacement of an isotropic medium propagates through the medium.

If the displacement u_I is irrotational, then u_I satisfies the constraint

$$\nabla \times u_I = 0 \tag{5.3.2}$$

and Eq. (5.3.1) becomes

$$\frac{\partial^2 u_I}{\partial t^2} = \frac{(\lambda + 2\mu)}{\rho}\nabla(\nabla \cdot u_I) \tag{5.3.3}$$

The vector identity

$$\nabla(\nabla \cdot u) = \nabla \times (\nabla \times u) + \nabla^2 u \tag{5.3.4}$$

for an irrotational vector is

$$\nabla(\nabla \cdot u_I) = \nabla^2 u_I \tag{5.3.5}$$

so that Eq. (5.3.1) becomes the wave equation

$$\frac{\partial^2 u_I}{\partial t^2} = v_I^2 \nabla^2 u_I \tag{5.3.6}$$

The velocity of wave propagation v_I for an irrotational displacement u_I is

$$v_I = \sqrt{\frac{(\lambda + 2\mu)}{\rho}} \tag{5.3.7}$$

A solution of Eq. (5.3.6) that is irrotational is the solution for a longitudinal wave propagating in the z-direction with amplitude u_0, frequency ω, and wave number k:

$$\boldsymbol{u}_I = u_0 e^{i(kz-\omega t)} \hat{k} \tag{5.3.8}$$

If the displacement $\boldsymbol{u}_S$ is solenoidal, then the displacement $\boldsymbol{u}_S$ satisfies the constraint

$$\nabla \cdot \boldsymbol{u}_S = 0 \tag{5.3.9}$$

and Eq. (5.3.1) becomes

$$\frac{\partial^2 \boldsymbol{u}_S}{\partial t^2} = -\frac{\mu}{\rho} \nabla \times (\nabla \times \boldsymbol{u}_S) \tag{5.3.10}$$

The vector identity in Eq. (5.3.4) reduces to

$$\nabla \times (\nabla \times \boldsymbol{u}_S) = -\nabla^2 \boldsymbol{u}_S \tag{5.3.11}$$

for a solenoidal vector. Substituting Eq. (5.3.11) into Eq. (5.3.10) yields the wave equation

$$\frac{\partial^2 \boldsymbol{u}_S}{\partial t^2} = v_S^2 \nabla^2 \boldsymbol{u}_S \tag{5.3.12}$$

The velocity of wave propagation v_S for a sinusoidal displacement $\boldsymbol{u}_S$ is

$$v_S = \sqrt{\frac{\mu}{\rho}} < v_I \tag{5.3.13}$$

A solution of Eq. (5.3.12) that is solenoidal is the solution for a transverse wave propagating in the z-direction:

$$\boldsymbol{u}_S = u_0 e^{i(kx-\omega t)} \hat{k} \tag{5.3.14}$$

Irrotational displacement $\boldsymbol{u}_I$ represents a longitudinal P (primary) wave, while the solenoidal displacement $\boldsymbol{u}_S$ represents a slower transverse S (secondary) wave. The S-wave is the shear wave, and the P-wave is the compressional wave. Other expressions for seismic velocities are presented later in this chapter.

It was noted at the beginning of the preceding derivation that Eq. (5.3.1) is based on the assumptions that the medium is elastic, homogeneous, and isotropic. In practice, these assumptions often yield useful results. Relaxing the assumptions by, for example, attempting to model an anisotropic system requires more complex analysis. The greater complexity is needed to mathematically model some systems found in nature, such as heterogeneous and fractured reservoirs. Naturally fractured reservoirs can give rise to both a fast shear wave and a slow shear wave. The additional data associated with anisotropic shear waves provide information about fracture orientation that is useful in reservoir characterization and can be used to provide more accurate images of the subsurface.

5.4 Acoustic Impedance and Reflection Coefficients

Seismic waves are displacements that propagate through the earth. The displacement can be caused by a mechanical source. These waves have many attributes, such as frequency content and transit time, but our focus here is on reflection amplitude. When the seismic wave encounters a reflecting surface, it is partially transmitted and partially reflected. A seismic reflection occurs at the interface between two regions with different acoustic impedances. Acoustic impedance Z is defined as the product of bulk density and compressional velocity:

$$Z = \rho_B V_P \qquad\qquad (5.4.1)$$

Bulk density for a rock–fluid system is

$$\rho_B = (1 - \phi)\rho_m + \phi\rho_f \qquad\qquad (5.4.2)$$

where

ϕ = porosity
ρ_m = rock matrix grain density
ρ_f = fluid density

Fluid density for an oil-water-gas system is

$$\rho_f = \rho_o S_o + \rho_w S_w + \rho_g S_g \qquad\qquad (5.4.3)$$

where

ρ_ℓ = fluid density of phase ℓ
S_ℓ = saturation of phase ℓ

Subscripts o, w, and g stand for oil, water, and gas, respectively.

The acoustic impedance Z is also called compressional impedance because of its dependence on compressional velocity. A similar definition can be made using shear velocity to define shear impedance as the product of bulk density and shear velocity:

$$Z_S = \rho_B V_S \qquad\qquad (5.4.4)$$

Figure 5.6 shows an incident wave as a bold arrow directed downward in Layer 1. The energy in the incident wave is split between the reflected wave in Layer 1

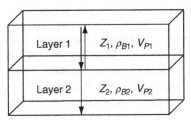

Figure 5.6 The seismic attributes for the reflection coefficient.

and the transmitted wave in Layer 2. The reflection coefficient RC at the interface between the two layers with acoustic impedances Z_1 and Z_2 in Figure 5.6 is given by

$$RC = \frac{Z_2 - Z_1}{Z_2 + Z_1} \tag{5.4.5}$$

Equation (5.4.5) assumes zero-offset—in other words, that the incident plane wave is propagating in a direction that is normal, or perpendicular, to a horizontal reflecting interface. Seismic images are indicators of acoustic contrasts because the reflection coefficient depends on the difference between acoustic impedances in two adjacent layers.

The seismic reflection coefficient RC at the interface between two formations with equal acoustic impedances is zero. Nonzero reflection coefficients occur when a wave encounters a difference between acoustic impedances. The acoustic impedance changes if there is a change in either bulk density or compressional velocity as the wave propagates from one medium to another. If the change in acoustic impedance is large enough, the reflected wave can be detected at the surface and the seismic method is said to have measured the contrasts in acoustic impedance. Similar remarks apply to the reflection of shear waves and the analysis of shear impedance.

The large density difference between an overlying water-bearing seal and an underlying gas-bearing reservoir can make it difficult to image the structure of a gas reservoir with compressional waves. The resulting impedance contrast is due to lithology changes that are relatively small compared to the change in impedance associated with fluid phase differences. One way to improve the image is to work with shear waves. An analysis of shear waves can yield information about reservoir structure for reservoirs containing gas because fluids have no resistance to shear and the shear modulus for a fluid is zero. The reflected shear waves are therefore imaging only the structure.

In some physical systems, such as naturally fractured reservoirs, the incident compressional wave splits into two reflected shear waves with different propagation velocities. An analysis of slow and fast reflected shear waves can provide information about the system that would not be available by analyzing only compressional wave information. Multicomponent seismology is an emerging technology that is designed to extract information from shear wave behavior.

5.4.1   Seismic Inversion

Seismic inversion is the process of transforming seismic measurements into quantitative estimates of rock properties that are needed for reservoir engineering calculations. Seismic inversion attempts to determine rock properties by correlating seismic attributes to rock properties. It is important to use physically meaningful correlations to guide the distribution of rock properties between wells. At the same time, seismic inversion should preserve rock properties at wells. Correlations are obtained by cross-plotting seismic attributes with groups of rock properties. Examples of cross-plots for a formation with permeability K and net thickness h_{net} include acoustic impedance versus porosity ϕ, seismic amplitude versus flow capacity (Kh_{net}) or rock quality (ϕKh_{net}), and seismic amplitude versus oil productive capacity ($S_o \phi Kh_{net}$) for a reservoir with oil saturation S_o.

One of the earliest examples of seismic inversion that included a field test of the technique was published by de Buyl et al. (1988b). They forecast rock properties using seismic inversion—that is, by finding a correlation between seismic measurements and rock properties. They then compared the forecasts with measurements that are obtained directly at wells. A similar comparison was made between measurements at wells and forecasts prepared using only well logs. The forecasts made with seismic inversion were at least as accurate as the forecasts made with well log data only and were more accurate in some cases.

5.5   Seismic Data Acquisition, Processing, and Interpretation

The primary role of geophysics in the oil and gas industry has historically been mapping structure. Structure typically refers to the controlling folds, faults, and dips of subsurface formations. It may also include stratigraphic features such as unconformities and pinchouts. Three steps are required to prepare a structural model using seismic measurements: data acquisition, data processing, and interpretation.

In reflection seismology, subsurface features are mapped by measuring the time it takes an acoustic signal to propagate from an energy source to a reflector and back to a receiver. Acoustic energy appears in the form of waves or wavetrains. Associated with each wave is a frequency. Observable seismic frequencies are typically in the range of 10 to 60 Hz.

Energy sources such as dynamite or weight-dropping equipment are used to initiate vibrations in the earth. The two most common types of vibrations are compressional (P-) waves and shear (S-) waves. P-waves are the waves that have been measured by traditional seismic methods. S-waves are insensitive to fluid properties and can be used to image structures that may be difficult or impossible to image using P-waves. For example, gas chimneys and gas caps disrupt P-wave continuity and make it difficult to observe structural features in the associated reservoir. By contrast, S-waves are virtually unaffected by fluid properties and may yield a structural image through the gas.

Receivers at the surface or in wellbores are used to detect the vibrations generated by controlled sources. The travel time from the source to the receiver is the primary information recorded by the receivers during the data acquisition phase of a seismic survey. Other information such as amplitude and signal attenuation can also be acquired and used in the next step: data processing.

Data processing is used to transform travel time images to depth images. This requires the conversion of time measurements to depths, which in turn depend on the velocity of propagation of the acoustic signal through the earth. Acoustic velocity contrast at the interface between two subsurface zones is indicated by the difference in acoustic impedance previously described. The contrast between acoustic impedances associated with two adjacent layers provides the best means of obtaining an acoustic image of the subsurface. One of the most effective methods of improving the signal-to-noise ratio of the measured travel times associated with reflections from subsurface layer interfaces is to use common depth point (CDP) data stacking.

CDP data stacking requires the superpositioning, or summing, of overlapping reflections from a number of source points at each subsurface point. The number of source points that generate overlapping signals is referred to as "fold." Thus, a 30-fold data gather uses data from 30 sources at each subsurface point. Superpositioning of vibrations from multiple sources results in constructive interference to enhance the signal and destructive interference to minimize the random noise. The result can be a significant enhancement of the signal-to-noise ratio.

Once a satisfactory measurement of travel time is obtained, the travel time must be converted to depths using the process of depth migration. The relationship between travel time and depth requires the preparation of a velocity model. The velocity model provides a distribution of velocity as a function of depth. A model of the acoustic velocity in each layer from the source to the reflector must be defined. The velocity distribution depends on the stratigraphic description of the subsurface geology.

The quality of velocity models can be improved by using vertical seismic profiles (VSP) or checkshots in wellbores. The checkshot survey is obtained by initiating a vibration at the surface and recording the seismic response in a borehole sonde. If the vibration source is vertically above the seismic receiver, the checkshot is a VSP with zero offset. If the source is not vertically above the receiver, the checkshot is a VSP with offset. A reverse VSP is obtained by placing the source in the wellbore and the receiver on the surface. The VSP and checkshot information can be compared with well logs and core samples to validate the velocity model used for the conversion of travel times to depths. Once the seismic image has been transformed from a time representation to a depth representation, it is ready for the third step in the seismic survey process: interpretation.

The subsurface structure obtained from seismic measurements is part of the observational information that is used to develop a geologic model. Dorn (1998) estimated the volume of data V to be interpreted from a 3-D seismic survey using the formula $V = L \times C \times S \times D$, where L is the number of lines, C is the number of cross-lines, S is the number of time samples, and D is the number of data types.

The number of lines and cross-lines can range from 2,000 to 3,000 each in a marine survey. The geologic model is an interpretation of the seismic data volume or seismic image as a function of depth. Optimum reservoir characterization is achieved by integrating seismic information with all other available information, such as regional geology and well information. Well information can include well logs, cores, and production or injection information.

Geologic models based on 3-D seismic surveys consist of millions of cells of information. The number of cells is related to the bin size, which depends on the number of shot points and receivers. The bin size of a 3-D seismic survey is the product of one-half the interval between receivers ΔR and the interval between shot points ΔS, so bin size $= (\Delta R/2) \times \Delta S$. Each cell is a bin and contains all of the available information, such as acoustic impedance and velocities, lithology, moduli, porosity, permeability, and net thickness. The number of cells in the geologic model depends on how many bins are needed to cover the volume of interest. Ideally, the set of cells should cover the entire subsurface volume of interest.

Computers are needed to manipulate the large volume of information in a 3-D seismic survey. Computer visualization is the most effective way to view the large volume of data. Several applications of 3-D seismic surveys to exploration and production cases are presented in Weimer and Davis (1996), and Biondi (2006) discusses 3-D seismic imaging techniques.

Most modern flow models are unable to handle the large data volume used to characterize the reservoir. Flow simulators are usually able to solve large grid systems, but the time it takes to complete the calculations using available computers is often considered prohibitive. Consequently, it is necessary to average the information in several cells to prepare a smaller set of data with larger cells. Cell size in a flow model is usually much larger than seismic bin size. The need to transform a large data set for a relatively fine grid to a smaller data set for a relatively coarse grid is referred to as the upscaling problem in reservoir flow modeling. The upscaling problem can be avoided if sufficient computer resources are available, but most practical problems require a coarsening of the geophysical data volume for use in reservoir flow models.

5.6 Seismic Resolution

The ability to distinguish two features that are very close together is the essence of resolution. This becomes important in seismic resolution when imaging a reflecting surface in either the vertical or horizontal directions. These reflecting surfaces can represent facies changes, fluid contacts, or any change in acoustic impedance that is relevant to reservoir characterization. The magnitude of seismic resolution will determine the usefulness of seismic surveys.

5.6.1 *Vertical Resolution*

The ability to distinguish two features in the vertical direction depends on interference effects and the wavelength associated with the seismic event. The dominant seismic wavelength at a reflector λ_d is the ratio of the interval velocity v_i to the dominant frequency f_d:

$$\lambda_d = v_i/f_d \qquad (5.6.1)$$

The dominant frequency is one over the dominant period of the seismic event. Thus, if the dominant period of the seismic event at a reflector is 50 milliseconds, or 0.050 seconds, the dominant frequency f_d is 20 Hz. If the local interval velocity is 14,000 feet per second, the corresponding dominant wavelength is 700 feet. An analysis of the interference of waves shows that the maximum vertical resolution δz_V is one-fourth the dominant wavelength λ_d:

$$\delta z_V = \frac{\lambda_d}{4} = \frac{v_i}{4f_d} \qquad (5.6.2)$$

An increase in the dominant frequency will cause a decrease in the dominant wavelength and a decrease in the maximum vertical resolution of the seismic measurement. In our example, this corresponds to a resolution of approximately 175 feet.

The relatively coarse vertical resolution available using surface seismic imaging has discouraged the routine application of reservoir geophysics in the reservoir management process. It is often necessary in reservoir management to identify thin features that cannot be identified by using surface seismic methods. Examples of difficult to image thin features include high-permeability intervals and fractures. Crosswell seismic imaging and VSP can be used to improve vertical resolution. A crosswell seismic image can be obtained by placing a vibration source in one well and a receiver in another well. Although crosswell seismic imaging is usually performed between vertical wells, advances in technology have made it possible to conduct crosswell seismic imaging in wells that may be deviated or horizontal. The resulting resolution for crosswell imaging is on the order of feet rather than tens of feet (Williams et al., 1997).

5.6.2 Lateral Resolution

The ability to distinguish two features in the lateral, or horizontal, direction depends on the reflection of a spherical wave from a flat surface. The maximum horizontal resolution δz_H is approximately equal to the radius r_F of the first-order Fresnel zone:

$$\delta z_H = r_F = \frac{v_i}{2} \sqrt{\frac{\Delta t}{f_d}} \qquad (5.6.3)$$

where Δt is the two-way travel time. It is instructive to show how to derive Eq. (5.6.3) as an illustration of how seismic resolution is determined.

The radius r of the first Fresnel zone is found from the Pythagorean relation

$$r^2 + z^2 = \left(z + \frac{\lambda_d}{4}\right)^2 \tag{5.6.4}$$

where

$z =$ the depth from the recording device to the reflecting interface
$\lambda_d =$ the dominant wavelength

Expanding the left-hand side of Eq. (5.6.4) and simplifying gives

$$r^2 = \frac{\lambda z}{2} + \frac{\lambda^2}{16} \tag{5.6.5}$$

Now we use $z = v_i \Delta t/2$ and $\lambda_d = v_i/f_d$ in Eq. (5.6.5) to find

$$r = \frac{v_i}{2}\left[\frac{\Delta t}{f_d} + \frac{1}{4f_d^2}\right]^{1/2} \tag{5.6.6}$$

Keeping only first-order terms in the dominant frequency gives Eq. (5.6.3).

For our vertical resolution example, if Δt is 0.6 seconds, the Fresnel radius r_F is about 1,200 feet. Thus, horizontal features, such as pinchouts, can only be resolved to approximately 1,200 feet. Improvements in horizontal resolution, corresponding to reductions in δz_H, can be made by increasing the dominant frequency.

5.6.3 Exploration versus Development Geophysics

The resolution associated with development geophysics tends to be more quantitative than the resolution associated with exploration geophysics. Wittick (2000) has pointed out that the difference in resolution is due to the role of calibration. Pennington (2001) drew a similar distinction between exploration geophysics and reservoir geophysics. Reservoir geophysics and development geophysics may be viewed as equivalent in the context of reservoir management. Seismic surveys in exploration geophysics acquire seismic data in areas where no wells have penetrated the target horizon. By contrast, well information should be available for calibrating seismic surveys in development geophysics. The lack of well–seismic ties—that is, correlations between well data and seismic data—in exploration geophysics results in seismic data that are not calibrated.

By contrast, development geophysical surveys are conducted in fields where wells have penetrated the target horizon. Development geophysics may have more seismic data than exploration geophysics, and it is possible to tie seismic lines to well information, such as well logs. The result is a set of seismic data that have been calibrated to "hard data" from the target horizons. The availability of well control data makes it possible to extract more detailed information from seismic data.

The calibration of seismic data can be used to improve the apparent resolution of the data and prepare a tool for estimating reservoir properties at interwell locations. Wittick (2000) identified two steps in the calibration process:

1. Correlate well logs, cores, and other well data with seismic attributes to determine how changes in a reservoir property alter the acoustic properties of the reservoir.
2. Use synthetic seismograms and seismic models to determine how changes in the acoustic properties of the reservoir affect seismic attributes.

When seismic data have been calibrated to measured reservoir properties, the data can be used to make reasonably accurate quantitative estimates of reservoir properties. De Buyl et al. (1988b) published one of the earliest examples of the seismic calibration process and provided a prediction of reservoir properties at two wells. Discussions of some of the first applications of reservoir geophysics can be found in Sheriff (1992). For more recent developments in reservoir geophysics, see Pennington (2007).

5.7 Stratigraphy

Finding the most likely locations for new fields is typically the domain of geologists and geophysicists. While some early fields were found by luck, collecting seismic data increased the probability of success. As mile after mile of seismic data were collected, a need arose for a model that allowed geoscientists to interpret some of the large-scale sedimentological features that were seen in the data. The following discusses two widely used interpretation techniques.

5.7.1 Seismic Stratigraphy

Structural features, like salt domes and anticlines, have been recognized from the beginning of seismic data acquisition. In the 1970s, geoscientists devised a way to interpret large packages of sediment being deposited in relation to sea level (Vail et al., 1977). The method, called seismic stratigraphy, allows a global framework of deposition to be developed by interpreting seismic lines that show how sediments were deposited in a given region. By interpreting the conditions under which a sedimentary rock formation was deposited, a geoscientist can infer the surrounding rock types and obtain a better understanding of the reservoir using seismic stratigraphy.

As an example, suppose a shoreline moves inland during periods of high sea levels, while during low sea levels the shoreline moves seaward. The inland movement of a shoreline is referred to as transgression, while the seaward movement of a shoreline is called regression. During a period of transgression, the landmass may be flooded by an encroaching body of water, such as a sea or an ocean. The resulting deposition of sediments is a transgressive sequence that often consists of sandstone on the bottom grading upward into mud and then limestone (Montgomery, 1990). By contrast, in a period of regression, the landmass rises relative to sea level, and the result is a receding shoreline with subsequent exposure of additional landmass to weathering. When the sea is inland, called a high stand, sediments are more likely to be deposited. A low stand yields conditions where erosion is taking place. The Valley Fill case study in this chapter is an application of this depositional concept.

5.7.2 Sequence Stratigraphy

A representation of the physical structure of the reservoir, the reservoir architecture, requires the specification of areal boundaries, elevations or structure tops, gross thickness, net-to-gross thickness, and, where appropriate, descriptions of faults and fractures. One of the most useful techniques for developing a model of reservoir architecture is sequence stratigraphy.

Sequence stratigraphy began in the 1960s and 1970s with the acquisition of high-quality offshore seismic data (Selley, 1998; Mulholland, 1998). Van Wagoner and colleagues (1988) formalized many of the definitions and concepts used in sequence stratigraphy. A sequence is a genetically related succession of strata that is bounded by unconformities or their correlative conformities. Following Hubbard and colleagues (1985), Selley (1998) presented the seismic sequence stratigraphic analysis in four steps:

1. Identify sequence boundaries.
2. Define sequence geometry and interpret the depositional environment.
3. Identify seismic reflector continuity and shape. Define the internal structure of the sequence.
4. Identify seismic reflector shapes and amplitude. Further refine the internal structure of the sequence.

Sequences are controlled by changes in relative sea level. Sequence boundaries are created by changes in eustatic sea level. Eustatic sea level is the global sea level relative to the center of the earth. The eustatic sea level changes when the volume of water in the oceans changes. The volume of ocean water depends on factors such as glaciation, ocean temperature, and groundwater volume. The accumulation of sediment within a sequence depends on changes in eustatic sea level, the rate of subsidence of the continental lithosphere relative to the asthenosphere due to tectonic forces (tectonic subsidence), and climatic effects.

Sequence stratigraphy is the study of rock relationships within a chronostratigraphic framework. The term "chronostratigraphy" refers to the chronological ordering of layers. A chronostratigraphic framework is defined by chronostratigraphic surfaces, where a chronostratigraphic surface is a surface with younger rocks above and older rocks below. Chronostratigraphic surfaces are useful time markers because they are correlatable over large lateral distances. Sequence boundaries are potentially correlatable chronostratigraphic surfaces.

Sequence stratigraphy may be applied at scales that are related to global changes of sea level (Selley, 1998; Mulholland, 1998; Coe, 2003). First-order eustatic cycles span at least 50 million years and apply to time-scales on the order of the fossil-bearing Phanerozoic age (see Table 3.1). Second-order eustatic cycles span from 3 to 50 million years and apply to time-scales on the order of geologic eras.

Third-order eustatic cycles span from 300,000 to 3 million years and apply to time-scales commonly associated with seismic stratigraphy.

The application of sequence stratigraphy to reservoir characterization has many advantages. Chronostratigraphic interpretation of seismic lines makes it possible for geophysicists to estimate the geologic age of seismically observed strata to at least the level of a geologic period; improve the accuracy of facies identification; identify probable source rock intervals and the location of probable reservoir facies; and develop both tectonic and sedimentation histories of new or poorly understood basins. Geologists can apply sequence stratigraphy to both clastic and carbonate systems. Sequence stratigraphy can be used to develop more accurate surfaces for mapping and correlating facies; predict reservoir, source, and sealing facies; identify stratigraphic traps; and project reservoir trends into areas with limited data.

Despite its many advantages, sequence stratigraphy is based on some fundamental assumptions that are subject to dispute (Selley, 1998; Coe, 2003). For example, one assumption of sequence stratigraphy is that the changes in sea level are global and can be correlated. Another assumption is that seismic reflectors are time horizons, even though seismic reflections occur at the interface between two different lithologies rather than changes in time. If the assumptions are not satisfied, the application of sequence stratigraphy can yield inaccurate results.

CS.5 Valley Fill Case Study: V_P/V_S Model

Traditionally, seismic data have provided information about the structure of a reservoir. The distribution of the ratio V_P/V_S of compressional to shear velocities is shown in Figure CS.5A for the top of the Valley Fill reservoir. The V_P/V_S ratio shows which parts of the area of interest are in the reservoir and which are not. The Valley Fill model in Figure CS.5A is interpreted as a meandering channel in an incised valley with regional dip. Region 1 (black bounded by white) is the region contained in the meandering channel, and Region 2 (grey) is the region outside of the meandering channel. The white area shows the boundary of the channel. The channel boundary is not sharply defined.

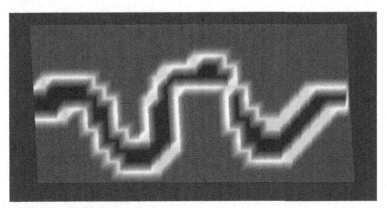

Figure CS.5A V_P/V_S distribution.

Exercises

5-1. Verify that Eqs. (5.3.8) and (5.3.14) are solutions of Eqs. (5.3.6) and (5.3.12), respectively.

5-2. What is the seismic reflection coefficient RC at the interface between two formations with equal acoustic impedances?

5-3A. Use Eq. (5.4.5) to estimate seismic reflection coefficient RC when impedance in Layer 2 is 10 percent greater than impedance in Layer 1.

5-3B. Use Eq. (5.4.5) to estimate seismic reflection coefficient RC when impedance in Layer 2 is 10 pecent less than impedance in Layer 1.

5-4. Use Eq. (5.6.2) to estimate the dominant frequency needed to achieve a vertical resolution of 10 feet for the following rock types and associated compressional velocities: tight gas sandstone (15,322 ft/s), high-porosity sandstone (12,467 ft/s), and clay (11,188 ft/s). Repeat the calculation for a vertical resolution of 100 feet.

5-5. Use Figure CS.5A to revise your structure map of the Valley Fill reservoir. Are your "dry holes" in the channel?

5-6. Use the area in Figure CS.5A and the gross thickness from Exercise 3-3 to estimate the bulk volume of the channel. Recall that the total area being studied is 6,000 ft $\times$ 3,000 ft $= 18 \times 10^6$ ft^2. *Hint:* Approximately 20 to 25 percent of the area in Figure CS.5A is channel.

5-7. Assume that the Valley Fill channel has an oil saturation of 70 percent. Use the bulk volume from Exercise 5-6, a porosity of 22 percent, and a net-to-gross ratio of 100 percent to estimate the amount of oil that is in the Valley Fill channel. Express the volume of oil in cubic feet and barrels.

5-8. Estimate bulk density for a water-saturated rock given the following data:

Parameter	Value
ρ_m (lb/cu ft)	166
ρ_w (lb/cu ft)	62.4
ϕ (frac.)	0.15

6 Petrophysics

The study of the mechanical and acoustical properties of reservoir rocks and fluids is the focus of petrophysics. Petrophysical information is valuable in well logging, time-lapse seismology, and reservoir modeling. This chapter defines petroelastic parameters, introduces elasticity theory, presents models that can be used to estimate petroelastic and geomechanical quantities, and discusses the importance of petrophysics in time-lapse seismology.

6.1   Elastic Constants

The behavior of an object when it is subjected to deforming forces is described by the theory of elasticity. Elasticity is the property of the object that causes it to resist deformation. The deforming force applied to one or more surfaces of the object is called "stress." Stress has the unit of pressure, or force per unit area, and is proportional to the force causing the deformation. The deformation of the object in response to the stress is called "strain." Strain is a dimensionless quantity that reflects the relative change in the shape of the object as a result of the applied stress. Figure 6.1 illustrates the relationship between stress and strain.

If the stress is not too great, object can return to its original shape when the stress is removed. In this case, Hooke's law states that stress is proportional to strain. The proportionality constant is called the elastic modulus, which is the ratio of stress to strain:

$$\text{Elastic modulus} = \frac{\text{stress}}{\text{strain}} \tag{6.1.1}$$

Dimensional analysis shows that the elastic modulus has the unit of pressure. The elasticity of a substance determines how effective the object is in regaining its original form.

Stresses are either one or a combination of three basic stresses: compressional stress, tensile stress, and shear stress. The corresponding strains are compressional strain, tensile strain, and shear strain, respectively. The corresponding elastic moduli are bulk modulus, Young's modulus, and shear modulus. The following discusses each of these moduli.

A compressional or volume stress is a deforming force that is applied to the entire surface of an object, as shown in Figure 6.2. A negative compressional stress is sometimes referred to as an "expansional" stress. Compressional stress is the ratio

Integrated Reservoir Asset Management. DOI: 10.1016/B978-0-12-382088-4.00006-2

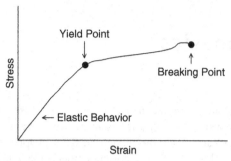

Figure 6.1 Stress–strain curve for an elastic solid.

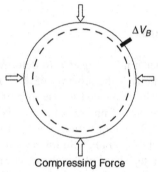

Figure 6.2 Compressional stress.

of the magnitude F of the force applied perpendicularly to the surface of the object divided by the surface area A of the object. The corresponding compressional strain is the ratio of the change in bulk volume ΔV_B of the object divided by the original bulk volume V_B of the object. The ratio of compressional stress to compressional strain is the bulk modulus K_B of the object; thus

$$K_B = -\frac{F/A}{\Delta V_B/V_B} = \frac{-\Delta P}{\Delta V_B/V_B} = -V_B\frac{\Delta P}{\Delta V_B} \tag{6.1.2}$$

where $-\Delta P$ is the pressure applied to the surface of the object. The negative sign is inserted to ensure that K_B is always a positive number.

Compressional stresses can change the volume and shape of the object. Bulk modulus measures the resistance of the object to compressional stresses. The smaller the change in volume ΔV_B for an applied stress ΔP, the larger the bulk modulus. The reciprocal of bulk modulus is bulk compressibility c_B; thus

$$c_B = -\frac{1}{V_B}\frac{\Delta V_B}{\Delta P} = \frac{1}{K_B} \tag{6.1.3}$$

Compressibility can be used to estimate the subsidence of a layered reservoir. Suppose the reservoir is divided into a sequence of permeable and impermeable layers. Assuming the area of each layer does not change, the bulk compressibility becomes $c_B \approx -\Delta h/h\Delta P$, where h is thickness and Δh is the change in thickness h. The sum of the changes in the thickness of each layer subject to pressure depletion is the subsidence of a formation.

Tensile stress is defined as the magnitude F of the force applied along an elastic rod divided by the cross-sectional area A of the rod in a direction that is perpendicular to the applied force. The corresponding tensile strain is the ratio of the change in length ΔL of the rod to the original length L. Young's modulus E is defined by the ratio

$$E = \frac{\text{tensile stress}}{\text{tensile strain}} = \frac{F/A}{\Delta L/L} \tag{6.1.4}$$

The value of Young's modulus increases as the deformation associated with a given tensile stress decreases. Young's modulus is a measure of the resistance of an object to a change in its length.

Suppose the rod is a bar. Tensile stresses can change the length of the bar by either stretching or compressing the bar, as shown in Figure 6.3. The width w also changes when the length L of the bar changes. Poisson's ratio ν is a measure of the relative magnitude of these changes. It is the ratio of the fractional change in width $\Delta w/w$ to the fractional change in length $\Delta L/L$; thus

$$\nu = \frac{\Delta w/w}{\Delta L/L} \tag{6.1.5}$$

If the rod is a cylinder instead of a bar, the width of the bar is replaced by the diameter of the cylinder in Equation (6.1.5).

Another important measure of solid deformation is shear modulus, also called rigidity modulus. As with the other moduli, it is the ratio of a stress to a strain. Shear stress is the ratio of the magnitude of the tangential force F to the area A of the face

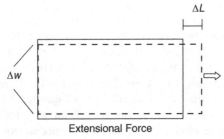

Figure 6.3 Tensile stress.

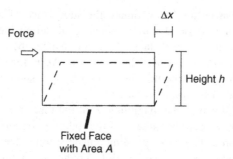

Figure 6.4 Shear stress.

of the object being sheared. Shear strain is the ratio of the horizontal distance Δx that the sheared face moves to the height h of the object. These terms are illustrated in Figure 6.4 for a simple geometric object. Notice that one face of the object is fixed so that the applied tangential force acts on the unrestrained surface of the object. In terms of these definitions, shear modulus μ is given by

$$\mu = \frac{\text{shear stress}}{\text{shear strain}} = \frac{F/A}{\Delta x/h} \tag{6.1.6}$$

Shear modulus is a measure of the resistance of an object to the movement of the plane of contact between two contiguous parts of the object in a direction parallel to the plane of contact.

Bulk modulus K_B and shear modulus μ are related to Young's modulus E and Poisson's ratio v by the expressions

$$K_B = \frac{E}{3(1 - 2v)} \tag{6.1.7}$$

and

$$\mu = \frac{E}{2 + 2v} \tag{6.1.8}$$

The values of Poisson's ratio v range from 0.05 for very hard, rigid rocks to about 0.45 for soft, poorly consolidated materials (Telford et al., 1990). The values of $\{E, K_B, \mu\}$ range from 1.4×10^6 to 28.9×10^6 psia (10 to 200 Gpa). Telford and colleagues noted that Young's modulus E usually has the largest value, while the shear modulus μ has the smallest value of the three quantities $\{E, K_B, \mu\}$. Young's modulus and Poisson's ratio can be obtained from well log or seismic survey measurements of seismic velocities. Calculation of moduli from seismic velocities is discussed next.

6.2 Elasticity Theory

An elastic body is a solid body that can be deformed by the application of an external force. The deformation of an elastic body may be described by analyzing the displacement of a particle in a general anisotropic medium from its initial position. Our discussion of elasticity will refer to mathematical objects known as tensors (e.g., Fanchi, 2006c).

A tensor in an n-dimensional space is a function whose value at a given point in space does not depend on the coordinate system used to cover the space. Examples of tensors include temperature, porosity, velocity, and permeability. Temperature and porosity are scalar tensors, or tensors of rank zero. Velocity is a tensor of rank one, and permeability is a tensor of rank two. A tensor of rank one in 3-D can be written as a 3×1 matrix or column vector with three rows and one column. A tensor of rank two in 3-D can be written as a 3×3 square matrix—that is, a matrix with three rows and three columns. The permeability tensor and the stress tensor may be written as 3×3 matrices.

In Figure 6.5, we let $\vec{r}$ denote the initial distance to a particle at point P_0 from the origin of the coordinate system and $\vec{s}$ denote the displacement of the particle to point P from P_0. Application of external forces changes the position of the particle from P_0 to P_1. The change in position P_1 relative to P_0 is expressed by the displacement

$$\vec{s} = \sum_{j=1}^{3} \hat{e}_j s_j \tag{6.2.1}$$

where

$\{\hat{e}_j\}$ = unit vectors $(\hat{i}, \hat{j}, \hat{k})$
$\{s_j\}$ = components of the displacement vector

The components $\{s_j\}$ are functions of space and time. If we write the spatial coordinates $\{x, y, z\}$ as $\{u_1, u_2, u_3\}$, the change in the displacement vector $\vec{s}$ with respect to the spatial coordinates is

$$\delta\vec{s} = \sum_{k=1}^{3}\sum_{j=1}^{3} \hat{e}_j \frac{\partial s_j}{\partial u_k} \delta u_k = \sum_{j=1}^{3} \hat{e}_j \left[\sum_{k=1}^{3} \frac{\partial s_j}{\partial u_k} \delta u_k \right] = \left(\delta\vec{u} \cdot \nabla \right) \vec{s} \tag{6.2.2}$$

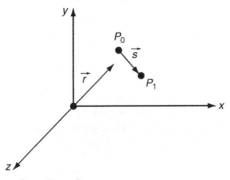

Figure 6.5 Displacement from P_0 to P_1.

The derivatives $\{\partial s_j/\partial u_k\}$ are components of a rank two tensor. We can rewrite $\vec{\delta s}$ in terms of symmetric and antisymmetric terms by writing the ith-component as

$$\delta s_i = \frac{1}{2}\left(\frac{\partial s_i}{\partial u_k} + \frac{\partial s_k}{\partial u_i}\right)\delta u_k - \frac{1}{2}\left(\frac{\partial s_k}{\partial u_i} - \frac{\partial s_i}{\partial u_k}\right)\delta u_k \qquad (6.2.3)$$

or

$$\delta s_i = \eta_{ik}\delta u_k - \xi_{ik}\delta u_k \qquad (6.2.4)$$

The antisymmetric terms $\{\xi_{ik}\}$ represent a rotation, while the symmetric terms $\{\eta_{ik}\}$ represent a distortion or strain.

Element ik of the strain tensor is

$$\eta_{ik} = \frac{1}{2}\left(\frac{\partial s_i}{\partial u_k} + \frac{\partial s_k}{\partial u_i}\right) \qquad (6.2.5)$$

The strain tensor is symmetric because $\eta_{ik} = \eta_{ki}$. The diagonal elements $i = k$ of $\{\eta_{ik}\}$ represent distortions, while the off-diagonal elements $i \neq k$ of $\{\eta_{ik}\}$ represent shear strains.

6.2.1   Hooke's Law

The strain tensor may be written as the matrix

$$\eta = \begin{bmatrix} \eta_{11} & \eta_{12} & \eta_{13} \\ \eta_{21} & \eta_{22} & \eta_{23} \\ \eta_{31} & \eta_{32} & \eta_{33} \end{bmatrix} \qquad (6.2.6)$$

Similarly, the stress tensor may be expressed as the matrix

$$\sigma = \begin{bmatrix} \sigma_{11} & \sigma_{12} & \sigma_{13} \\ \sigma_{21} & \sigma_{22} & \sigma_{23} \\ \sigma_{31} & \sigma_{32} & \sigma_{33} \end{bmatrix} \qquad (6.2.7)$$

The elements of the stress tensor have units of pressure—namely, force per unit area. Normal stresses are given by the diagonal elements $\{\sigma_{ik} \text{ for } i = k\}$, and tensional stresses are given by off-diagonal elements $\{\sigma_{ik} \text{ for } i \neq k\}$. Like the strain tensor $\{\eta_{ik}\}$, the stress tensor $\{\sigma_{ik}\}$ is symmetric—that is, $\sigma_{ik} = \sigma_{ki}$.

If the deformation is small, Hooke's law states that strain $\{\eta_{ik}\}$ is proportional to stress $\{\sigma_{ik}\}$. Hooke's law lets us express stress in terms of strain as

$$\sigma_{ij} = \sum_{k=1}^{3} \sum_{l=1}^{3} c_{ijkl} \, \eta_{kl} \qquad (6.2.8)$$

where the 81 elements of the stress tensor $\{c_{ijkl}\}$ are called elastic constants. Alternatively, we can write strain in terms of stress by using the expression

$$\eta_{kl} = \sum_{i=1}^{3} \sum_{j=1}^{3} \gamma_{ijkl} \, \sigma_{ij} \qquad (6.2.9)$$

where the 81 elements of $\{\gamma_{ijkl}\}$ are called elastic compliances.

Symmetries may be used to reduce the number of nonzero elastic constants. If the elastic body is isotropic, the elastic constants have only two independent components λ and μ. The components λ and μ are called Lamé constants. The stress–strain tensor for an isotropic, elastic body may be written in terms of the Lamé constants as

$$c_{ijkl} = \lambda \delta_{ij} \delta_{kl} + \mu \big(\delta_{ik} \delta_{jl} + \delta_{il} \delta_{jk} \big) \qquad (6.2.10)$$

where δ is the Kronecker delta function. The Kronecker delta function $\delta_{\alpha\beta} = 1$ if $\alpha = \beta$, and $\delta_{\alpha\beta} = 0$ if $\alpha \neq \beta$.

The Lamé constant μ is called the shear modulus or rigidity and relates the element of the stress tensor to the strain tensor by

$$\sigma_{ij} = \mu \eta_{ij} \quad \text{for} \quad i \neq j \qquad (6.2.11)$$

In addition to the preceding relationship, the Lamé constants λ and μ are related to Young's modulus and Poisson's ratio.

6.2.2 Effective Vertical Stress

We illustrate stress tensor concepts by calculating effective vertical stress (Badri et al., 2000). The effective stress tensor σ_{ij}^{*} for a porous medium is the difference between the total stress tensor S_{ij} and the pore pressure P; thus

$$\sigma_{ij}^{*} = S_{ij} - P\delta_{ij} \qquad (6.2.12)$$

The vertical component of $\left\{ \sigma_{ij}^{*} \right\}$ is

$$\sigma_{33}^{*} = S_{33} - P \qquad (6.2.13)$$

Pore pressure P is internal pressure, and confining stress S_{33} is external pressure. In the case of universal compaction in the vertical direction, the vertical component of total stress S_{33} at depth h relative to a reference depth, or datum, h_{ref} represents the weight of the fluids and the formation in the interval $h_{ref} \leq z \leq h$. If the density ρ of the fluid-filled medium is known as a function of depth for this interval, then S_{33} is given by the integral

$$S_{33} = g \int_{h_{ref}}^{h} \rho(z)dz \tag{6.2.14}$$

and g is the acceleration of gravity. Formation density may be obtained from a density log. If density is approximately constant $(z) \approx \rho_c$ in the interval $h_{ref} \leq z \leq h$, then

$$S_{33} = \rho_c g(h - h_{ref}) \tag{6.2.15}$$

The pore pressure may be estimated using pressure transient analysis or seismic velocities.

6.3 The Petroelastic Model

Seismic attributes can be related to one another using a petroelastic model. One example of a petroelastic model, the IFM (Integrated Flow Model) petroelastic model (Fanchi, 2009), is presented here.

6.3.1 Key Assumptions for the Petroelastic Model

The formulation of the petroelastic model is based on the key assumptions: (1) the seismic environment is elastic; and (2) temperature changes do not significantly affect shear modulus, dry frame bulk modulus, and the associated grain bulk modulus, or grain density. The dry frame bulk modulus refers to the bulk deformation of a dry, porous material when subjected to a confining pressure (Mavko et al., 1998; Batzle, 2006). The term "dry" means no fluid is present in the pore space, including gas. If gas is present in the pore space, then the material is gas-saturated.

The elasticity assumption allows us to use Hooke's law. This should be a reasonable assumption in regions where rock failure does not occur over the range of pressure and temperature that can be encountered during the life of a reservoir. The temperature assumption should be valid over the temperature range of interest—namely, original reservoir temperature to temperatures associated with recovery processes such as thermal EOR processes. Temperature effects associated with changes in fluid properties should be used in the calculation of seismic attributes when appropriate.

In addition to the preceding assumptions, we observe that reservoir conditions (e.g., pressure, temperature, saturation, and fluid flow) do not depend on the calculation of seismic attributes. Instead, seismic attributes depend on reservoir conditions. That means that the petroelastic model can be accessed after flow calculations are completed in a flow simulator using a one-way coupling scheme.

6.3.2   IFM Petroelastic Model Velocities

The petroelastic model is used to estimate seismic attributes: compressional velocity (V_P), shear velocity (V_S), and associated acoustic impedances (Z_P, Z_S):

$$V_P = \sqrt{\frac{S^*}{\rho^*}}, \quad S^* = K^* + \frac{4\mu^*}{3} \tag{6.3.1}$$

$$V_s = \sqrt{\frac{\mu^*}{\rho^*}} \tag{6.3.2}$$

$$Z_P = \rho^* V_P, \quad Z_S = \rho^* V_S \tag{6.3.3}$$

where

S^* = stiffness
K^* = bulk modulus
μ^* = shear modulus
ρ^* = bulk density

The bulk density is calculated as

$$\rho^* = (1 - \phi)\rho_m + \phi\rho_f \tag{6.3.4}$$

where

ϕ = porosity
ρ_m = rock matrix grain density
ρ_f = fluid density

Fluid density is given by

$$\rho_f = \rho_o S_o + \rho_w S_w + \rho_g S_g \tag{6.3.5}$$

where

ρ_ℓ = fluid density of phase ℓ
S_ℓ = saturation of phase ℓ

Subscripts o, w, and g stand for oil, water, and gas, respectively.

Measurements of compressional and shear velocities may be reported in terms of the ratio V_P/V_S. The advantage of working with the V_P/V_S ratio is the elimination of bulk density, which may be a poorly known quantity. This can be seen by taking the ratio of Eqs. (6.3.1) and (6.3.2) to find

$$\frac{V_P}{V_S} = \sqrt{\frac{K^* + \frac{4\mu^*}{3}}{\mu^*}} = \sqrt{\frac{4}{3} + \frac{K^*}{\mu^*}} \tag{6.3.6}$$

Noting that moduli K^* and μ^* are greater than zero, we find that $V_P/V_S \geq \sqrt{4/3}$ from Eq. (6.3.6).

When velocities, particularly shear velocities, are not available, estimates can be made from general correlations. An example of a correlation between acoustic velocities in sandstone, porosity, and clay content C is given by Castagna and colleagues (1985):

$$V_P = 5.81 - 9.42\phi - 2.21C \tag{6.3.7}$$

and

$$V_S = 3.89 - 7.07\phi - 2.04C \tag{6.3.8}$$

The units are km/s for acoustic velocities, while clay volume fraction C and porosity ϕ are fractions. Notice the linear dependence of acoustic velocity on porosity.

6.3.3 IFM Petroelastic Model Moduli

Saturated bulk modulus K^* is modeled as a form of Gassmann's equation (Gassmann, 1951)—namely,

$$K^* = K_{IFM} + \frac{\left[1 - \frac{K_{IFM}}{K_m}\right]^2}{\frac{\phi}{K_f} + \frac{1 - \phi}{K_m} - \frac{K_{IFM}}{K_m^2}} \tag{6.3.9}$$

where

K_{IFM} = dry frame bulk modulus
K_m = rock matrix grain bulk modulus
K_f = fluid bulk modulus

Gassmann's equation represents perfect coupling between the pore fluid and the solid skeleton of the porous medium (Schön, 1996). This relation is strictly valid only for isotropic, homogeneous rock composed of a single mineral. Gassmann's equation is widely used in rock physics because of its relative simplicity and limited data requirement. Equation (6.3.9) has the form of Gassmann's equation, but the parameters in Eq. (6.3.9), including the moduli, are allowed to change as reservoir conditions change.

Fluid bulk modulus is the inverse of fluid compressibility C_f:

$$K_f = 1/C_f \qquad (6.3.10)$$

Fluid compressibility is the saturation weighted average of phase compressibility— that is,

$$C_f = c_o S_o + c_w S_w + c_g S_g \qquad (6.3.11)$$

The moduli K_{IFM} and K_m are modeled using the following equations:

$$K_{IFM} = A_0 + A_1 P_e^{e_1} + A_2 \phi + A_3 \phi^2 + A_4 \phi P_e^{e_2} + A_5 \sqrt{C} \qquad (6.3.12)$$

and

$$K_m = A_0 + A_1 P_e^{e_1} + A_5 \sqrt{C} \qquad (6.3.13)$$

where

$A_0, A_1, A_2, A_3, A_4, A_5, e_1,$ and e_2 = model coefficients
P_e = effective pressure
C = clay content volume fraction

The shear modulus is calculated using

$$\mu^* = B_0 + B_1 P_e^{\varepsilon_1} + B_2 \phi + B_3 \phi^2 + B_4 \phi P_e^{\varepsilon_2} + B_5 \sqrt{C} \qquad (6.3.14)$$

where $B_0, B_1, B_2, B_3, B_4, B_5, \varepsilon_1,$ and ε_2 are model coefficients.

The coefficients in Eqs. (6.3.12) and (6.3.14) are determined by fitting the equations to measurements of moduli. If moduli are linear functions of porosity with no dependence on effective pressure or clay content, the coefficients A_0, B_0 are positive; A_1, B_1 are negative; and the remaining coefficients are zero. In general, the value of each model coefficient depends on the set of data being fit by the model. For example, Murphy and colleagues (1993) present the bulk and the shear moduli for quartz sandstone that have a linear dependence on porosity, and Castagna and colleagues (1985) present bulk and shear moduli for clastic silicate rocks that have a quadratic dependence on porosity. In these cases, many of the coefficients are set to zero. A more complex example that depends on porosity, effective pressure, and clay content is the set of data presented by Han (1986) for well-consolidated Gulf Coast sandstones. The match of Han's data by Eberhart-Phillips (1989) has been converted to the petroelastic model format described here (Fanchi, 2003).

Equations (6.3.1) and (6.3.2) can be rearranged to express moduli in terms of measured acoustic velocities; thus bulk modulus of rock saturated with pore fluid is

$$K^* = \rho_B \left[V_P^2 - \frac{4}{3} V_S^2 \right] \tag{6.3.15}$$

and effective shear modulus is

$$\mu^* = \rho_B V_S^2 \tag{6.3.16}$$

Measurements of shear velocity and compressional velocity at zero porosity yield an effective bulk modulus that equals grain modulus K_m, thus

$$K_m = K_{dry} = K_{sat} = K^* \text{ at } \phi = 0 \tag{6.3.17}$$

This result can be derived using Gassmann's equation and recognizing that bulk modulus equals grain modulus as porosity goes to zero.

Dry frame bulk modulus can be calculated from measured acoustic velocities, fluid properties, matrix density, and grain modulus using the equation

$$K_{dry} = \frac{K_{sat} \left(\dfrac{\phi}{K_f} + \dfrac{1-\phi}{K_m} \right) - 1}{\dfrac{\phi}{K_f} + \dfrac{1-\phi}{K_m} - \dfrac{2}{K_m} + \dfrac{K_{sat}}{K_m^2}} \tag{6.3.18}$$

where we write saturated bulk modulus as $K_{sat} = K^*$ and dry frame bulk modulus as $K_{dry} = K_{IFM}$ to emphasize their physical content. Equation (6.3.18) was derived by solving Gassmann's equation for dry frame bulk modulus and calculating saturated bulk modulus using Eq. (6.3.9).

6.3.4 IFM Petroelastic Model Effective Pressure

Effective pressure P_e for the petroelastic model is calculated as

$$P_e = P_{con} - \alpha P \tag{6.3.19}$$

where

    P_{con} = confining pressure
    α = Biot coefficient correction factor
    P = pore pressure

The Biot coefficient is the ratio of the change in pore volume to the change in bulk volume of a porous material in a dry or drained state (Mavko et al., 1998). The choice of Biot coefficient α depends on rock characteristics and the measurement procedure. For more discussion of the Biot coefficient, see Mavko et al. (1998) or Tiab and Donaldson (1996). The Geertsma-Skempton correction is a reasonable choice for the Biot coefficient unless there is reason to select one of the other following options:

Option 1: No correction

$$\alpha = 1 \tag{6.3.20}$$

Option 2: Geertsma-Skempton correction

$$\alpha = 1 - \left(\frac{K_{IFM}}{K_m}\right) \tag{6.3.21}$$

Option 3: Suklje's modification to the Geertsma-Skempton correction

$$\alpha = 1 - (1 - \phi)\left(\frac{K_{IFM}}{K_m}\right) \tag{6.3.22}$$

Option 4: Modified Geertsma-Skempton correction using effective bulk modulus

$$\alpha = 1 - (1 - \phi)\left(\frac{K^*}{K_m}\right) \tag{6.3.23}$$

Confining pressure P_{con} may be estimated from an average overburden gradient, γ_{OB}, as

$$P_{con} = \gamma_{OB} z \tag{6.3.24}$$

where z is depth.

6.3.5 IFM Petroelastic Model Grain Density

Rock matrix grain density, ρ_m, is calculated as a quadratic function of clay content volume fraction C; thus

$$\rho_m = \kappa_0 + \kappa_1 C + \kappa_2 C^2 \tag{6.3.25}$$

where κ_0, κ_1, κ_2 are model coefficients. If there is no known functional dependence on C, then set $\rho_m = \kappa_0, \kappa_1 = \kappa_2 = 0$.

6.4 The Geomechanical Model

A geomechanical model can be constructed using the petroelastic model presented previously to estimate several geomechanical parameters: Poisson's ratio (ν), Young's modulus (E), Uniaxial compaction (Δh), and Horizontal stress (σ_H). Dynamic Poisson's ratio (ν_d) is calculated from compressional (P-wave) velocity and shear (S-wave) velocity:

$$\nu_d = \frac{0.5 V_P^2 - V_S^2}{V_P^2 - V_S^2} \tag{6.4.1}$$

Static Poisson's ratio (v_s) is calculated from dynamic Poisson's ratio using the algorithm based on results in Wang (2000):

$$v_s = av_d^b + c \qquad (6.4.2)$$

where a, b and c are dimensionless coefficients. Coefficients a, b are linear functions of $\log(P_e)$ where P_e is effective pressure; thus

$$a = a_1 + a_2 \log(P_e) \qquad (6.4.3)$$

and

$$b = b_1 + b_2 \log(P_e) \qquad (6.4.4)$$

If there is no known functional dependence on P_e, static Poisson's ratio can be set equal to dynamic Poisson's ratio by specifying $a = 1, b = 1, c = 0$.

Dynamic Young's modulus (E_d) is calculated from shear modulus in Eq. (6.3.11) and dynamic Poisson's ratio in Eq. (6.4.1):

$$E_d = 2(1 + v_d)\mu^* \qquad (6.4.5)$$

Young's modulus has the same unit as shear modulus. Static Young's modulus (E_s) is calculated from dynamic Young's modulus using the algorithm based on results in Wang (2000):

$$E_s = d'E_d^{b'} + c' \qquad (6.4.6)$$

where a', b' are dimensionless coefficients, and c' has the same unit as shear modulus. Coefficients a', b' are linear functions of $\log(P_e)$; thus

$$a' = a_1' + a_2' \log(P_e) \qquad (6.4.7)$$

and

$$b' = b_1' + b_2' \log(P_e) \qquad (6.4.8)$$

If there is no known functional dependence on P_e, static Young's modulus can be set equal to dynamic Young's modulus by specifying $a' = 1, b' = 1, c' = 0$.

Conventional geomechanical models include the compacting reservoir deformation effects shown in the upper half of Figure 6.6—namely, surface extension, compression, and reservoir compaction. The compaction model sketched in part (b) of

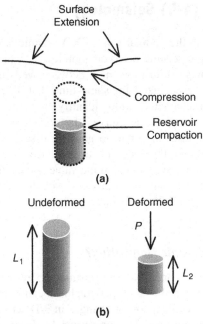

Figure 6.6 Schematic of reservoir compaction features: (a) compacting reservoir deformation and (b) approximate as uniaxial compaction.
Source: Adapted from Fanchi, 2006a.

Figure 6.6 approximates all of these effects as uniaxial compaction of the reservoir, which is estimated from static Poisson's ratio as

$$\Delta h = \frac{1}{3}\left[\frac{1+\nu_s}{1-\nu_s}\right]\phi\,c_\phi h_{net}(P-P_{init}) \qquad (6.4.9)$$

where

ϕ = porosity
c_ϕ = porosity compressibility
h_{net} = net thickness
P_{init} = initial pore pressure
P = pore pressure

Horizontal stress is estimated from pore pressure P, confining pressure P_{con}, static Poisson's ratio ν_s, and Biot coefficient α as

$$\sigma_H = \frac{\nu_s}{1-\nu_s}(P_{con}-\alpha P)+\alpha P \qquad (6.4.10)$$

In this model, vertical stress can be approximated as confining pressure P_{con}.

6.5 Time-Lapse (4-D) Seismology

Time-lapse seismology is the comparison of 3-D seismic surveys at two or more points in time. Time-lapse seismic reservoir monitoring, which is often called 4-D seismic, allows us to image fluid movement between wells in many systems. The oil and gas industry has recognized for some time that 4-D seismic can improve the quality of reservoir characterization, identify movement of fluid interfaces, and help operators locate bypassed reserves. Improved reservoir characterization and fluid monitoring are also important to geological sequestration of greenhouse gases. The primary purpose of this section is to introduce the concept of 4-D seismic monitoring. Several authors have discussed the value and challenges of 4-D seismic monitoring in the reservoir management process, including Calvert (2005), Lumley (2001, 2004), Fanchi et al. (1999), and Jack (1998).

6.5.1 What Is 4-D Seismic Monitoring?

Time-lapse seismology compares one 3-D seismic survey with one or more repeat 3-D seismic surveys taken in the same geographic location at different times. 4-D seismic monitoring is the comparison of changes in 3-D seismic surveys as a function of the fourth dimension: time. By comparing the differences in measurements of properties, such as travel times, reflection amplitudes, and seismic velocities, changes in the elasticity of the subsurface can be monitored over time.

There are two principal elastic parameters that affect seismic waves: the bulk modulus and the shear (or rigidity) modulus. The bulk modulus is related to rock and fluid compressibility. Shear (S-) waves are affected by bulk density and shear modulus, while compressional (P-) waves are affected by bulk density and both bulk and shear moduli. The combination of bulk and shear moduli used in the calculation of P-wave velocities is called stiffness and is a measure of the rock frame stiffness and pore fluid stiffness. Stiffness is quantitatively defined in Eq. (6.3.1).

Reservoir elasticity is affected by lithology, fluid content, and variations in pore pressure. Seismic velocity (V), attenuation (Q), and reflectivity measurements contain information on the fluid distribution in the reservoir. For example, the ratio of compressional velocity to shear velocity (V_P/V_S) is dependent on bulk modulus, shear modulus, and the density of the rock. Bulk modulus and density are functions of porosity and fluid content in the pore space. Seismic monitoring of changes in reservoir elasticity can be linked to properties associated with the movement of fluids in the reservoir. This link yields information that can be used to improve the validity of fluid flow models and the reservoir management decisions that rely on flow model forecasts.

Technological innovations have made it possible to probe the elastic properties of a reservoir by recording P- and S-waves as they pass through the reservoir. The ability to monitor both P- and S-waves is referred to as multicomponent imaging. Time-lapse, multicomponent seismology is a tool for sensing changes to bulk rock properties and fluid properties as a function of time. An example of a time-lapse, multicomponent survey is the 4-D, 3-C seismic survey. One vertical and two horizontal velocity components are recorded during this survey. The recording procedure is similar to that of earthquake seismologists and facilitates the combined recording of P- and S-waves.

Seismic anisotropy is a measure of the fine-scale structure in the reservoir. An anisotropic reservoir exhibits differences in properties as a function of spatial orientation. Horizontal permeability, for example, will be greater in one direction than in another. In an anisotropic medium, a shear wave splits into two orthogonally polarized components (S1 and S2). The S1 wave is faster, and its velocity and attenuation are affected by lithology, porosity, and pore saturants. By contrast, the S2 wave is slower, and its velocity and attenuation are affected by reservoir features such as fractures. The different dependencies of the S1 and S2 waves provide information for determining dynamic reservoir properties such as permeability, porosity, and fluid saturations. Multicomponent seismic studies have been especially useful in the characterization of reservoir rock properties, including lithology, porosity, and fractures.

6.5.2 Forward Modeling

Forward modeling is the calculation of seismic attributes from fundamental reservoir properties. These attributes are needed to understand changes in seismic variables as a function of time, which is the basis for time-lapse (4-D) seismic analysis. Rock properties include rock frame elastic properties such as bulk modulus and shear modulus. The moduli are used to calculate elastic stiffness. Other rock properties include porosity, shale volume, and grain density. Rock properties can be obtained from well logs, laboratory measurements of core properties, and correlations.

Distributions of pressure (P), temperature (T), fluid composition (Z), and saturation (S) can be obtained from flow simulators. The P-T-Z-S distributions are related to pore fluid properties such as phase densities and compressibilities. Seismic attributes are calculated from rock properties, pore fluid properties, and P-T-Z-S distributions in the rock physics model. The rock physics model is an algorithm for calculating seismic attributes such as compressional velocity, shear velocity, acoustic impedance, and the reflection coefficient. The petroelastic model just discussed is an example of a rock physics model.

One procedure for calculating seismic attributes in use today is a sequential procedure that retains the separation between flow model results and forward modeling results. The results of the flow model are exported to a file that can be imported by the forward modeling program. Sequential workflow steps can be performed as a sequence of independent workflow actions using separate software packages, or they can be performed in a software package that is designed to simplify the data transfer process. An example of the latter process is to use a software package to control the execution of the flow simulator and the forward modeling program. Separation of flow and petroelastic calculations in a sequential process is relatively inefficient, and care must be exercised to avoid the use of inconsistent data sets.

An alternative to the forward modeling process is the integrated flow model. An integrated flow model is a flow simulator that includes a petroelastic model. A prototype integrated flow model (IFLO) is used as the simulator in this book.

6.5.3 Guidelines for Applying 4-D Seismic Monitoring

Wang (1997) identified several criteria that can be used to identify good candidate reservoirs for time-lapse seismic reservoir monitoring. One criterion is the magnitude of the bulk and the shear moduli. The moduli characterize the elastic frame of a reservoir. Reservoirs with relatively small moduli have weak elastic frames and are good candidates for time-lapse seismic monitoring. Examples include reservoirs with unconsolidated sands or fractured reservoirs. Wang also observed that reservoirs subject to large compressibility changes in either the rock or the pore fluids can exhibit a significant seismic response over time. Some examples include reservoirs in which the gas phase is either appearing or disappearing.

Integrated flow model studies have found that 4-D seismic should be most effective in regions where the gas phase is appearing or disappearing. This observation is in agreement with Wang's criteria and has been substantiated in the field. For example, Kelamis and colleagues (1997) noted the importance of similar gas saturation behavior in their study of a clastic oil reservoir with a large gas cap in the Arabian Gulf. The contrast between fluid compressibilities is greatest when one survey images only liquid, while a subsequent survey images both liquid and gas.

Very small amounts of gas—as low as one saturation percent—are sufficient to change the total compressibility of a reservoir system by an order of magnitude or more. The change in fluid compressibility with time can generate observable differences in seismic response when comparing one survey with another. The differences in seismic response appear in measurements of attributes like compressional and shear velocities, and dynamic Poisson's ratio.

The necessity to compare two surveys that have been conducted at different points in time over the same region leads to the issue of survey repeatability. The signal in a 4-D seismic survey is the magnitude of the change in acoustic response of the reservoir between two surveys taken over the same region at different times. Detection of the signal requires that the differences between time-lapse surveys should be due to actual reservoir changes and not to differences in data acquisition, processing, or interpretation.

Fluid phase behavior, such as the appearance or disappearance of free gas saturation during the life of the field, can be used to schedule sequential 3-D seismic surveys. From a reservoir management perspective, the scheduling of time-lapse seismic surveys should optimize the acquisition of reservoir engineering information.

A common factor in most screening criteria is the identification of significant changes to properties that directly influence seismic response. For example, if a reservoir is subjected to large pressure or temperature changes, the resulting change in petroelastic properties can lead to an observable change in seismic response. Case studies substantiate the screening criteria outlined here. They demonstrate that a 4-D seismic survey can be an effective tool for reservoir management. It is important to keep in mind, however, Lumley and Behrens's warning that "this new technology is not a panacea but rather an exciting emerging technology that requires very careful analysis to be useful" (1997, p. 998)

CS.6 Valley Fill Case Study: Bulk Moduli

The area of interest in the Valley Fill case study is the area containing the productive wells shown in Figure CS.1A. This area has an average effective porosity of 22 percent and an average permeability of 150 md. The core from the productive interval was water-wet sandstone.

No measurements of moduli were available, so correlations published by Murphy and colleagues (1993) for quartz sandstone and shown in Figure CS.6A were used to determine grain, bulk, and shear moduli for sandstone with a porosity of 22 percent. These moduli were then used to compute shear and compressional velocities. The bulk frame modulus is 16.2 GPa (2.35 × 10^6 psia); the grain modulus is 36.0 GPa (6.22 × 10^6 psia); the shear modulus is 18.5 GPa (2.68 × 10^6 psia); and the grain density is 2.67 g/cc (167 lbf/cu ft). The unit "lbf" refers to pounds of force. These petroelastic properties were applied throughout the area of interest.

Initial water saturation in cores from the productive channel ranged from 30 to 100 percent. Cores from wells outside of the productive channel were water saturated.

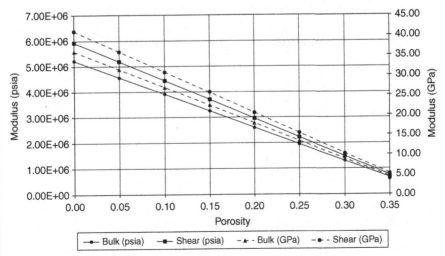

Figure CS.6A Moduli.
Source: Adapted from Murphy et al., 1993.

Exercises

6-1. Assume the compressional velocity in Eq. (6.3.1) is equivalent to the wave velocity in Eq. (5.2.7). Furthermore, assume $\rho^* = \rho$ and $\mu^* = \mu$. Show that $K^* = (3\lambda + 2\mu)/3$.

6-2. Use Figure CS.6A to estimate the bulk modulus and shear modulus of sand with 22 percent porosity. What is the grain modulus?

6-3. Calculate Young's modulus and Poisson's ratio using the bulk and shear moduli in Exercise 6-2 and Eqs. (6.1.7) and (6.1.8).

6-4. Suppose $\rho(z) = a + bz^n$ where a, b, n are constants. Calculate σ_{33}^* using Eq. (6.2.14).

6-5. Derive Eq. (6.3.18) from Gassmann's equation for bulk modulus in Eq. (6.3.9).

6-6. Derive Eq. (6.4.1) from Eqs. (6.3.15), (6.3.16), (6.1.7), and (6.1.8).

6-7. Effective bulk modulus K^* can be written in terms of porosity ϕ, dry rock bulk modulus K_B, grain modulus K_G, and fluid modulus K_F as

$$K^* = K_B + \frac{[1-b]^2}{\dfrac{\phi}{K_F} + \dfrac{1-\phi-b}{K_G}}, \quad b = \frac{K_B}{K_G}$$

What is the effective bulk modulus K^* if $K_B = K_G$?

6-8A. Complete the following table. *Hint:* First calculate

$$b = \frac{K_B}{K_G} = \frac{1 - \dfrac{K^*}{K_F}\phi - \dfrac{K^*}{K_G}(1-\phi)}{2 - \left[\dfrac{K^*}{K_G} + (1-\phi) + \dfrac{K_G}{K_F}\phi\right]}$$

and then use $K_B = bK_G$ to calculate bulk modulus.

Parameter	Value
V_P (ft/s)	18,736
V_S (ft/s)	10,036
V_P/V_S	
K^* (psi)	7.30×10^6
K_G (psi)	9.70×10^6
K_B (psi)	
K_F (psi, brine)	2.97×10^5
ϕ (frac.)	0.15

6-8B. Estimate bulk compressibility and fluid compressibility from the data in the previous table. *Hint:* Compressibility is the reciprocal of modulus.

7 Well Logging

Well logs are often acquired after a well has been drilled. They provide valuable information about the formation within a few feet of the wellbore. The properties of the formation are inferred from the well log response. This chapter describes the different types of logs that are available, explains why combinations of logs are used, and discusses techniques and limitations of well log interpretation. Well logging technology continues to improve, so this discussion focuses on the more common logs. A thorough review of well logs is beyond the scope of this book, and the interested reader should consult Bassiouni (1994), Selley (1998), several articles in Holstein (2007), or contact appropriate service companies for details of specific logs.

7.1 Drilling and Well Logging

The giga scale is comparable in size to the reservoir but is too coarse to provide the detail needed to design a reservoir management plan. The mega scale is the scale at which we begin to integrate well log and well test data into a working model of the reservoir. Several types of logs can contribute to the mega scale level of reservoir characterization. Many of these logs are discussed in this chapter in the context of formation evaluation after a brief discussion of drilling methods. Modern drilling techniques are reviewed in several articles in Mitchell (2007).

7.1.1 Drilling Methods

As noted in Chapter 1, the first method of drilling for oil in the modern era was cable-tool drilling. This method was replaced by rotary drilling in the mid-1800s. Rotary drilling uses a rotating drill bit with nozzles for shooting drilling mud into the wellbore. Drilling mud is a combination of clay and water. The rotary drill bit allows the circulation of drilling mud through the drill pipe and back up the annulus between the tubing and wellbore walls. It is used to cool the drill bit, counter formation pressure, and carry drill cuttings up the wellbore to the surface for removal from the system. A modern drilling rig is shown in Figure 7.1. The rotary table is set in the drill floor.

Once a hole has been drilled, it is necessary to complete the well. A modern wellbore is shown in Figure 7.2. Steel casing and cement are typically used to isolate shallow formations from fluid in the wellbore. Cement is injected through the drill pipe and into the annulus between the drill pipe and the borehole wall. Cement is used to form a barrier between the borehole and adjacent formations. Leaks in the

Integrated Reservoir Asset Management. DOI: 10.1016/B978-0-12-382088-4.00007-4

Figure 7.1 A drilling rig.

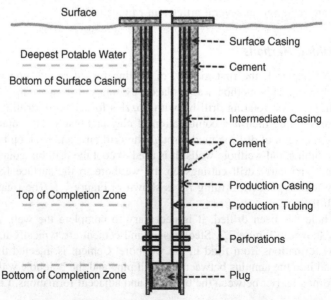

Figure 7.2 A wellbore diagram for a vertical well.

cement can result in fluid movement between the productive interval and other formations. This can affect production and injection of fluids and make it more difficult to understand the petroleum system. Cement bond logs can be used to check the integrity of the cement.

Wellbores may be completed using a variety of techniques. A common completion technique is to place tubing adjacent to the productive interval and then use a perforation gun to shoot perforations through the tubing walls, as illustrated in Figure 7.2. Reservoir fluid can flow into the tubing through the perforation holes. A plug is set at the bottom of the completion zone shown in Figure 7.2 to prevent fluid from flowing into the borehole below the completion zone. In open-hole completions, tubing is set above the productive interval and the fluid is allowed to fill the open wellbore until it can flow into the tubing and up the wellbore. A wire screen can be placed in the wellbore adjacent to the productive interval in reservoirs where rock fragments or grains can move into the wellbore. This can happen, for example, when an unconsolidated formation is being produced.

Figure 7.2 shows a vertical well. Completion techniques are modified when the wellbore is deviated or horizontal. The type of drilling mud and the type of completion can influence the selection of a well log or suite of well logs that will provide optimum information about the productive interval. Additional details can be found in the references.

7.1.2 Well Logging Principles

Formation evaluation is the acquisition of data and quantification of parameters needed for drilling, production, reservoir characterization, and reservoir engineering. The data may be determined by direct measurement or by remote sensing. Data from direct measurements include drill cuttings, core samples, fluid properties, and production testing. Data obtained by remote sensing range from electromagnetic and sonic wave signals to the detection of elementary particles and the monitoring of pressure changes in the reservoir. Many well logs use remote sensing techniques.

Well logs are obtained by running a tool into the wellbore. The tool can detect physical properties such as temperature, diameter of the wellbore, electrical current, radioactivity, or sonic reflections. Logging tools are designed to function best in certain types of environments. The environment depends on a variety of factors, including temperature, lithology, and fluid content. The theoretical analysis of log signals is often based on the assumption that the formation is infinite in extent with homogeneous and isotropic properties. Tool performance will not be optimal in other environments.

The measurement environment is usually represented by an idealized representation that includes several zones. Figure 7.3 illustrates the distribution of drilling fluids in the vicinity of a newly drilled wellbore. Four zones are identified near the wellbore: the mud cake, the flushed zone, the zone of transition, and the uninvaded zone. The pressure difference between the borehole and the formation

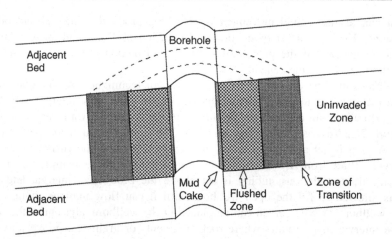

Figure 7.3 Schematic of invasion zones.

can force drilling mud into the permeable formation. Large particles in the drilling mud will be filtered out at the sandface and create a mud cake. The remaining filtered liquid is called the mud filtrate. The pressure differential forces the mud filtrate into the formation. The mud cake can reduce the productivity or the injectivity of a well. A measure of the degree of wellbore damage caused by the mud cake is the "skin" of the well.

Pore space in the flushed zone has been completely swept by mud filtrate from the drilling operation. Beyond the flushed zone is the zone of transition between the flushed zone and the uninvaded zone. The volume of pore space that is swept by the mud filtrate ranges from completely swept at the interface between the flushed zone and the zone of transition to completely unswept at the interface between the zone of transition and the uninvaded zone. The flushed zone and the zone of transition are considered invaded zones because original *in situ* fluids have been displaced by fluids from the borehole. The uninvaded zone is the part of the formation that has not been altered from its original state by the drilling operation. Reservoir rock and fluid properties in the uninvaded zone are needed for accurate reservoir characterization and are the objective of near wellbore remote sensing tools.

7.1.3 Depth of Investigation

The depth of investigation is a measure of the volume of the formation that is primarily responsible for the well log signal. If we assume that the formation has a uniform cylindrical shape for a formation with thickness h and porosity ϕ, then the volume investigated is $\phi r^2 h$, where the radius r is the depth of investigation into the formation. The depth of investigation of a well log is usually characterized as shallow, medium, or deep. Depth of investigation can range from a few inches to several feet.

7.2   Direct Measurement Logs

Several logs, as described in the following subsections, provide direct information about the borehole environment.

7.2.1   The Driller's Log and Mud Logs

The driller's log is a record of observations made by the driller during the course of drilling the borehole. The driller's log includes such information as rate of penetration of the drill bit, rock and fluid types encountered as a function of borehole depth, and any other information that is considered worth recording by the driller. The rate of penetration can be used to distinguish between hard and soft formations. In addition to the driller's log, mud logs are recorded while the well is being drilled.

Drilling mud is circulated in the wellbore to remove drill cuttings and control pressure. Mud logs provide measurements of parameters such as the rate of penetration, the detection of natural gas, and the types of cuttings that are encountered at various depths along the wellbore. Combining this information with rock type observations and correlations with other logs provides information that can be used to identify formations.

7.2.2   Caliper Logs

Ideally, the borehole shape will be a uniform cylinder with the diameter of the drill bit used to drill the borehole. In practice, the diameter and expected cylindrical shape of the borehole may differ substantially from a uniform cylinder with the diameter of the drill bit. The actual borehole shape depends on the type of formation, and it can be measured with a caliper log.

7.2.3   Sample Logs

Sample logs are collections of physical samples of cores and cuttings that are extracted from the borehole. "Cuttings" are bits of rock that have been produced by the action of the drill bit during the drilling operation. Cuttings are returned to the surface by the circulation of drilling mud. Once at the surface, they can be analyzed to indicate the lithology of the drill bit environment. Cuttings are not always representative of the current position of the drill bit because bits of rock from shallower formations may be broken off by the circulating mud and brought to the surface with the cuttings. Nevertheless, cuttings are the first solid evidence of lithology obtained at the well site.

Core samples are obtained by a coring procedure in which the drill bit is replaced by a coring bit. Since coring requires a separate bit, the amount of core depends on the cost and relative benefit of acquiring core from a particular borehole. Not all wells will be cored. Those wells that are cored are usually cored over a limited depth interval, such as the depth interval associated with the productive formation. Cores can be used for a variety of reservoir characterization purposes because they provide a small sample of the reservoir. Analyses of core samples should recognize that some characteristics of the core may change during the extraction process because of changes in temperature, pressure, and borehole fluid invasion. Native state coring may be used to obtain core samples at conditions that are close to original *in situ* conditions, but the added expense limits the applicability of native state coring to special situations.

7.3 Lithology Logs

Lithology logs are used to determine rock type. Most reservoirs are found in sedimentary rocks and are either clastics or carbonates, as shown in Figure 7.4. Clastics include sandstone, conglomerate, shaly sand, and shale. Carbonates include limestone, dolomite, and evaporite. Next, we consider two types of lithology logs: the SP log and the gamma ray log.

7.3.1 Spontaneous Potential Logs

The spontaneous potential (SP) log is an electric log that provides information about permeable beds. It records the direct current (DC) voltage difference, or electrical potential, between two electrodes as a function of depth. One of the electrodes has a fixed position, and the other electrode is allowed to move. The spontaneous potential E_{SP} can be estimated from the empirical relationship (Selley, 1998):

$$E_{SP} = \kappa \log \frac{R_{mf}}{R_w} = (a_1 + a_2 T) \log \frac{R_{mf}}{R_w} \qquad (7.3.1)$$

where

κ = a linear function of temperature T with parameters a_1 and a_2
R_{mf} = the resistivity of the mud filtrate
R_w = the resistivity of formation water

The temperature dependence of resistivity must be taken into consideration when using Eq. (7.3.1) for E_{SP}.

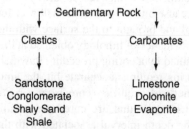

Figure 7.4 Common reservoir rock types.

The electrical potential is associated with the presence of ions in the vicinity of the well. The number of ions in *in situ* fluids depends on the type of fluid occupying the pore space. *In situ* brine usually contains several inorganic salts that ionize in water. The potential difference is a function of the difference in salinity between borehole fluid and formation water. If the formation water salinity is greater than the mud filtrate salinity, the SP will deflect in a negative direction. A small SP log deflection implies that the formation does not contain high-salinity fluids, such as impermeable shales. A large SP log deflection implies that the formation does contain high-salinity fluids, such as permeable beds. If the SP log encounters shale, the small SP deflection associated with the shale beds is used to establish a "shale baseline." An increase in the SP deflection relative to the shale baseline is often used as an indicator of more permeable beds.

SP logs can be used for a variety of purposes. They can detect permeable beds and indicate their thickness. An examination of the SP log yields information about both net and gross thickness. The shape of the SP log can yield information about the depositional environment. Quantitative estimates of formation water resistivity and formation shaliness can be determined from the SP log.

SP logs do, however, have limitations. If the resistivity of formation water equals the resistivity of the mud filtrate, no SP exists. Corrections must be applied to the SP log to account for thin beds and large borehole diameters.

7.3.2 Gamma Ray Logs

Gamma rays are photons (particles of light) with energies ranging from 10^4 ev (electron volts) to 10^7 ev. Gamma ray logs are used to detect *in situ* radioactivity from naturally occurring radioactive materials such as potassium, thorium, and uranium. In general, shale contains more radioactive materials than other rock types. Consequently, the production of gamma rays by radioactive decay is greater in the presence of shale. A high gamma ray response implies the presence of shales, while a low gamma ray response implies the presence of clean sands or carbonates.

7.4 Porosity Logs

Well logs that can be used to obtain porosity include density logs, acoustic logs, and neutron logs. Each of these logs is described in the following subsections.

7.4.1 Density Logs

The formation density log is useful for determining the porosity of a logged interval. The porosity is given by

$$\phi = \frac{\rho_{ma} - \rho_b}{\rho_{ma} - \rho_f} \qquad (7.4.1)$$

where

    ϕ = porosity
    ρ_{ma} = density of dry rock
    ρ_b = bulk density recorded by the log
    ρ_f = fluid density

Density logs rely on Compton scattering to determine bulk density. In Compton scattering, a gamma ray photon transfers some of its energy to an electron in the inner orbital of an atom. The loss of energy of the photon changes the frequency of the gamma ray photon and can cause the ejection of the electron from the atom. The density log measures electron density by detecting gamma rays that undergo Compton scattering. The intensity of scattered gamma rays is proportional to electron density. Electron density is the number of electrons in a volume of the formation and is proportional to bulk density. If the measurement of gamma rays is large, it implies a large electron density and correspondingly large bulk density.

Density logs are useful for determining hydrocarbon density and for detecting hydrocarbon gas with low density compared to rock matrix or liquid densities. A low density implies high hydrocarbon gas content, while a high density suggests low hydrocarbon gas content.

7.4.2 Acoustic Logs

Acoustic or sonic logs provide another technique for measuring porosity. The acoustic log emits a sound wave and measures the transit time associated with the propagation of the sound wave through the medium to a recorder. The speed of sound depends on the density of the medium, and the density of the medium depends on the relative volumes of rock and pore space. In particular, the bulk density ρ_B of the medium is

$$\rho_B = (1 - \phi)\rho_{ma} + \phi\rho_f \tag{7.4.2}$$

where ϕ is porosity, and ρ_{ma} and ρ_f are rock matrix and fluid densities, respectively.

The bulk volume of a formation with large pore space contains more fluid in the bulk volume than a formation with small pore space. Since fluid density is typically several times smaller than matrix density, the bulk density of a system with relatively large pore space is less than the bulk density of a system with relatively small pore space.

The speed of sound v through the formation is approximately given by Wyllie's time average equation

$$\frac{1}{v} = \frac{\phi}{v_f} + \frac{(1 - \phi)}{v_{ma}} \tag{7.4.3}$$

where

    v_f = the speed of sound in interstitial fluids
    v_{ma} = the speed of sound in the matrix

Wyllie's equation can be written in terms of the interval transit time Δt as

$$\Delta t = \phi \Delta t_f + (1 - \phi)\Delta t_{ma} \qquad (7.4.4)$$

since the speed of sound is inversely proportional to transit time and the subscripts f and ma again denote fluid and matrix. The transit time Δt is the time measured by the acoustic log. A long transit time implies a slow speed of sound propagation. The presence of hydrocarbons increases the interval transit time. Solving the transit time form of Wyllie's equation for porosity gives

$$\phi = \frac{\Delta t - \Delta t_{ma}}{\Delta t_f - \Delta t_{ma}} \qquad (7.4.5)$$

Corrections for compaction can be made to Wyllie's equation for unconsolidated sands. The compaction factor is an empirical factor that can be estimated from the interval transit time for adjacent shale.

If we note that sound waves propagate faster in rock matrix than in fluid, we can make some general observations about acoustic log response. A long travel time implies slow speed of sound propagation and large pore space. Conversely, a short transit time implies a high speed of sound propagation and small pore space.

Acoustic logs are used for a variety of purposes in addition to porosity determination. For example, an acoustic log can be compared directly to seismic measurements. This comparison makes it possible to calibrate giga scale seismic measurements to mega scale measurements. Calibration with an acoustic log can significantly improve the accuracy of the acoustic velocity model used to convert seismic travel times to depths.

7.4.3 Neutron Logs

The neutron log can be used to determine the hydrogen content of the logged interval by counting captured gamma rays or neutrons counted at a detector. Fast neutrons emitted by a radioactive source in the neutron log are slowed by collisions to thermal energies. The ensuing thermal neutrons are captured by nuclei, which then emit detectable gamma rays. Hydrogen nuclei are especially effective because hydrogen has a relatively large capture cross-section for thermal neutrons. The neutron log response indicates the concentration of hydrogen in the fluid-filled pore space.

Oil and water have about the same concentration of hydrogen, while gas has a relatively low hydrogen concentration. The presence of a significant amount of hydrogen will appear as a large gamma ray response. A small response suggests a low hydrogen concentration. If we note that natural gas such as methane (CH_4) has a small hydrogen concentration relative to other molecules, such as water (H_2O), a small gamma ray response can be interpreted as indicating the presence of gas.

7.5 Resistivity Logs

Resistivity logs are the oldest type of well log. They were first applied to formation evaluation in 1927 by Conrad and Marcel Schlumberger and Henri Doll. We consider some electrical properties of rocks and then discuss specific resistivity logs. Resistivity is proportional to electrical resistance R_e. If we consider a conductor of length L and cross-sectional area A, resistivity of the conductor R_c is given by

$$R_c = R_e \frac{A}{L} \qquad\qquad (7.5.1)$$

The resistivity R_0 of a porous material saturated with an ionic solution is equal to the resistivity R_w of the ionic solution times the formation resistivity factor F of the porous material; thus

$$R_0 = FR_w \qquad\qquad (7.5.2)$$

The formation resistivity factor F, which is sometimes referred to as the formation factor, can be estimated from an empirical relationship between formation resistivity factor F and porosity ϕ. The empirical relationship is

$$F = a\phi^{-m} \qquad\qquad (7.5.3)$$

where the cementation exponent m varies from 1.14 to 2.52, and the tortuosity factor a varies from 0.35 to 4.78 (Bassiouni, 1994) for sands. Figure 7.5 shows a semilogarithmic plot of formation resistivity factor $F(a,m)$ versus porosity for sands. Parameters a and m depend on pore geometry; a depends on tortuosity, and m depends on the degree of consolidation of the rock.

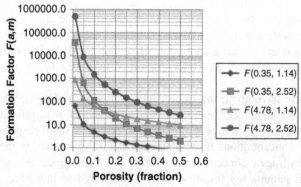

Figure 7.5 Formation resistivity factor for sands.

If a porous medium is partially saturated by an electrically conducting, wetting phase with saturation S_w, the formation resistivity R_t of the partially saturated medium is given by Archie's empirical relationship

$$R_t = R_0 S_w^{-n} \qquad (7.5.4)$$

where n is called the saturation exponent. Note that $R_t = R_0$ when the porous medium is completely saturated by the wetting phase—that is, at $S_w = 1$. Combining the preceding relationships and solving for wetting phase saturation in terms of resistivities and formation resistivity factor give

$$S_w = \left(\frac{FR_w}{R_t} \right)^{1/n} \qquad (7.5.5)$$

Equation (7.5.5) is Archie's equation for wetting phase saturation, which is usually water saturation.

Resistivity logs measure electrical resistivity in the borehole. The resistivity of formation fluid depends on the concentration of ionized particles in the fluid. Formation water is brine with a substantial concentration of ionizing molecules, such as sodium chloride and potassium chloride. The ions in solution decrease the resistivity of the fluids in the formation. Hydrocarbons, on the other hand, generally do not contain comparable levels of ions and therefore have a larger electrical resistivity. Resistivity differences between brine and hydrocarbons make it possible to use resistivity logs to distinguish between brine and hydrocarbon fluids. The ability of a rock to support an electric current depends primarily on fluid content in pore space because rock grains are usually nonconductive. Thus, high resistivity suggests the presence of hydrocarbons, while low resistivity implies the presence of brine. We consider two types of resistivity logs: electrode and induction logs.

7.5.1 Electrode Logs

Electrodes in the electrode tool are connected to a generator. Electrical current passes from the electrodes through the borehole fluid, into the formation, and finally to a remote reference electrode. Depth of investigation is controlled by the spacing of electrodes and, more recently, by focusing electrical current using logs such as the spherically focused log. The observed current gives information about the resistivity of the formation. Electrode logs must be used with conductive ("salty") mud to allow a current to pass through the borehole fluid.

7.5.2 Induction Logs

Induction logs measure formation conductivity, which is the inverse of resistivity. The conductivity is induced by a focused magnetic field. Transmitting coils in the tool emit a high-frequency alternating current (AC) signal that creates an alternating magnetic field in the formation. Induced secondary currents are created in the formation by the alternating magnetic field. The secondary currents create new magnetic fields that are recorded by receiver coils. The transmitting and receiving coils are on opposite ends of the tool.

The induction log provides information that is proportional to conductivity. Induction logs are most accurate when used with nonconductive or low-conductivity mud. They can therefore be used with "fresh" mud, or boreholes filled with air or oil. Induction logs can be used for fluid type evaluation and to identify coals or other non-conducting materials.

7.6   Other Types of Logs

New logs are being developed as technology evolves and the needs of industry change. Some of these logs do not fit into the major categories presented in previous sections. We will discuss some of them now.

7.6.1   Photoelectric Logs

In the photoelectric effect, a low-energy gamma photon that collides with an atom can transfer all of its energy to an inner orbital electron and cause the ejection of the electron from the atom. The amount of energy required to eject the electron depends on the atomic number. Some of the gamma photon energy will be used to eject the electron, and the rest is transferred as kinetic energy. Photoelectric logs measure the absorption of low-energy gamma photons by atoms in the formation. They provide information about the atomic number of atoms in the formation that can be used to infer the composition of a formation.

7.6.2   Dipmeter Logs and Borehole Imaging

The dipmeter log is a tool for measuring the direction of dip associated with beds that are adjacent to the borehole. Three or four microresistivity measurements are recorded simultaneously. The measurements are made on different sides of the borehole and then recombined by computer to infer bedding dip. Similar information can be obtained using acoustic borehole imaging tools. Borehole imaging measurements can be used for thin-bed evaluation, fracture identification and analysis, structural and stratigraphic dip analysis, sedimentary facies analysis, and textural analysis.

7.6.3   MWD and LWD

Most logs are run in open holes or cased holes. Some logs may be run while the well is still being drilled and the drill pipe is present in the borehole. These logs are referred to as measurement while drilling (MWD) or logging while drilling (LWD) logs. Logs that can be run during drilling include resistivity logs, gamma ray logs, density logs, neutron logs, and electrode logs. These logs are especially useful in long, directionally drilled wells for providing real-time information on borehole environment during the drilling operation.

7.7 Reservoir Characterization Issues

Several issues arise in the application of well logs to reservoir characterization. The major applications of common log types are summarized in Table 7.1. Some of the more common issues are discussed in this section.

7.7.1 Well Log Legacy

Well logging technology has developed over many decades. The improvement in technology from one generation of well logs to another has been remarkable. This is both an asset and a liability. It is an asset for new fields where the most modern technology can be applied, but it is a liability for old fields where old logs must be combined with new logs in the analysis of a field. Care must be taken when working with logs from different generations of technology to be sure that analytical techniques are appropriate for each well log.

7.7.2 Cutoffs

The volume of reserves V_R is the product of pore volume, hydrocarbon saturation S_h, and recovery factor R_F. Writing pore volume as the product of porosity ϕ, area A, and net thickness h, V_R is given by

$$V_R = \phi A h S_h R_F \qquad (7.7.1)$$

Porosity and saturation are usually obtained from well logs. The estimate of these parameters is often accompanied by the specification of porosity and saturation cutoffs. A cutoff specifies the minimum value of the parameter that is considered a part of the productive formation. Cutoffs may be used for permeability in addition to porosity and saturation.

Table 7.1 Major Applications of Common Log Types

Log Type	Lithology	Hydrocarbons	Porosity	Dip
Electric				
SP	X			
Resistivity	X	X		
Radioactive				
Gamma Ray	X			
Neutron		X	X	
Density		X	X	
Sonix	X	X	X	
Dipmeter				X

Source: Adapted from Selley, 1998.

7.7.3   Cross-Plots

Cross-plots of well log data may be used to determine such factors as porosity, lithology, and gas saturation. A cross-plot is a plot of one well log parameter against another. For example, the Pickett cross-plot is a cross-plot of porosity versus resistivity. It is motivated by Archie's equation for wetting phase saturation expressed in the form

$$R_t = \phi^{-m} R_w S_w^{-n} \qquad (7.7.2)$$

where the coefficient $a = 1$ in the empirical relationship for formation resistivity factor F. Taking the logarithm of both sides of Eq. (7.7.2) gives

$$\log R_t = -m \log \phi + \log R_w - n \log S_w \qquad (7.7.3)$$

It can be seen from this equation that a log-log plot of porosity and resistivity should give a straight line with slope m and an intercept that depends on R_w and S_w.

Another example of a useful cross-plot is the plot of neutron log porosity versus density log porosity. The resulting cross-plot can be used to provide an estimate of the shaliness (shale content) of the formation. The porosity log estimate of shale content can then be used to validate the usually more reliable estimate of shale content obtained from gamma ray logs.

7.7.4   Correlation between Wells

Two of the most important uses of well logs are the determination of formation extent and the continuity of formations between logged wells. Correlations between wells are used to define formations and productive intervals. An example of a correlation technique is the fence diagram. A fence diagram is prepared by aligning well logs in their proper spatial position and then drawing lines between the well logs that show the stratigraphic correlation. Fence diagrams illustrate correlations between wells and can show formation pinchouts, unconformities, and other geologic discontinuities.

7.7.5   Log Suites

Modern logging techniques combine logging tools to obtain a more reliable representation of formation properties. A combination of well logging tools is usually needed to minimize ambiguity in log interpretation. For example, the combination neutron-density log is a combination log that consists of both neutron log and density log measurements. Possible gas-producing zones can be identified by the log traces of the combination neutron-density log. The presence of gas increases the density log porosity and decreases the neutron log porosity. If a sonic log is added to the log suite, quantitative information about lithology can be estimated using cross-plots, and the log suite can be used to calibrate seismic data.

Sonic log interpretation depends on lithology. In particular, the interval transit time in carbonates depends on the relative amount of primary and secondary porosity. Primary porosity is associated with the matrix, and secondary porosity is associated with features such as fractures and *vugs*. Subtracting sonic porosity from total porosity recorded using neutron or density logs gives an estimate of secondary porosity.

One more log, the gamma ray log, is usually added to the suite of logs used to evaluate gas-bearing formations. The gamma ray log measures natural radioactivity in a formation. It provides a measurement of shale content and can be used for identifying lithologies, correlating zones, and correcting porosity log results in formations containing shale.

CS.7 Valley Fill Case Study: Well Logs

Synthetic well logs are presented in Figure CS.7A for Wells 7, 3, and 9 as the figure is viewed from left to right. Again moving from left to right for each well, Figure CS.7A shows the spontaneous potential (SP) log, seismic reflection coefficients (RC), and the resistivity (Res.) log. Depth is measured in feet. The SP log and seismic reflection coefficients can be used to estimate regional dip by estimating the dip of the productive interval. The resistivity log shows that the upper part of the productive interval in Well 3 has a higher resistivity than the other wells. This information is consistent with the observation that Well 3 is an oil producer, while Wells 7 and 9 are not. Furthermore, the depth at which a change in

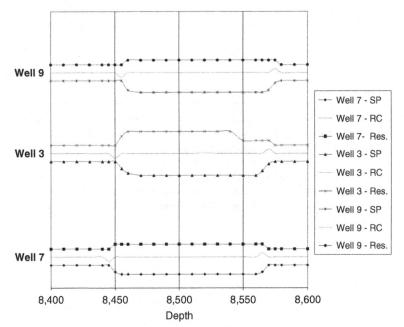

Figure CS.7A Well logs for Valley Fill reservoir.

resistivity takes place in Well 3 from a higher to a lower value within the productive interval denoted by the spontaneous potential log is an indication of the depth of the oil–water contact in the productive interval. This combination of well logs supports the Valley Fill interpretation.

Exercises

7-1. A well is drilled to a depth of 6,000 feet, and the wellbore is filled with drilling mud. What is the hydrostatic pressure of the drilling mud at 6,000 feet if the density of the drilling mud is 10 lb/gal, which corresponds to a pressure gradient of 0.53 psi/ft?

7-2. Suppose a density log is making measurements in a formation where the density of sandstone is 2.65 g/cc and the fluid density of brine is 1.05 g/cc. Use Eq. (7.4.12) to plot porosity versus bulk density for the density log. Assume bulk density ranges from 2 g/cc to 2.65 g/cc.

7-3. Suppose an acoustic log is making measurements in a water-saturated limestone formation. The limestone matrix has a transit time of 44 μs/ft = 44×10^{-6} s/ft corresponding to an acoustic velocity of 23,000 ft/s. The brine transit time is 189 μs/ft = 189×10^{-6} s/ft corresponding to an acoustic velocity of 5,300 ft/s. Use Eq. (7.4.5) to plot porosity versus log transit time for the acoustic log. Assume that log transit time ranges from 44 μs/ft to 104 μs/ft.

7-4. Fill in the following table using Eq. (7.5.3) to calculate the formation resistivity factor for a resistivity log.

	Sand		Carbonate	
a	0.81	0.81	1	1
m	2	2	2	2
Porosity	0.1	0.2	0.1	0.2
F				

7-5. Density log signals and acoustic log signals can be written as $x_{log} = (1 - \phi)x_{ma} + \phi x_f$, where ϕ is porosity and the well log signal x_{log} is the sum of the signal $(1 - \phi)x_{ma}$ associated with the matrix and the signal ϕx_f associated with the fluid. Solve $x_{log} = (1 - \phi)x_{ma} + \phi x_f$ for ϕ in terms of well log signals.

7-6. Estimate the regional dip using the well logs in Figure CS.7A. *Hint:* Determine the dip angle using the RC trace of all three wells to define the top of the formation.

7-7. Estimate the oil-water contact using the resistivity log for Well 3 shown in Figure CS.7A.

8 Well Testing

Well tests can provide information about reservoir structure and the expected flow performance of the reservoir. They help refine the operator's understanding of the field and often motivate changes in the way the well or the field is managed. Additional information about well testing can be found in Matthews and Russell (1967), Earlougher (1977), Sabet (1991), Economides et al. (1994), Horne (1995), Dake (2001), Lee (2007), and Miller and Holstein (2007). This chapter discusses a variety of well tests that can be performed on production and injection wells, vertical and deviated/horizontal wells, and combinations of wells. It also describes the information that can be obtained from the well tests and shows how to use well test information in reservoir management.

8.1 Pressure Transient Testing

Pressure transient testing uses changes in measurable pressure performance to infer reservoir parameters such as flow capacity, average reservoir pressure in the drainage area, reservoir size, boundary and fault locations, wellbore damage and stimulation, and well deliverability. The information from well tests can be combined with data from other sources to obtain additional reservoir parameters. For example, the estimate of flow capacity from a pressure transient test can be combined with a well log estimate of net pay to determine effective permeability in the volume of the reservoir investigated by the well test. A summary of parameters that can be determined by well tests is presented later in this chapter.

Pressure changes in a well test are induced in a system by changing the flow rate of one or more wells and recording the variation in pressure as a function of time using pressure gauges. Analysis of the pressure response provides information about reservoir flow capacity in the radius of investigation of the well. The pressure response is analyzed by plotting pressure and its derivative as a function of time. Analytical techniques depend on the type of test run, the reservoir geology, and the flowing fluids. A selection of the most common tests is discussed in this chapter to illustrate the procedures and type of information that can be obtained from pressure transient testing of wells.

A pressure response can be elicited from a well by changing the well's flow rate. The pressure response at the well passes through three stages: early-time wellbore-dominated response (transient state); late-time, boundary-dominated response (pseudo-steady state); and intermediate-time infinite-acting response during a transitional

Integrated Reservoir Asset Management. DOI: 10.1016/B978-0-12-382088-4.00008-6

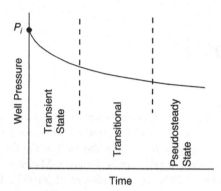

Figure 8.1 Pressure decline at a well.

stage between the early- and late-time responses. The three stages are sketched in Figure 8.1 for a well with initial pressure P_i. The infinite-acting response often behaves as radial flow.

The flow rate of a well can be changed by either increasing or decreasing the rate. The four basic types of transient tests for production or injection wells are flow rate increases or decreases in a production well, and flow rate increases or decreases in an injection well. Pressure buildup and drawdown tests are run on production wells. The buildup test measures pressure increases after a flowing well has been shut in. By contrast, a drawdown test measures pressure decline as the production rate of a well is increased. A similar set of tests can be run on injection wells. The falloff test measures pressure decline after an injection well is shut in, while injectivity tests measure the pressure increase as the injection rate is increased. These tests can be run on gas, oil, and water wells.

Many well tests are based on a discontinuous change in flow rate. For example, in a pressure buildup test, the test is conducted by shutting in a producing well. Incremental changes from one rate to another, rather than abrupt changes between flow and no-flow conditions, can be used to provide additional information. Deliverability tests in gas wells and pressure pulse testing are examples of tests that use incremental rate changes to generate observable pressure variations. Gas well deliverability tests tell us about the flow capacity of gas wells. Pressure pulse tests provide evidence of pressure communication between wells.

8.2   Oil Well Pressure Transient Testing

The diffusivity equation is the starting point for understanding the behavior of pressure transient tests. The diffusivity equation for a single-phase liquid with fluid pressure P is

$$\frac{\partial^2 P}{\partial r_D^2} + \frac{1}{r_D}\frac{\partial P}{\partial r_D} = \frac{\partial P}{\partial t_D} \qquad (8.2.1)$$

where dimensionless radius r_D and dimensionless time t_D are defined as

$$r_D \equiv \frac{r}{r_w}; \ t_D \equiv 0.000264 \frac{kt}{\phi(\mu c_T)_i r_w^2} \tag{8.2.2}$$

in terms of the variables

t = time (hr)
r = radial distance from well (ft)
k = permeability (md)
ϕ = porosity (fraction)
μ = viscosity of liquid (cp)
c_T = total compressibility (psia^{-1})
r_w = wellbore radius (ft)

The subscript i indicates that viscosity μ and total compressibility c_T are evaluated at initial pressure. The group of parameters $k/\phi\mu c_T$ with porosity ϕ and total compressibility c_T occurs frequently in the context of the diffusivity equation and is called the diffusivity coefficient.

The diffusivity equation was derived for a set of assumptions that must be noted to avoid indiscriminate, and inappropriate, applications of its solutions. The diffusivity equation is based on the assumption that the reservoir is homogeneous and isotropic in porosity, permeability, and thickness. The production well is assumed to be completed through the entire thickness of formation to ensure that radial flow is occurring in the formation. The fluid flowing into the well is assumed to be a single-phase with constant fluid viscosity. The fluid is considered slightly compressible with a constant compressibility. The total system compressibility is the sum of rock compressibility c_r and fluid phase compressibilities times phase saturations—thus, $c_T = c_r + c_o S_o + c_w S_w + c_g S_g$ for a system with oil, water, and gas phases. The diffusivity equation does not include gravity forces, which corresponds to the assumption that gravity effects are negligible. These assumptions lead to Eq. (8.2.1). Although the following discussion is presented in terms of oil wells, it is applicable to any system that satisfies the preceding assumptions, such as a water production well.

Solutions of the diffusivity equation assume that the well is a line source with a constant flow rate at the well. The solutions depend on the imposed boundary conditions. For a reservoir that acts like it does not have an outer boundary, we have the solution for an infinite-acting reservoir:

$$P_D(r_D, t_D) = -\frac{1}{2}\left[Ei\left(\frac{r_D^2}{4t_D}\right)\right] \tag{8.2.3}$$

where the term $Ei(\ldots)$ is the exponential integral

$$-Ei(-x) = \int_x^\infty \frac{e^{-u}}{u} du \tag{8.2.4}$$

Dimensionless pressure P_D in oil field units is

$$P_D = \frac{KH}{141.2QB\mu}\left(P_i - P_{wf}\right) \tag{8.2.5}$$

where

K = permeability (md)
H = thickness (ft)
P_i = initial reservoir pressure (psia)
P_{wf} = well flowing pressure (psia)
Q = flow rate (STB/d)
B = formation volume factor (RB/STB)
μ = viscosity (cp)

Equation (8.2.3) is valid throughout the reservoir, including at the wellbore, where the dimensionless radius $r_D = 1$. In many cases, the inequality $t_D/r_D^2 > 10$ is valid at $r_D = 1$, so that the exponential integral solution can be approximated by

$$P_{wD} = \frac{1}{2}(\ln t_D + 0.80907) + S \tag{8.2.6}$$

where S is called skin. A positive value of skin S represents well damage, while a negative value represents a stimulated well. In oil field units and transforming the logarithm to base 10, Eq. (8.2.6) becomes

$$P_{wf} = P_i - 162.6\frac{QB\mu}{KH}\left(\log t + \log\frac{K}{\phi\mu c_T r_w^2} + 0.8686S - 3.2274\right) \tag{8.2.7}$$

8.2.1 Flow Regimes

Flow regimes are associated with different boundary conditions. Three flow regimes are usually identified: steady state, pseudosteady state, and transient state. The flow regime depends on the boundary condition, and it can be identified by the rate of change in pressure with time. The steady state flow regime corresponds to a system in which the mass flow rate is constant everywhere, and pressure is constant with respect to time ($dP/dt = 0$). An example of a system that exhibits steady state flow is a reservoir connected to an infinite-acting aquifer. The corresponding boundary condition is referred to as the constant pressure boundary.

The pseudosteady state regime applies to a system in which both the wellbore and the average reservoir pressure change with time. Pressure changes at a constant rate (dP/dt = constant) in the pseudosteady state regime. The system behaves like a closed system, and there is no fluid movement across boundaries. An example of this type of reservoir is a reservoir with closed boundaries and no fluid encroachment from sources such as aquifers or leaking faults.

The final regime is the transient state, which is the flow regime in which pressure changes as a function of time ($dP/dt = f(t)$). In this case, there are no restrictions on fluid movement. Identifying the appropriate method for analyzing a pressure transient test depends on correctly identifying flow regime.

8.2.2 Diagnostic Analysis of Pressure Transient Tests

Pressure transient test analysis begins with a diagnostic analysis. The first step in the diagnostic analysis is to plot the logarithm of pressure versus the logarithm of time. The behavior of pressure on this plot determines the flow regime and the corresponding analytical techniques. The steady state flow regime is characterized by pressures that do not change with time ($dP/dt = 0$). In this case, pressure P equals constant α so that $P = \alpha$ and α is independent of time. The resulting time derivative satisfies the time derivative criterion ($dP/dt = 0$). The steady state flow regime appears as a horizontal line on a diagnostic plot of log P versus log t. An example of the steady state flow regime is the late-time pressure response for infinite-acting radial flow in which dimensionless pressure has the form $P_D \ln(r_e/r_w)$, where r_e is the drainage radius of the well and r_w is the wellbore radius.

The pseudosteady state flow regime is characterized by pressures that change at a constant rate ($dP/dt = $ constant). In this case, pressure P is proportional to time such that $P = \beta t$, where β is independent of time. The resulting time derivative satisfies the time derivative criterion $dP/dt = \beta$. The pseudosteady state flow regime appears as a straight line with a unit slope on a diagnostic plot of log P versus log t. This can be seen by noting that the logarithm of $P = \beta t$ is log $P = $ log $\beta +$ log t. The geometric characteristics of the pseudosteady state flow regime on a diagnostic plot are determined by comparing log $P = $ log $\beta + $ log t to the equation for a straight line $y = mx + b$, with x the independent variable, y the dependent variable, m the slope, and b the y-axis intercept. The relationships between the pseudosteady state flow regime case and the straight line are $x = $ log t, $y = $ log P, $m = 1$, $b = $ log β. An example of the pseudosteady state flow regime is the pressure response for a closed system with area A in which dimensionless pressure is related to dimensionless time by $P_D = 2\pi t_D (r_w^2/A) + \dots$ with the logarithm given by log $P_D = $ log $t_D + \dots$.

The transient state flow regime is characterized by pressures that change as a function of time. Suppose $f(t)$ is a function of time such that pressure $P = f(t)$. The time derivative in this case has the general form $dP/dt(t)/dt$. The time derivative in the transient state flow regime must satisfy the inequality $dP/dt \neq $ constant. If dP/dt is a nonzero constant, then we are in the pseudosteady state flow regime, and if $dP/dt = 0$, then we are in the steady state flow regime. A diagnostic plot of log P versus log t should exhibit a nonlinear relationship in the transient state flow regime. The transient state flow regime can be seen in the early-time pressure response for infinite-acting radial flow in which dimensionless pressure P_D is proportional to $\ln t_D /2$. A semilogarithmic plot of P_D versus $\ln t_D$ yields a line with slope one-half.

8.2.3 The Superposition Principle

Pressure transient test analyses rest upon an important assumption called the super-position principle. The superposition principle asserts that the total pressure change at a point in the reservoir is a linear sum of the changes in pressure due to each well in the reservoir. The assumption of linear superposition is a theoretical consequence of the linear diffusivity equation that is substantiated by field performance. The superposition principle implies that a pressure disturbance will propagate through the reservoir even if the source of the disturbance changes or disappears. A quantitative expression of the superposition principle is

$$\Delta P_{TOTAL} = \Delta P_{Well\ A} + \Delta P_{Well\ B} + \Delta P_{Well\ C} + \dots \tag{8.2.8}$$

where the pressure change at a well is given by

$$\Delta P_{Well} = P_{Well} - P_{initial} \tag{8.2.9}$$

and $P_{initial}$ is the initial reservoir pressure.

8.2.4 The Pressure Buildup Test

The concepts of pressure transient testing can be clarified by considering a particular application: the pressure buildup (PBU) test. The PBU test is performed by first flowing the well at a stabilized rate Q for a stabilized flow period t_F. In practice, the stabilized rate Q is the last rate prior to shutting in the well, and the flow time t_F is given by

$$t_F = \frac{\text{cumulative production } [STB]}{Q[STB/D]} \tag{8.2.10}$$

The well is shut-in for a duration Δt after the stabilized flow period while reservoir pressure is recorded as a function of time. The results of the PBU test are then analyzed using the superposition principle.

The superposition principle is applied to the PBU test by recognizing that the shut-in condition is equivalent to adding the pressure response for two wells producing at different rates and at different times from the same location. From this perspective, the original well flows at rate Q for the entire period of the test. The shut-in condition is represented by introducing an image well, which is an imaginary well that is introduced for mathematical purposes. The addition of the pressure response due to an image well at the same location as the actual well but producing at rate $-Q$ beginning at time t_F gives the shut-in condition. The rates for the PBU test are shown in Figure 8.2, and the PBU analytical procedure is outlined following.

It is useful to recall at this point that dimensionless time t_D is given by

$$t_D = \frac{0.000264Kt}{\phi\mu c_T r_w^2} \tag{8.2.11}$$

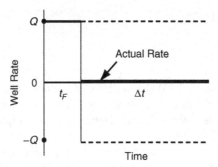

Figure 8.2 Rates for the PBU test.

Dimensionless pressure P_D is a function of dimensionless time. The solution of the diffusivity equation in terms of dimensionless pressure and dimensionless time may be written in the form

$$P_{ws} = P_i - 141.2 \frac{QB\mu}{KH} [P_D(t_D) + \text{constant}] \qquad (8.2.12)$$

where

P_{ws} = shut-in pressure (psia)
P_i = initial pressure (psia)
Q = stabilized flow rate (STB/D)

Dimensionless pressure changes linearly with the logarithm of time during the infinite-acting time period; thus

$$P_D = \frac{1}{2} [\ln(t_D) + \text{constant}] \qquad (8.2.13)$$

Combining Eqs. (8.2.12) and (8.2.13) for the PBU case with the superposition principle eliminates the unspecified constant in these equations and gives

$$P_{ws} = P_i - 141.2 \frac{QB\mu}{KH} [P_D([t_F + \Delta t]_D) - P_D(\Delta t_D)] \qquad (8.2.14)$$

The subscript D attached to a variable indicates that the dimensionless form of the variable is being used. The dimensionless pressure in Eq. (8.2.14) is replaced by Eq. (8.2.13) to yield the simplified form

$$P_{ws} = P_i - m\log(t_H) \qquad (8.2.15)$$

where t_H is Horner time

$$t_H = \frac{t_F + \Delta t}{\Delta t} \qquad (8.2.16)$$

and the variable m is given by

$$m \equiv 162.6 \frac{QB\mu}{KH} \qquad (8.2.17)$$

It can be seen from Eq. (8.2.15) that a plot of P_{ws} versus the logarithm of Horner time will give a straight line with slope $-m$. This semilogarithmic plot is called the Horner plot, and the analysis is the Horner analysis. Figure 8.3 illustrates the main features of the Horner plot.

The early-time behavior of Figure 8.3 is at the right-hand side of the figure and proceeds to later times on the left-hand side of the figure. The rapid buildup of pressure at early times corresponding to Horner time ≈ 25 or greater is the wellbore storage effect, which is discussed in more detail following. The data for the infinite-acting period comprises the straight line in the Horner time range from ~ 25 to ~ 4. For Horner times less than 4, the slope of the infinite-acting line changes because the reservoir boundary is beginning to influence the shut-in pressure response.

A straight line drawn through the infinite-acting period of the Horner plot, such as the one in Figure 8.3, has a slope $-m$, where m has units of psia/log cycle. The value of m is approximately 0.8 psia/log cycle for the PBU data shown in Figure 8.3. Once the value of m is known, it is possible to rearrange Eq. (8.2.17) to obtain an estimate of flow capacity in the region of investigation of the well test; thus

$$KH \equiv 162.6 \frac{QB\mu}{m} \qquad (8.2.18)$$

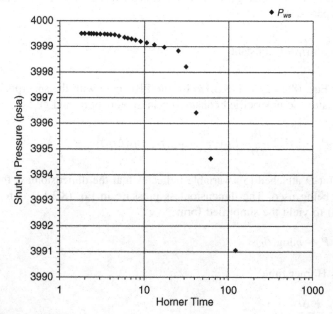

Figure 8.3 A Horner plot.

An estimate of net formation thickness H can then be used with Eq. (8.2.18) to estimate an effective permeability. Net formation thickness is usually obtained from well log analysis. Notice that permeability obtained from well testing represents a larger sample volume than laboratory measurements of permeability.

Dimensionless skin factor S can be estimated from the Horner plot. The procedure is to extrapolate the straight line drawn through the infinite-acting period to the Horner time at a shut-in time $\Delta t = 1$ hour. The corresponding skin factor is

$$S = 1.151\left[\frac{P_{1hr} - P_{wf}(t_F)}{m} - \log\frac{Kt_F}{(t_F + 1)\phi\mu c_T r_w^2} + 3.2274\right] \qquad (8.2.19)$$

where the well flowing pressure $P_{wf}(t_F)$ at the end of the stabilized flow period t_F is required for the calculation.

Skin factor S uses actual well flowing pressure, which is usually less than that predicted by the radial flow diffusivity equation. The additional pressure drop is proportional to rate and can be viewed as a zone of reduced permeability, or "skin," in the vicinity of the well. The pressure drop associated with skin is

$$(\Delta P)_{skin} = S\left(\frac{141.2QB \int \mu}{KH}\right) \qquad (8.2.20)$$

Actual pressure drop is the sum of line source pressure drop and skin pressure drop: $(\Delta P)_{actual} = (\Delta P)_{line\ source} + (\Delta P)_{skin}$. Skin usually ranges from $-5 < S < 50$. A positive skin corresponds to damage or reduction in effective permeability near the well, while a negative skin represents stimulation.

8.2.5 Wellbore Storage

The pressure response to a change in flow rate is made more complicated by wellbore storage. The effect of the wellbore's finite volume on pressure response is called the "wellbore storage effect." The wellbore pressure drops when the well is first open to flow, as shown in Figure 8.4. Initial fluid production includes expansion of fluid in the wellbore as a result of pressure decline. Wellbore storage is the effect of the finite wellbore volume on well flow response when the well flow rate changes. Wellbore storage prevents the flow rate at the sandface from instantaneously responding to a change in flow rate at the surface.

Early-time pressure data is dominated by wellbore storage and is related to the flow test duration t by $P = (\kappa/C)t$, where the proportionality constant κ and wellbore storage coefficient C are independent of flow time t. The corresponding time derivative is $dP/dt = \kappa/C$. A diagnostic plot of $\log P$ versus $\log t$ gives a straight line with unit slope; thus $P = (\kappa/C)t$ implies $\log P = \log t + \log(\kappa/C)$. The straight line with unit slope was also encountered in the pseudosteady state flow regime, but in this case the wellbore storage straight line appears in the period of time immediately following the rate change.

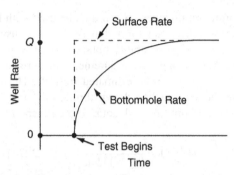

Figure 8.4 The effect of wellbore storage on flow rate.

Wellbore storage is known by many names, including afterflow, afterproduction, afterinjection, and wellbore unloading or loading. Two of the most common types of wellbore storage are compressive storage and changing liquid level. Compressive storage is due to the compression or expansion of fluid in the wellbore when the wellbore is completely full of a single-phase fluid. The wellbore storage coefficient C is estimated from the compressibility relation

$$C = V_w c \tag{8.2.21}$$

where

    C = wellbore storage coefficient (bbl/psia)
    V_w = total wellbore volume (bbl)
    c = average compressibility of wellbore fluid (psia^{-1})

By contrast, changing liquid level occurs in production wells on pump or gas-lift, and in injection wells taking fluid on vacuum. The wellbore storage coefficient C for changing liquid level is estimated by

$$C = \frac{V_u}{\left(\dfrac{\rho}{144}\dfrac{g}{g_c}\right)} \tag{8.2.22}$$

where

    C = wellbore storage coefficient (bbl/psia)
    V_u = wellbore volume per unit length (bbl/ft)
    ρ = fluid density (lb/ft^3)
    $\rho/144$ = hydrostatic gradient of fluid (psia/ft)
    g = acceleration of gravity (ft/sec^2)
    g_c = unit conversion factor (32.17 lb$_m$ ft/(lb$_f$ sec^2))

Wellbore storage from the changing liquid level is usually much larger than compressive storage.

One procedure for determining wellbore storage is to prepare the usual log–log diagnostic plot of pressure versus time. Early-time data on the diagnostic plot are dominated by wellbore storage and should exhibit a unit slope straight line. The analytical procedure begins with a plot of pressure change ΔP versus shut-in time Δt on a log–log graph. Draw a unit slope through the early-time data and calculate

$$C = \frac{QB}{24}\left(\frac{\Delta t}{\Delta P}\right)_{\text{on the unit slope}} \tag{8.2.23}$$

where

    C = wellbore storage coefficient (bbl/psia)
    Q = flow rate (STB/day)
    B = formation volume factor (RB/STB)
    t = time (hr)
    P = pressure (psia)

Notice that the wellbore storage coefficient is expressed in terms of reservoir barrels of volume per psia of pressure change. The value of the wellbore storage coefficient for the data shown in Figure 8.3 is approximately 0.6 bbl/psia.

8.2.6 Interpreting Pressure Transient Tests

Several "rules of thumb" can be used to interpret pressure transient tests. We begin with the preparation of a diagnostic plot that includes both pressure versus time and the derivative of pressure with respect to time. The "pressure derivative" referred to in the literature is actually the calculation of the product of shut-in time Δt and the derivative of pressure with respect to shut-in time. Thus, $\Delta t\, dP/d(\Delta t)$ or $dP/d\ln(\Delta t)$. This quantity, which we refer to here as the pressure derivative to be consistent with the literature, is a more sensitive indicator of reservoir characteristics than the pressure response. A procedure for numerically calculating the pressure derivative is discussed by Horne (1995). The interpretations in Table 8.1 refer to the slope of the pressure derivative curve on a log–log diagnostic plot.

Pressure transient test interpretation is aided by information from other disciplines, such as the current understanding of reservoir structure and knowledge of fluid phase behavior. As with other reservoir characterization techniques, it is

Table 8.1 Rules of Thumb for Interpreting Pressure Transient Tests

Effect	Slope on Log–Log Plot
Wellbore Storage	Unit slope (rise 1/run 1)
Spherical Flow	Negative half slope (drop 1/run 2)
Radial Flow	Zero slope (horizontal)
Linear Flow	Positive half slope (rise 1/run 2)
Bilinear Flow	Positive quarter slope (rise 1/run 4)

important to seek a concept of the reservoir that is consistent with data from all sources. The selected model should be consistent with all available geoscience and engineering data.

8.2.7 Radius of Investigation

The radius of investigation is the distance the pressure transient moves away from the wellbore in the time interval following the change in flow rate. It may be estimated from a Horner analysis for a PBU test using the equation

$$r_i = 0.029 \sqrt{\frac{K \Delta t}{\phi \mu c_T}} \qquad\qquad (8.2.24)$$

where

r_i = radius of investigation (ft)
Δt = shut-in time (hr)
K = permeability (md)
ϕ = porosity (fraction)
μ = viscosity (cp)
c_T = total compressibility (1/psia)

Equation (8.2.24) assumes radial flow, steady state conditions, an infinite-acting reservoir, and single-phase flow.

Although approximate, the radius of investigation can indicate the distance to reservoir features that cause the slope of the pressure transient response to change. For example, the change in slope at late times in Figure 8.3 indicates that a no-flow boundary has been reached. By substituting the time when the infinite-acting period changes to a late-time boundary effect, we can estimate the distance from the tested well to the boundary to be about 400 feet. The radius of investigation can also provide information about no-flow barriers such as sealing faults or permeability pinchouts. This type of information should be compared with the geological concept of the reservoir and geophysical indications of structural discontinuities. The most accurate characterization of the reservoir is usually the one that provides a realization of the reservoir that is consistent with all of the data that is available from every discipline.

8.3   Gas Well Pressure Transient Testing

The analysis of gas well testing is based on many of the same principles as those for oil well testing. The procedure is to change the flow of a well and record the pressure variation with time. The primary purposes of gas well testing are to determine reservoir characteristics using pressure transient testing and to forecast gas well deliverability using gas deliverability tests. Our focus here is on pressure transient testing. The following discussion uses conversion factors assuming standard temperature and pressure are $60°F = 520°R$ and 14.7 psia.

8.3.1 Diffusivity Equation

A diffusivity equation can be derived for gas wells just as it can for oil wells. The diffusivity equation for gas wells differs, however, in the treatment of nonlinear fluid properties. In particular, the diffusivity equation for a single-phase gas flow is expressed in terms of the real gas pseudopressure $m(P)$ (psia^2/cp); thus

$$\frac{\partial^2 m(P)}{\partial r_D^2} + \frac{1}{r_D}\frac{\partial m(P)}{\partial r_D} = \frac{\partial m(P)}{\partial t_D} \tag{8.3.1}$$

Real gas pseudopressure $m(P)$ is defined as

$$m(P) = 2 \int_{P_{ref}}^{P} \frac{P'}{\mu Z} dP' \tag{8.3.2}$$

where

P_{ref} = reference pressure (psia)
P' = dummy variable of integration (psia)
Z = real gas compressibility factor (fraction)
μ = gas viscosity (cp)

Dimensionless radius and dimensionless time are defined as

$$r_D \equiv \frac{r}{r_w}; \quad t_D \equiv 0.000264\frac{Kt}{\phi(\mu c_T)_i r_w^2} \tag{8.3.3}$$

and

t = time (hr)
r = radial distance from well (ft)
K = permeability (md)
ϕ = porosity (fraction)
μ = gas viscosity (cp)
c_T = total compressibility (psia^{-1})
r_w = wellbore radius (ft)

The subscript i in Eq. (8.3.3) indicates that μ and c_T are evaluated at initial pressure.

The general solution of the diffusivity equation for gas wells is

$$\Delta m(P) \equiv m(P_i) - m(P_{wf}) = 1422\frac{qT}{KH}\left[m_D(t_D) + S + D|q|\right] \tag{8.3.4}$$

where

m_D = dimensionless real gas pseudopressure
S = skin (dimensionless)
D = non-Darcy flow coefficient $((Mscf/D)^{-1})$
q = surface flow rate of gas $(Mscf/D)$
T = reservoir temperature $(°R)$
K = permeability (md)
H = formation thickness (ft)

Analysis of gas well pressure transient testing is directly analogous to the analysis of the oil well pressure transient testing with pressure being replaced by real gas pseudopressure.

We illustrate gas well test analysis by considering a pressure buildup test. The test begins with the flow of gas from the well at a stabilized rate for a duration t_F. Pressure is then recorded as a function of shut-in time Δt when the well is shut in after the stabilized flow period. The superposition principle is applied in the usual way to find dimensionless pseudopressure m_D as a function of dimensionless time. The solution of the real gas diffusivity equation is written in the form

$$m(P_{ws}) = m(P_i) + 1422 \frac{qT}{KH} [m_D(t_D) + \text{constant}] \qquad (8.3.5)$$

where P_{ws} is well shut-in pressure.

Dimensionless pseudopressure increases linearly with the logarithm of time during the infinite-acting period. Using this observation in Eq. (8.3.5) gives

$$m(P_{ws}) = m(P_i) + \frac{1637qT}{KH} \left[\log\left(\frac{Kt}{\phi(\mu c_T)r_w^2} \right) - 3.23 + 0.869S' \right] \qquad (8.3.6)$$

$$S' \equiv S + D|q|$$

The skin factor S' for a gas well includes the usual skin factor S plus a factor that is proportional to gas rate q with proportionality constant D. Turbulent gas flow is represented by the rate-dependent part of S'. Combining these equations for the pressure buildup case gives

$$m(P_{ws}) = m(P_i) + 1637 \frac{qT}{KH} [m_D([t_F + \Delta t]_D) - m_D(\Delta t_D)] \qquad (8.3.7)$$

or

$$m(P_{ws}) = m(P_i) + 1637 \frac{qT}{KH} \log\left(\frac{t_F + \Delta t}{\Delta t} \right) \qquad (8.3.8)$$

8.3.2 The Horner Analysis

Equation (8.3.8) is the gas analog of the Horner equation for a single-phase oil well. It is based on the usual assumptions: radial flow, steady state conditions, infinite-acting reservoir, and single-phase flow. The procedure for analyzing gas well tests proceeds as in the liquid case:

1. Plot $m(P_{ws})$ versus Horner time $(t_F + \Delta t)/\Delta t$.
2. Estimate flow capacity from slope m:

$$KH = 1637\frac{qT}{m} \tag{8.3.9}$$

Do not confuse slope m with real gas pseudopressure $m(P)$.

3. Estimate skin using extrapolated shut-in pressure at 1 hour; thus

$$S' = 1.151\left[\frac{m(P_{1hr}) - m(P_{wf})}{m} - \log\left(\frac{K}{\phi\mu c_T r_w^2}\right) + 3.23\right] \tag{8.3.10}$$

8.3.3 Pressure Drawdown Analysis

The preceding procedure applies to pressure buildup tests in gas wells. An analogous procedure can be used to analyze pressure drawdown tests in gas wells. The change in real gas pseudopressure is

$$\begin{aligned} \Delta m(P) &= m(P_i) - m(P_{wf}) \\ &= 1637\frac{qT}{KH}\left[\log\left(\frac{Kt}{\phi\mu c_T r_w^2}\right) - 3.23 + 0.869S'\right] \\ S' &\equiv S + D|q| \end{aligned} \tag{8.3.11}$$

where t is the time the well is flowing at rate q. The analytical procedure for obtaining meaningful results consists of the following steps:

1. Plot $\Delta m(P)$ versus $\log t$.
2. Estimate flow capacity from slope m:

$$KH = 1637\frac{qT}{m} \tag{8.3.12}$$

Do not confuse slope m with real gas pseudopressure $m(P)$.

3. Estimate skin from

$$S' = 1.515\left[\frac{m(P_{1hr}) - m(P_{wf})}{m} - \log\left(\frac{K}{\phi c_T r_w^2}\right) + 3.23\right] \tag{8.3.13}$$

8.3.4 The Two-Rate Test

If the gas flow rate is large, it may be necessary to perform the two-rate test. In this test, two single-rate tests at two different flow rates $\{q_1, q_2\}$ are conducted. Results from the single-rate tests provide enough information to determine two values of skin $\{S_1', S_2'\}$. The resulting two equations for skin may be solved for the two unknowns S and D in the expression $S' = S + D|q|$.

8.3.5 The Reservoir Limits Test

The drawdown test provides information about the limits of a reservoir. The reservoir limits test requires that pressure drawdown be continued until pseudosteady state (PSS) flow is achieved. The beginning of PSS flow is given by the stabilization time t_s:

$$t_s \approx 380 \frac{\phi \mu c_T A}{K} \tag{8.3.14}$$

where stabilization time, t_s, is in hours and A is the drainage area in square feet. Drainage area depends on drainage radius, which is uncertain. The drainage area for a radial system may be approximated as $A = \pi r_e^2$, with r_e the drainage radius. The uncertainty in drainage radius will introduce uncertainty in the estimate of stabilization time. Consequently, the onset of PSS flow is only an approximation.

The pseudopressure equation for PSS flow has the straight line form

$$m(P_{wf}) = m't + m(P_{\text{intercept}}) \tag{8.3.15}$$

when $m(P_{wf})$ is plotted against t. The quantity m' is the slope of the line. It has the form

$$m' = -2.356 \frac{qT}{\phi(\mu c_T)HA} \tag{8.3.16}$$

for a reservoir with thickness H and temperature T. The constant $m(P_{\text{intercept}})$ is the time-independent intercept of the infinite-acting straight line. Given the slope m', we can estimate the drainage volume V_d (cu ft) as

$$V_d = \phi HA = -2.360 \frac{qT}{m'(\mu c_T)} \tag{8.3.17}$$

8.3.6 The Radius of Investigation

The radius of investigation for a pressure transient test in a gas well is an estimate of the distance the pressure transient moves away from the wellbore in a specified time. It may be estimated for a gas well using

$$r_i = 0.0325\sqrt{\frac{K\Delta t}{\phi\mu c_T}}$$ (8.3.18)

for $r_i \le r_e$, where

    r_e = drainage radius (ft)
    r_i = radius of investigation (ft)
    Δt = shut-in time (hr)
    K = permeability (md)
    ϕ = porosity (fraction)
    μ = viscosity (cp)
    c_T = total compressibility (1/psia)

A comparison of Eq. (8.3.18) with Eq. (8.2.24) shows that the radius of investigation for gas wells has the same functional dependence as the radius of investigation for oil wells, but the coefficient is larger for gas than for oil.

Before leaving this section, it is worth noting the dependence of stabilization time and radius of investigation on permeability. Tight gas reservoirs have very low permeabilities (less than 1 md). Consequently, the stabilization time for a tight gas reservoir is very long when compared to the stabilization time of conventional reservoirs with permeabilities greater than 1 md. Similarly, the radius of investigation of a tight gas reservoir is relatively small when compared to the radius investigation of a conventional reservoir. It is often too expensive to conduct a pressure transient test in a tight gas well until the reservoir boundary is reached because the stabilization time is too long.

8.4 Well Test Capabilities

Many types of well tests are available that can provide information about properties of interest to reservoir characterization. Several are listed in Table 8.2 (Kamal et al., 1995), and Lee (2007) includes a discussion of well tests in horizontal wells, hydraulically fractured wells, and wells in naturally fractured reservoirs. Table 8.2 illustrates the type of mega scale information that can be obtained from well test data. Many tests are performed on a single well, while others require changing rates or monitoring pressures in two or more wells. The wells in many tests can be either vertical or horizontal. Table 8.2 identifies the properties that can be determined by each test and notes the time in the life of the project when the test is most likely to be run. It is usually necessary to run a variety of well tests as the project matures. These tests help refine the operator's understanding of the field and often motivate changes in the way the well or the field is operated.

Although it is beyond the scope of this book to discuss all these tests in detail, we have attempted to convey a sense of how well testing can contribute to the reservoir characterization process. Pressure transient testing provides information

Table 8.2 Reservoir Properties Obtainable from Transient Tests

Type of Test	Properties	Development Stage
Drill stem tests	Reservoir behavior Permeability Skin Fracture length Reservoir pressure Reservoir limit Boundaries	Exploration and appraisal wells
Repeat-formation tests/ Multiple-formation tests	Pressure profile	Exploration and appraisal wells
Drawdown tests	Reservoir behavior Permeability Skin Fracture length Reservoir limit Boundaries	Primary, secondary, and enhanced recovery
Buildup tests	Reservoir behavior Permeability Skin Fracture length Reservoir pressure Reservoir limit Boundaries	Primary, secondary, and enhanced recovery
Step-rate tests	Formation parting pressure Permeability Skin	Secondary and enhanced recovery
Falloff tests	Mobility in various banks Skin Reservoir pressure Fracture length Location of front Boundaries	Secondary and enhanced recovery
Interference and pulse tests	Well communication Porosity Interwell permeability Vertical permeability	Primary, secondary, and enhanced recovery
Layered reservoir tests	Layer properties Horizontal permeability Vertical permeability Skin Average layer pressure Outer boundaries	Throughout reservoir life

about individual well performance, wellbore damage, reservoir pressure, and reservoir fluid flow capacity. Perhaps even more important from a reservoir characterization perspective, pressure transient testing can be used to estimate the distance to reservoir boundaries, structural discontinuities, and communication among wells. The reader should consult the literature for more information about specific tests.

CS.8 Valley Fill Case Study: Well Pressures

Well tests provide information about reservoir continuity, flow capacity, and pressure distribution. Figure 8.3 is a typical Horner plot for wells in the Valley Fill study. Table CS.8A is a tabulation of shut-in pressure versus time data for the pressure buildup test. The PBU test used a production period of 24 hours at a stabilized flow rate of 100 STB/D in a well with inner radius of 0.25 feet.

The slope of the infinite-acting period corresponds to a permeability of approximately 150 md for a formation thickness of 120 feet. For lack of better data, the permeability distribution is assumed to be isotropic with horizontal permeability equal to 150 md. Vertical permeability is assumed to be one-tenth the horizontal permeability. No direct measurements of permeability anisotropy or vertical permeability were made.

Table CS.8B shows the initial pressures and associated datum depths at well locations prior to production. The pressures were obtained from a pressure distribution determined using a reservoir flow model that was matched to well performance. They could have been measured in the field using a drill stem test (DST) or a repeat formation test (RFT).

Table CS.8A Pressure Buildup Test Data

Shut-In Time (hrs)	Shut-In Pressure (psia)	Shut-In Time (hrs)	Shut-In Pressure (psia)
0.2	3,991.05	6.0	3,999.42
0.4	3,994.63	7.0	3,999.46
0.6	3,996.42	8.0	3,999.47
0.8	3,998.21	9.0	3,999.48
1.0	3,998.84	10.0	3,999.49
1.5	3,998.98	12.0	3,999.49
2.0	3,999.08	14.0	3,999.50
2.5	3,999.15	16.0	3,999.50
3.0	3,999.21	18.0	3,999.51
3.5	3,999.26	20.0	3,999.51
4.0	3,999.30	25.0	3,999.51
4.5	3,999.34	30.0	3,999.52
5.0	3,999.37		

Table CS.8B Initial Pressures in the Valley Fill Reservoir

Well	Pressure (psia)	Datum Depth (ft)
1	4,006	8,543
2	4,003	8,538
3	4,012	8,558
4	4,008	8,548
5	4,014	8,563
6	4,008	8,548

Exercises

8-1. Plot Horner time $t_H = (t_F + \Delta t)/\Delta t$ as a function of shut-in time in the range 1 hr $\leq \Delta t \leq$ 300 hours for flowing times of 10, 100, and 1,000 hours. Is Horner time largest at small shut-in time or large shut-in time?

8-2. Use Eq. (8.2.2) to calculate dimensionless time for a well that has produced 10,000 STB/D of dry oil for 15 days. Other data are as follows:

$K = 90$ md
$\phi = 0.17$
$\mu = 13.2$ cp
$c_T = 20.0 \times 10^{-6}$ psi^{-1}
$r_w = 0.5$ ft

8-3. Suppose the following physical properties apply to a drill stem test (DST) in an oil well:

$K = $ permeability $= 150$ md
$\phi = $ porosity $= 0.20$
$\mu = $ viscosity $= 1.0$ cp
$c_T = $ total compressibility $= 10 \times 10^{-6}$ psia^{-1}

Calculate the radius of investigation at shut-in times of 0.5 day, 1 day, and 2 days using Eq. (8.2.24).

8-4. Suppose the following physical properties apply to a pressure transient test on a gas well:

$K = $ permeability $= 1.07$ md
$\phi = $ porosity $= 0.14$
$\mu = $ viscosity $= 0.0159$ cp
$c_T = $ total compressibility $= 5.42 \times 10^{-4}$ psia^{-1}

Calculate the radius of investigation at shut-in times of 0.5 day, 1 day, and 2 days using Eq. (8.3.18).

8-5. Prepare a Horner plot of the PBU data in Table CS.8A. Find the slope of the infinite-acting straight line and use it to estimate the permeability of the producing formation.

8-6. A gas reservoir contains two layers of rock. Layer 1 has permeability $= 1.07$ md, and Layer 2 has permeability $= 150$ md. Use Eq. (8.3.14) to estimate drawdown stabilization time in each layer for a well with the following physical properties:

$r_e = $ drainage radius $= 100$ ft
$\phi = $ porosity $= 0.14$
$\mu = $ viscosity $= 0.0159$ cp
$c_T = $ total compressibility $= 5.42 \times 10^{-4}$ psia^{-1}
Area is estimated as $A = \pi r_e^2$.

9 Production Evaluation Techniques

The ability to extract fluids, or "produce," from a subsurface reservoir is one of the key objectives of reservoir management. Production analysis methods include empirical and analytical techniques. Empirical techniques fit a curve to production data, while analytical techniques match a physics-based mathematical model to production data. An empirical technique is decline curve analysis, and an analytical technique is material balance analysis. Semi-analytic techniques combine empirical relationships and physics-based mathematical models. Empirical techniques are widely used to prepare reserves estimates and yield information that can help differentiate among reservoir realizations.

Analytic and semi-analytic methods can be used to identify flow regimes and characterize reservoirs. Production evaluation techniques are used for a variety of reservoir management purposes, such as estimating the original size and shape of the reservoir, identifying flow paths, and forecasting reservoir performance. In this chapter the production evaluation techniques decline curve analysis, gas well deliverability, material balance analysis, production performance monitoring, and tracer production are discussed.

9.1 Decline Curve Analysis

Decline curve analysis is the study of the relationship between oil flow rate q and time t for producing wells (Arps, 1945; Economides et al., 1994; Towler, 2002). In the early 1900s, production analysts observed that future production could be predicted by fitting an exponential equation to historical decline rates. The exponential equation worked well for many oil reservoirs in production at the time, but it did not adequately represent the behavior of some producing wells in depletion drive reservoirs. A better fit was obtained using a hyperbolic decline equation for these wells. Assuming constant flowing pressure, a general equation for the empirical exponential and hyperbolic relationships used in decline curve analysis is

$$\frac{dq}{dt} = -aq^{n+1} \tag{9.1.1}$$

where a and n are empirically determined constants. The empirical constant n ranges from 0 to 1.

Solutions to Eq. (9.1.1) show the expected decline in flow rate as the production time increases. Fitting an equation in the form of Eq. (9.1.1) to flow rate data is

Integrated Reservoir Asset Management. DOI: 10.1016/B978-0-12-382088-4.00009-8

referred to as decline curve analysis. Three decline curves have been identified based on the value of n. The *exponential decline* curve corresponds to $n = 0$ and has the solution

$$q = q_i e^{-at} \tag{9.1.2}$$

where q_i is initial rate and a is a factor that is determined by fitting Eq. (9.1.2) to well or field data.

The *hyperbolic decline* curve corresponds to a value of n in the range $0 < n < 1$. The rate solution has the form

$$q^{-n} = nat + q_i^{-n} \tag{9.1.3}$$

where q_i is initial rate and a is a factor that is determined by fitting Eq. (9.1.3) to well or field data.

The *harmonic decline* curve corresponds to $n = 1$. The rate solution is equivalent to Eq. (9.1.3) with $n = 1$; thus

$$q^{-1} = nat + q_i^{-1} \tag{9.1.4}$$

where q_i is initial rate and a is a factor that is determined by fitting Eq. (9.1.4) to well or field data.

Decline curves are fit to actual data by plotting the logarithm of observed rates versus time t. The semilog plot yields the following exponential decline equation:

$$\ln q = \ln q_i - at \tag{9.1.5}$$

Equation (9.1.5) has the form $y = mx + b$ for a straight line with slope m and intercept b. In the case of exponential decline, time t corresponds to the independent variable x, $\ln q$ corresponds to the dependent variable y, $\ln q_i$ is the intercept b, and $-a$ is the slope m of the straight line. Cumulative production for decline curve analysis is the integral of the rate from the initial rate q_i at time $t = 0$ to the rate q at time t. For example, the cumulative production for the exponential decline case is

$$N_p = \int_0^t q\,dt = \frac{q_i - q}{a} \tag{9.1.6}$$

The decline factor a is for the exponential decline case and is found by rearranging Eq. (9.1.5); thus

$$a = -\frac{1}{t} \ln \frac{q}{q_i} \tag{9.1.7}$$

Oil reserves can be estimated by fitting a decline curve to oil production and then extrapolating the decline curve to an abandonment rate. The difference between current production and production at abandonment is an estimate of reserves.

9.2 Gas Well Deliverability

Gas well deliverability tests are used to predict the flow rate of a gas well during reservoir depletion (Canadian Energy Resources Conservation Board, 1975; Beggs, 1984; Ahmed, 2000; Lee, 2007). Measurements of reservoir pressures and their corresponding flow rates are obtained during the test and are subsequently analyzed. Selection of a deliverability test depends on the length of time needed to stabilize the rate of pressure decline. Pressure is stabilized when pressure fluctuations are minimized. An estimate of stabilization time is given by

$$t_s = 1{,}000 \frac{\phi \mu r_e^2}{kP_r} \tag{9.2.1}$$

where

t_s = stabilization time (hr)
P_r = stabilized reservoir pressure (psia)
μ = gas viscosity at P_r (cp)
ϕ = porosity (fraction)
k = effective permeability (md)
r_e = outer radius of drainage area (ft)

Stabilized reservoir pressure P_r is obtained by shutting in the well until reservoir pressure stabilizes.

Single well gas deliverability tests include the conventional backpressure test, the isochronal test, and the modified isochronal test. Each flow rate for the conventional backpressure test should be maintained until the rate of pressure decline stabilizes. As reservoir permeability decreases, t_s increases. The stabilization time for the backpressure test may be excessive for some reservoirs, such as low-permeability reservoirs. In this case, the isochronal test may be used. The isochronal test consists of several flow periods of equal duration. Each flow period begins at static reservoir conditions and uses a different flow rate. The final flow rate is maintained (or extended) until a stabilized pressure is reached.

Although the isochronal test is faster than the conventional backpressure test, production time is lost while the well is shut in until the reservoir returns to static conditions between each flow period. An alternative approach, which is faster but less accurate, is the modified isochronal test. In this case, the shut-in period is the same duration as the flow period, and unstabilized shut-in pressures must be used to evaluate test results.

The applicability of a Simplified Backpressure Analysis (SBA) or a Laminar-Inertial-Turbulent (LIT) analysis depends on the gas flow regime. Lee (2007) pointed out that the SBA method is valid at low pressures (<2,000 psia), while the LIT (pseudopressure) method is valid for all pressures.

9.2.1 The Simplified Backpressure Analysis Method

The backpressure equation is

$$q_{sc} = C(P_r^2 - P_{wf}^2)^n \equiv C(\Delta P^2)^n \tag{9.2.2}$$

where

  C and n = empirical parameters
  q_{sc} = gas flow rate at standard conditions (MMSCFD)
  P_r = stabilized reservoir pressure (psia)
  P_{wf} = flowing wellbore pressure (psia)

Taking the logarithm of the backpressure equation for measurement i yields

$$\log q_{sci} = n \log(\Delta P^2)_i + \log C \tag{9.2.3}$$

If we plot $\log q_{sci}$ versus $\log(\Delta P^2)_i$, we obtain the equation for a straight line where n is the slope and $\log C$ is the intercept. The absolute open flow (AOF) of the well is the rate corresponding to $P_{wf} = 0$. Rates can be calculated from the backpressure equation using the values of C and n determined by a least squares fit of test data.

9.2.2 The Laminar-Inertial-Turbulent Method

The LIT method is applicable to any gas deliverability test. Test data are fit to the LIT equation

$$m(P_r) - m(P_{wf}) \equiv \Delta m = aq_{sc} + bq_{sc}^2 \tag{9.2.4}$$

where

  q_{sc} = gas flow rate at standard conditions (MMSCFD)
  $m(P_r)$ = pseudopressure corresponding to P_r (psia^2/cp)
  $m(P_{wf})$ = pseudopressure corresponding to P_{wf} (psia^2/cp)
  aq_{sc} = laminar flow
  bq_{sc} = inertial and turbulent flow

The LIT equation is second order in rate. Dividing Eq. (9.2.4) for measurement i by q_{sci} lets us write the LIT equation as

$$\frac{\Delta m_i}{q_{sci}} = bq_{sci} + a \tag{9.2.5}$$

Plotting $\Delta m_i/q_{sci}$ versus q_{sci} gives a straight line with slope b and intercept a. Values of a and b can be determined by a least squares fit of test data, and flow rate can be calculated by solving Eq. (9.2.4) for q_{sc} using the quadratic formula

$$q_{sc} = \frac{-a + \sqrt{a^2 + 4b\Delta m}}{2b}$$ (9.2.6)

Absolute open flow occurs when $m(P_{wf}) = 0$.

9.3 Material Balance

The law of conservation of mass is the basis of material balance calculations. Material balance is an accounting of material entering or leaving a system. The calculation treats the reservoir as a large tank of material and uses quantities that can be measured to determine the amount of a material that cannot be directly measured. Measurable quantities include cumulative fluid production and injection volumes for oil, water, and gas phases; accurate reservoir pressure measurements; and fluid property data from samples of produced fluids.

Material balance calculations may be used for several purposes. They provide an independent method of estimating the volume of oil, water, and gas in a reservoir for comparison with volumetric estimates. The magnitude of various factors in the material balance equation indicates the relative contribution of different drive mechanisms at work in the reservoir. Material balance can be used to predict future reservoir performance and aid in estimating cumulative recovery efficiency. More discussion of these topics can be found in Craft et al. (1991), Ahmed (2000), Dake (2001), Towler (2002), Walsh and Lake (2003), Lee (2007), and Miller and Holstein (2007). The form of the material balance equation depends on whether the reservoir is predominately an oil reservoir or a gas reservoir. Each of these cases is considered separately.

9.3.1 Oil Reservoir Material Balance

The general material balance equation for an oil reservoir is the Schilthuis material balance equation (1961):

$$
\begin{aligned}
N(B_t - B_{ti}) + NmB_{ti}\left(\frac{B_{gc} - B_{gi}}{B_{gi}}\right) &+ N\frac{B_{ti}S_{wio}}{1 - S_{wio}}\left(\frac{B_{tw} - B_{twi}}{B_{twi}}\right) \\
+ N\frac{mB_{ti}S_{wig}}{1 - S_{wig}}\left(\frac{B_{tw} - B_{twi}}{B_{twi}}\right) &+ N\left(\frac{1}{1 - S_{wio}} + \frac{m}{1 - S_{wig}}\right)B_{ti}c_f\Delta P \\
= N_pB_o - N_pR_{so}B_g &\\
+ \left[G_{ps}B_g + G_{pc}B_{gc} - G_iB_g'\right] &- (W_e + W_i - W_p)B_w
\end{aligned}
$$ (9.3.1)

This equation is derived by conserving volume and is referred to as the volumetric material balance by Dake (2001). All of the terms in Eq. (9.3.1) are defined in Table 9.1. The unit of each quantity is presented in square brackets in the Nomenclature table at the front of the book.

Table 9.1 Nomenclature for Equation (9.3.1)

B_g	Gas formation volume factor (FVF) (RB/SCF)
B_{gc}	Gas cap FVF (RB/SCF)
B_g'	Injected gas FVF (RB/SCF)
B_o	Oil FVF (RB/STB)
B_t	$B_o + (R_{si} - R_{so})B_g$ = composite oil FVF (RB/STB)
B_{tw}	$B_w + (R_{swi} - R_{sw})B_g$ = composite water FVF (RB/STB)
c_f	Formation (rock) compressibility (1/psia)
G	Initial gas in place (SCF)
G_i	Cumulative gas injected (SCF)
G_{pc}	Cumulative gas cap gas produced (SCF)
G_{ps}	Cumulative solution gas produced as evolved gas (SCF)
m	Ratio of gas reservoir volume to oil reservoir volume
N	Initial oil in place (STB)
N_p	Cumulative oil produced (STB)
R_{so}	Solution gas–oil ratio (SCF/STB)
R_{si}	Initial solution gas–oil ratio (SCF/STB)
R_{sw}	Solution gas–water ratio (SCF/STB)
R_{swi}	Initial solution gas–water ratio (SCF/STB)
S_g	Gas saturation (fraction)
S_o	Oil saturation (fraction)
S_w	Water saturation (fraction)
S_{wi}	Initial water saturation (fraction)
S_{wig}	Initial water saturation in gas cap (fraction)
S_{wio}	Initial water saturation in oil zone (fraction)
W_e	Cumulative water influx (STB)
W_i	Cumulative water injected (STB)
W_p	cumulative water produced (STB)
ΔP	$P_i - P$ = reservoir pressure change (psia)
P_i	Initial reservoir pressure (psia)
P	Reservoir pressure corresponding to cumulative fluid times (psia)

The physical significance of the terms in Eq. (9.3.1) can be displayed by first defining the terms

$$D_o = B_t - B_{ti}$$

$$D_{go} = mB_{ti}\left(\frac{B_{gc} - B_{gi}}{B_{gi}}\right)$$

$$D_w = \frac{B_{ti}S_{wio}}{1 - S_{wio}}\left(\frac{B_{tw} - B_{twi}}{B_{twi}}\right)$$

$$D_{gw} = \frac{mB_{ti}S_{wig}}{1 - S_{wig}} \left(\frac{B_{tw} - B_{twi}}{B_{twi}} \right)$$

$$D_r = \left(\frac{1}{1 - S_{wio}} + \frac{m}{1 - S_{wig}} \right) B_{ti}c_f \Delta P \tag{9.3.2}$$

Substituting Eq. (9.3.2) in Eq. (9.3.1) gives the general material balance equation in the form

$$N\left[D_o + D_{go} + D_w + D_{gw} + D_r\right] = N_pB_o - N_pR_{so}B_g$$
$$+ \left[G_{ps}B_g + G_{pc}B_{gc} - G_iB'_g\right] - \left(W_e + W_i - W_p\right)B_w \tag{9.3.3}$$

The terms on the right-hand side of Eq. (9.3.3) represent fluid production and injection, while the terms on the left-hand side represent volume changes. The physical significance of each term is summarized in Table 9.2.

Equation (9.3.1) is considered a general material balance equation because it can be applied to an oil reservoir with a gas cap and an aquifer. The derivation of the material balance equation is based on several assumptions: the system is in pressure equilibrium; the system is isothermal; available fluid property data are representative of reservoir fluids; production data is reliable; and gravity segregation of phases can be neglected.

If we rearrange the terms in the general material balance equation for an oil reservoir, Eq. (9.3.1), we can estimate the relative importance of different drive

Table 9.2 Physical Significance of Material Balance Terms

Term	Physical Significance
ND_o	Change in volume of initial oil and associated gas
ND_{go}	Change in volume of free gas
$N(D_w + D_{gw})$	Change in volume of initial connate water
ND_r	Change in formation pore volume
N_pB_o	Cumulative oil production
$N_pR_{so}B_g$	Cumulative gas produced in solution with oil
$G_{ps}B_g$	Cumulative solution gas produced as evolved gas
$G_{pc}B_{gc}$	Cumulative gas cap gas production
$G_iB'_g$	Cumulative gas injection
W_eB_w	Cumulative water influx
W_iB_w	Cumulative water injection
W_pB_w	Cumulative water production

Table 9.3 Drive Indices from the Schilthuis Material Balance Equation

Drive	Index
Solution Gas	$I_{sg} = ND_o/D_{HC}$
Gas Cap	$I_{gc} = ND_{go}/D_{HC}$
Water	$I_w = [(W_e - W_p)B_w]/D_{HC}$
Injected Fluids	$I_i = [W_iB_w + G_iB'_g]/D_{HC}$
Connate Water and Rock Expansion	$I_e = [N(D_w + D_{gw}) + ND_r]/D_{HC}$

mechanisms. Table 9.3 gives the indices representing different drives relative to the hydrocarbon production D_{HC} defined by

$$D_{HC} = N_pB_o + G_{pc}B_{gc} + [G_{ps} - N_pR_{so}]B_g \tag{9.3.4}$$

The sum of the drive indices equals one; thus

$$I_{sg} + I_{gc} + I_w + I_i + I_e = 1 \tag{9.3.5}$$

Equation (9.3.5) can be derived by rearranging Eq. (9.3.1). A comparison of the magnitudes of the drive indices indicates which drive is dominating the performance of the reservoir. If the sum of the drive indices in Eq. (9.3.5) does not equal one based on available data, Pletcher cautioned that the drive indices should not be normalized to one because this may obscure the usefulness of the drive indices and "lead to a false sense of security" (2002, p. 49).

9.3.2 The Gas Reservoir Material Balance

The general material balance equation for a gas reservoir can be derived from Eq. (9.3.1) by first recognizing that the relationship

$$GB_{gi} = NmB_{ti} \tag{9.3.6}$$

defines original gas in place G. Substituting Eq. (9.3.6) into Eq. (9.3.1) gives the general material balance equation

$$
\begin{aligned}
N(B_t - B_{ti}) &+ GB_{gi}\left(\frac{B_{gc} - B_{gi}}{B_{gi}}\right) + N\frac{B_{ti}S_{wio}}{1 - S_{wio}}\left(\frac{B_{tw} - B_{twi}}{B_{twi}}\right) \\
&+ G\frac{B_{gi}S_{wig}}{1 - S_{wig}}\left(\frac{B_{tw} - B_{twi}}{B_{twi}}\right) + \left(\frac{NB_{ti}}{1 - S_{wio}} + \frac{GB_{gi}}{1 - S_{wig}}\right)c_f\Delta P \\
&= N_pB_o + \left[G_{ps}B_g + G_{pc}B_{gc} - G_iB'_g\right] - N_pR_{so}B_g - (W_e + W_i - W_p)B_w
\end{aligned}
\tag{9.3.7}
$$

Equation (9.3.7) is further simplified by recognizing that the material balance for a gas reservoir does not include oil in place, so that $N = 0$ and $N_p = 0$. The resulting material balance equation is

$$
GB_{gi}\left(\frac{B_{gc} - B_{gi}}{B_{gi}}\right) + G\frac{B_{gi}S_{wig}}{1 - S_{wig}}\left(\frac{B_{tw} - B_{twi}}{B_{twi}}\right) + \left(\frac{GB_{gi}}{1 - S_{wig}}\right)c_f \Delta P
$$
$$
= \left[G_{pc}B_{gc} - G_iB_g'\right] - \left(W_e + W_i - W_p\right)B_w
$$
(9.3.8)

Water and formation compressibility are relatively small compared to gas compressibility. Consequently, Eq. (9.3.8) is often written in the following simplified form

$$
GB_{gi}\left(\frac{B_{gc} - B_{gi}}{B_{gi}}\right) = \left[G_{pc}B_{gc} - G_iB_g'\right] - \left(W_e + W_i - W_p\right)B_w
$$
(9.3.9)

9.4 Production Performance Ratios and Drive Mechanisms

The ratio of one produced fluid phase to another provides important information for understanding the dynamic behavior of a reservoir. Let Q_o, Q_w, Q_g be oil, water, and gas production rates, respectively. Frequently used produced fluid ratios include gas–oil ratio (GOR):

$$
GOR = Q_g/Q_o
$$
(9.4.1)

gas–water ratio (GWR):

$$
GWR = Q_g/Q_w
$$
(9.4.2)

and water–oil ratio (WOR):

$$
WOR = Q_w/Q_o
$$
(9.4.3)

One more produced fluid ratio is water cut, which is the water production rate divided by the sum of the oil and water production rates:

$$
WCT = Q_w/(Q_o + Q_w)
$$
(9.4.4)

The dominant produced fluid ratio depends on the mechanisms that are active in the reservoir.

Petroleum is initially produced by natural forces called drive mechanisms. The most common drive mechanisms are water drive, solution or dissolved gas drive, and gas cap drive. The most efficient drive mechanism for producing oil is water drive. Water from a subsurface source displaces the oil during the production process. The useful life of natural aquifer energy can be prolonged by balancing oil withdrawal with the rate of water influx. Water drive recovery ranges from 35 to 75 percent of the original oil in place (Ahmed, 2000).

In a solution gas drive process, gas dissolved in oil at reservoir pressure and temperature is liberated as the reservoir pressure declines during primary production. In addition to oil flow from high-pressure regions to lower pressure in the wellbore, gas expansion and movement carry some of the oil to the production wells. Solution gas drive recovery ranges from 5 to 30 percent of the original oil in place (Ahmed, 2000).

If an oil reservoir is in communication with free gas, the free gas is a gas cap that can provide energy to the production process. The resulting drive is referred to as gas cap drive. Oil production through wells that are completed in the oil zone causes a decline in reservoir pressure that allows gas to expand and displace oil-to-well completions. When the gas–oil contact reaches the uppermost perforations, large volumes of gas will be produced with the oil. Gas cap drive recovery ranges from 20 to 40 percent of the original oil in place (Ahmed, 2000). Higher recoveries are possible in steeply dipping reservoirs when flow capacity is sufficient to allow good oil drainage to downstructure production wells.

The recovery factor for gas reservoirs is usually much greater than that for oil reservoirs. Recovery factors as high as 80 to 90 percent of the original gas in place are economically possible for volumetric gas reservoirs undergoing depletion without water drive (Ahmed, 2000). The low density and high mobility of gas relative to oil are primarily responsible for the relatively large gas reservoir recovery factors. Recovery from water drive gas reservoirs tends to be lower, however, because gas is trapped as water encroaches into the gas zone. Typical recoveries range from 50 to 70 percent of the original gas in place (Ahmed, 2000).

Primary production from many reservoirs occurs under conditions when two or more of these drive mechanisms are active simultaneously. This type of drive is referred to as a combination drive. Field production performance depends on which mechanism is dominant at different points during the primary production period. As time evolves, the dominant mechanism may change. Detailed studies using reservoir simulators are often the only cost-effective means of determining the dominant mechanism and predicting the behavior of combination drive reservoirs.

The production performance of certain properties, such as reservoir pressure and GOR, can be used to identify the drive mechanism if a sufficient reservoir production history exists. For example, solution gas drive reservoirs show a significant increase in GOR followed by a decline in GOR as available gas is produced. By contrast, reservoir pressure declines more slowly in water drive reservoirs than in other reservoirs. Water drive is usually the most effective oil recovery mechanism if enough water is available to balance hydrocarbon withdrawal.

A number of injection processes have been developed to supplement natural reservoir energy and significantly increase oil and gas recovery. Water drive can be an effective means of displacing oil to production wells regardless of the source of water. Thus, if a reservoir does not have significant aquifer support, injection wells can be used to supplement existing natural resources. Water breakthrough is typically observed as a significant increase in produced WOR. If the volume of water injected is large enough, the reservoir pressure can increase, and free gas in the reservoir can be dissolved into the reservoir oil phase. The mechanism of free gas being forced to dissolve in the oil phase because water injection is causing an increase in reservoir pressure can appear as a reduction in GOR associated with an increase in reservoir pressure.

Another injection fluid is gas. The composition of injected gas can range from reinjection of produced gas or enriched gas to injection of inorganic gases such as carbon dioxide, nitrogen, or even air. The displacement efficiency of each gas depends on its interaction with reservoir rock and fluids. As in water injection, gas breakthrough can often be recognized when a significant increase in produced GOR occurs.

9.5 Production Stages

The production life of a reservoir begins when fluid is withdrawn from the reservoir. Production can begin immediately after the discovery well is drilled or several years later after several delineation wells have been drilled. Delineation wells are used to define the reservoir boundaries, while the development wells are used to optimize resource recovery. Optimization criteria are defined by management and should take into account the relevant governmental regulations. The optimization criteria may change during the life of the reservoir for a variety of reasons, including changes in technology, economic factors, and new information obtained during various stages of reservoir production: primary, secondary, and alternative. The stages of reservoir production are described in the following subsections.

9.5.1 Primary Production

Primary production is ordinarily the first stage of production. It relies entirely on natural energy sources. Production can result from the expansion of *in situ* fluids as pressure declines during primary reservoir depletion. Alternatively, production can occur when an *in situ* fluid such as oil is replaced by another fluid, such as water, natural gas, or air. The natural forces involved in primary production are called reservoir drives. The most common reservoir drives for oil reservoirs are water drive, solution or dissolved gas drive, and gas cap drive.

Gravity drainage is the least common of the primary production mechanisms. In this case, oil moves downstructure to a producing well. Downstructure movement of oil in an oil–water system is the result of a pressure gradient that favors downstructure oil flow over oil movement upstructure due to gravity segregation. Gravity drainage can be effective when it works. It is most likely to occur in shallow, highly permeable, steeply dipping reservoirs.

To extract oil from wells, it is sometimes necessary to provide energy using different types of pumps or to inject gas to increase the buoyancy of the gas–oil mixture. The earliest pumps used the same wooden beams that were used for cable-tool drilling. Oil companies developed central pumping power in the 1880s. Central pumping power used a prime mover—a power source—to pump several wells. In the 1920s, demand for the replacement of on-site rigs led to the use of a beam pumping system for pumping wells. A beam pumping system is a self-contained unit that is mounted at the surface of each well and operates a pump in the hole. More modern techniques include gas-lift and electric submersible pumps.

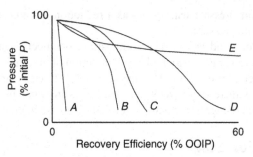

A : Liquid and rock expansion
B : Solution gas drive
C : Gas cap expansion
D : Gravity drainage
E : Water influx

Figure 9.1 A comparison of primary production mechanisms.

A schematic comparison of primary production mechanisms on reservoir pressure and recovery efficiency is shown in Figure 9.1. In many cases, two or more drive mechanisms are functioning simultaneously. The behavior of the field depends on which mechanism is most important at various times during the life of the field. The best way to predict the behavior of such fields is by using computer-based reservoir flow models.

Although the preceding discussion refers to oil reservoirs, similar comments apply to gas reservoirs. Water drive and gas expansion with reservoir pressure depletion are the most-common drives for gas reservoirs. Gas reservoir recovery can be as high as 70 to 90 percent of original gas in place (OGIP) because of the high mobility of gas relative to oil mobility.

Gas storage reservoirs have a different life cycle than gas reservoirs that are being depleted. Gas storage reservoirs are used as a warehouse for gas. If the gas is used as a fuel for power plants, it will also need to be periodically produced and replenished. The performance attributes of a gas storage reservoir are verification of inventory, assurance of deliverability, and containment against migration (Tek, 1996). The gas inventory consists of working gas and cushion gas. Gas deliverability must be sufficient to account for swings in demand, which can vary daily and seasonally. Gas containment is needed to conserve the amount of stored gas.

9.5.2 Secondary Production

Primary depletion is usually not sufficient to optimize recovery from an oil reservoir. Oil recovery can be both accelerated and increased by supplemental natural reservoir energy. The supplemental energy is provided using an external energy source, such as water injection or gas injection. The injection of water or natural gas may be referred to as pressure maintenance or secondary production. The latter term arose because injection usually follows a period of primary pressure depletion and is therefore the second production method used in a field. Many modern reservoirs incorporate pressure maintenance early in the production life of the field, sometimes from the beginning of production. In this case the reservoir is not subjected to a conventional primary production phase. The term "pressure maintenance" more accurately describes the reservoir management strategy for these fields than the term "secondary production."

9.5.3 Alternative Classifications

Both primary and secondary recovery processes are designed to produce oil by injecting fluids that do not mix with the oil. The result is an immiscible displacement method. Additional methods may be used to improve oil recovery efficiency by reducing residual oil saturation. Residual oil saturation is the saturation of oil that remains following displacement of oil by another fluid. Multiphase displacement concepts are discussed in more detail in Chapter 10. The reduction of residual oil saturation requires a change in such factors as interfacial tension (IFT) or wettability. Methods designed to reduce residual oil saturation have been referred to in the literature as tertiary production, enhanced oil recovery, and improved oil recovery. The term "tertiary production" was originally used to identify the third stage of the production life of the field, which typically occurred after waterflooding. The third stage of oil production would involve a process that was designed to mobilize waterflood residual oil.

Tertiary production processes were designed to improve displacement efficiency by injecting fluids or heat. They were referred to as enhanced recovery processes. It was learned, however, that some fields would perform better if the enhanced recovery process was implemented before the third stage in the life of the field. In addition, it was found that enhanced recovery processes were often more expensive than just drilling more wells in a denser pattern.

The drilling of wells to reduce well spacing and to increase well density is called "infill drilling." The birth of this term was coincident with the birth of the term "improved recovery," which includes both enhanced oil recovery and infill drilling. Some major improved recovery processes are waterflooding, gasflooding, chemical flooding, and thermal recovery. They are discussed in more detail in Chapter 15.

9.6 Tracer Tests

Tracers are chemicals that are injected into one part of the reservoir and monitored in other parts. They provide information that can be used in a variety of ways. This section discusses the kind of information that can be obtained from tracer testing, identifies several types of tracers, and outlines tracer test procedures. We then look at some simple yet practical tracer calculations in the next section. Dietz (1987), Loder (2000), and Dugstad (2007) provide additional information about tracer technology.

9.6.1 Tracer Test Information

Tracers can provide a wealth of information about reservoir flow paths, which are the paths that fluids follow when moving from one location to another. A flow path can include uniform movement through a permeable formation or channeling through a fracture or a high-permeability zone. Tracers can provide a direct connection between one point in the reservoir (a specific injection well) and another point in the reservoir (a specific production well). Not only can the path be defined, but the time it takes the tracer to traverse the path can be determined as well.

Directional flow trends can be identified by observing differences in tracer arrival times at production wells. Directional flow trends indicate the presence of flow anisotropy that needs to be identified in reservoir characterization. Different arrival times suggest preferred flow paths between tracer injection wells and producing wells. Changes in rates can alter areal sweep and influence arrival times. Rapid interwell communication can be identified if the tracer breakthrough time is short. This may imply channeling or a high-permeability streak. If an injection well does intersect channels or high-permeability streaks, vertical conformance treatments may be used to alter permeability.

Tracers may be used to estimate the fraction of the region of interest that has been flooded, or swept, by injected fluids. Estimates of sweep efficiency require knowledge of injection and production rates and well patterns. This knowledge can lead to improved estimates of reservoir pore volume and flow paths.

If tracer breakthrough occurs after injecting a small volume, it is possible to infer the existence of a channel, fracture, or high-permeability streak. By contrast, if tracer breakthrough does not occur until a large volume of tracer has been injected, the contacted reservoir is exhibiting a thief zone or a more uniform sweep. Thief zones can be natural or manmade. If thief zones are discounted, sweep can be estimated from injected volume. A measurement of tracer production volume can help distinguish a thief zone from uniform sweep because a thief zone tends to remove tracer from the produced fluids.

Flow barriers may be delineated by delayed tracer response. Although delayed tracer response could imply the existence of a thief zone, the presence of flow barriers, such as faults or shale breaks, could also delay the appearance of tracer at a producer. Again, the amount of tracer produced could be used to distinguish between the possibilities of flow barriers or thief zones. Alternatively, formation evaluation using well logs and geologic correlations would provide an independent interpretation that could help evaluate the meaning of delayed tracer response. For example, a fence diagram based on well logs may show that the thief zone interpretation is unlikely. In this case, delayed tracer response would be an indication of a flow barrier. It is possible that the flow barrier could be detected using seismic surveys.

The sequential application of tracer surveys can be used to evaluate the sweep improvement processes. If a tracer survey is conducted before and after applying a sweep improvement process, the difference in surveys can be used to infer the success of the process. Examples of sweep improvement processes include waterflooding, gasflooding, and steamflooding. If the survey after the process is the same as the survey before the process, it is likely that the process was ineffective. On the other hand, substantial changes in the tracer survey performance would suggest that the sweep improvement process had an effect, although not necessarily a positive one. The determination of incremental recovery is one measure that can be used in conjunction with sequential tracer surveys to assess the merit of the sweep improvement process.

Tracers may also be used to evaluate the flow of two fluids with different mobilities, such as water and polymer. One tracer design procedure is to inject a different tracer with each fluid and then compare breakthrough times for each tracer. The tracers should have similar characteristics for the comparison to be meaningful. For example, the breakthrough time should not be affected by the loss of one tracer due to biodegradation or radioactive decay while the other tracer is neither biodegradable nor radioactive. If the breakthrough times of the two tracers are not the same, it can be inferred that the tracers followed different flow paths through the reservoir. This implies that the different displacement fluids exhibited different flow characteristics and contacted (or swept) different parts of the reservoir.

9.6.2 Ideal Tracer Features

One way to assess the quality of a real tracer is to compare it to an ideal tracer. According to Greenkorn (1962), the ideal tracer would "follow the fluid of interest exactly, traveling at the same velocity as the fluid front." In addition, the ideal tracer would consist of easily detectable material, it would not interact with rock or oil, it would be cheap, and it would be free of environmental hazards.

9.6.3 Realistic Tracers for Aqueous Systems

The appropriate tracer to use in a system depends on the fluid phases in the system. In most field applications, tracers are expected to move through the aqueous phase. We focus here on a variety of realistic tracers for aqueous systems. Although none of the tracers exhibit ideal tracer behavior, several realistic tracers can provide valuable information when they are used properly.

Radioisotopes such as tritium (an isotope of hydrogen with two neutrons and one proton in the nucleus) are often used as tracers. Radioisotopes are easily detectable in small concentrations. They have insignificant absorption losses on rock, but they must be handled by licensed personnel and are closely regulated. Wilson and Frederick (1999) discuss some of the environmental issues in more detail.

Fluorescent dyes are also easily detectable in small concentrations. They should be used only when rapid communication is expected—for example, when breakthrough time is expected within a few days. Fluorescent dyes can be applied to the identification of flow paths through thief zones or natural fractures.

Several water-soluble salts can be used as tracers. Some of the examples of water-soluble salts are ammonium thiocyanate (NH_4SCN), ammonium nitrate (NH_4NO_3), sodium or potassium bromide (NaBr, KBr), sodium or potassium iodide (NaI, KI), and sodium chloride (NaCl). These salts are detectable by noting significant changes in the ionic content of produced water. Since many water-soluble salts exist *in situ*, especially sodium chloride, it is important to know the background levels of the salt. Significant deviations from the background level can then be interpreted as the arrival of an injected salt tracer.

Water-soluble alcohols can be inexpensive tracers. Some water-soluble alcohols that are appropriate for application as tracers are 2-propanol (IPA-isopropyl alcohol), methanol (MeOH), and ethanol (EtOH).

9.6.4 Perfluorocarbon Tracers

One of the more useful modern tracers is perfluorocarbon. Perfluorocarbon tracers, or PFTs, are fully fluorinated alkyl substituted cycloalkanes, and their advantages are similar to radioactive tracers (Senum and Fajer, 1992). PFTs have negligible background in atmospheric and subsurface environments. This means that small quantities of PFT may be used and readily detected. A desirable consequence of their chemical composition is the observation that chemically different PFTs may be simultaneously deployed, sampled, and analyzed. PFTs are environmentally safe because they are nontoxic, nonreactive, and nonflammable.

Perfluorocarbon tracers do, however, have some disadvantages. Their flow rate relative to the bulk flow rate of the fluid may be retarded by interphase mass transfer between the aqueous and hydrocarbon phases. The interphase mass transfer complicates the analysis of PFT travel time when compared with the analysis of the travel time associated with tracers that move at the bulk flow rate of the aqueous phase.

9.6.5 Practical Concerns

Several practical concerns must be considered when using tracers. They include some significant governmental and environmental issues in addition to technical issues. For example, environmentally safe methods for disposing of radioisotopes must be part of any prudent reservoir management plan. Similarly, care must be taken to prevent leaks of potentially toxic tracers such as bromides, iodides, and alcohols into potable aquifers. A tracer design should include a check of local government regulations.

Different issues arise from a technical perspective. A tracer survey requires the determination of background levels of the tracer in the region of interest. The claim that a tracer is being produced requires a demonstration that the produced tracer is significantly greater than background levels of tracer. This is especially applicable to radioisotopes and water-soluble salts. If the tracer is a water-soluble alcohol, the tracer survey must recognize that the water-soluble alcohols are susceptible to biodegradation. One way to combat the loss of water-soluble alcohols is to inject bactericides with the alcohols. An alcohol like 2-propanol is soluble in some oils. This can lead to the additional retention of 2-propanol and larger than expected losses or delays in appearance of the tracer if solubility is not included in the tracer design.

High background levels of a tracer can make it difficult to detect, especially if the deviation of the background level from an average background level is large. This is especially significant with naturally occurring inorganic salts or naturally occurring radioactive materials (NORM).

9.7 Tracer Test Design

Tracer test design begins with the identification of the pilot or pattern area (Terry et al., 1981; Dugstad, 2007). Determining the volume of the tracer that must be used depends on the volume of the system through which the tracer must pass. An estimate of the volume in the tracer test region can be obtained from volumetrics, material balance, or reservoir model studies. The volume of the region of interest is needed to estimate tracer breakthrough times and the amount of tracer needed for the study, as illustrated following.

A sample estimate of the mass of tracer m_T needed for injection is obtained by calculating

$$
\begin{aligned}
m_T &= (\text{water volume})(\text{density})(\text{concentration}) \\
&= (\pi r^2 h \phi S_w) \rho_T c_T
\end{aligned}
\tag{9.7.1}
$$

where

m_T = mass of tracer needed for injection (lbm)
r = distance between injector and producer (ft)
h = formation thickness (ft)
ϕ = porosity (fraction)
S_w = water saturation (fraction)
ρ_T = density of tracer solution (lbm/ft^3)
c_T = desired tracer concentration at producer (fraction)

A typical safety factor ranges from two to five times the estimated mass m_T to account for uncertainties in reservoir flow parameters.

One of the most important parameters in Eq. (9.7.1) is the desired tracer concentration. This concentration can be estimated by viewing the motion of a tracer molecule in the porous medium as a random walk process analogous to Brownian motion. According to this view (Scheidegger, 1954; Collins, 1961; Greenkorn, 1962), the tracer molecule obeys the convection–dispersion (C–D) equation for homogeneous reservoirs in N_d spatial dimensions—namely,

$$
\sum_{i=1}^{N_d} \left[D_i \frac{\partial^2 C}{\partial x_i^2} - v_i \frac{\partial C}{\partial x_i} \right] = \phi \frac{\partial C}{\partial t}
\tag{9.7.2}
$$

where $\{x_i, v_i, D_i\}$ are the components of position, Darcy velocity, and dispersion along the ith axis. Darcy velocity and dispersion are assumed to be constants in this analysis. Equation (9.7.2) has the solution

$$
C\left(\vec{x}, t\right) = C_0 \prod_{i=1}^{N_d} \frac{\exp\left[-\frac{1}{2D_i t}\left(x_i - \frac{v_i}{\phi} t\right)^2 \right]}{\sqrt{2\pi D_i t}}
\tag{9.7.3}
$$

where the constant C_0 is used to normalize the concentration. If length is in feet and time is in days, the velocity components are in feet/day, and the dispersion components are in feet^2/day. Although we refer to D_i as dispersion, it can also be used to approximate diffusion.

Equation (9.7.3) is a more general solution than we need for our tracer design estimate. It is sufficient for our purposes to consider the C–D equation in a horizontal plane. A horizontal plane has two space dimensions, so $N_d = 2$. The resulting C–D equation is obtained from Eq. (9.7.2) and has the form

$$D_1 \frac{\partial^2 C}{\partial x_1^2} + D_2 \frac{\partial^2 C}{\partial x_2^2} - v_1 \frac{\partial C}{\partial x_1} - v_2 \frac{\partial C}{\partial x_2} = \phi \frac{\partial C}{\partial t} \qquad (9.7.4)$$

where the subscripts denote orthogonal Cartesian coordinates in the horizontal plane. The solution of Eq. (9.7.4) is

$$C(x_1, x_2, t) = C_0 \frac{\exp\left[-\frac{1}{2D_1 t}\left(x_1 - \frac{v_1}{\phi}t\right)^2 - \frac{1}{2D_2 t}\left(x_2 - \frac{v_2}{\phi}t\right)^2 \right]}{\sqrt{(2\pi)^2 D_1 D_2 t^2}} \qquad (9.7.5)$$

If fluid flow is in the x_1 direction, then D_1 is the longitudinal dispersion D_L, and D_2 is the transverse dispersion D_T. As an illustration, assume the reservoir is isotropic, so that $D_L = D_T = D$. Equation (9.7.5) becomes

$$C(r, t) = \frac{C_0}{2\pi D t} \exp\left[-\frac{1}{2Dt}\left(r - \frac{v_r}{\phi}t \right)^2 \right] \qquad (9.7.6)$$

where

r = radial distance from injector (ft)
t = time after injection (days)
D = areal dispersion (ft^2/day)
v_r = Darcy velocity of frontal advance in radial direction (ft/day)

If the tracer is radioactive, Eq. (9.7.6) can be modified to account for the radioactive half-life of the tracer by including an exponential decay term:

$$C(r, t) = \frac{C_0}{2\pi D t} \exp\left[-\frac{1}{2Dt}\left(r - \frac{v_r}{\phi}t \right)^2 - \frac{0.693t}{t_H} \right] \qquad (9.7.7)$$

where

t_H = tracer half-life (days)

Equations (9.7.6) and (9.7.7) express tracer concentration as a function of radial position and time.

Flow path information within the region of interest can be used to predict tracer breakthrough locations. Actual tracer performance provides information about the quality of the reservoir characterization that was used to make the pretest estimates. The discussion that follows outlines the procedure for performing a tracer test.

The reservoir should undergo waterflooding long enough to fill any void space in the volume of the reservoir that is going to be swept by the tracer prior to the tracer test. This minimizes the potential loss of a tracer in the reservoir. Once the reservoir has been "pressured up," tracers should be injected as rapidly as possible. Rapid tracer injection forms a slug of tracer that can propagate through the reservoir. However, the injection rate should be designed so that the solubility of tracer in *in situ* water is not exceeded. If the solubility of a tracer is exceeded, the tracer can be lost by oversaturating the system. Tracer solubility can be determined by testing the solubility of the tracer in samples of *in situ* water. Table 9.4 presents the solubilities for some selected tracers (Terry et al., 1981).

It is necessary to implement a sampling program for detecting a tracer at producing wells in the region of interest. The sampling frequency should balance two factors: sample often enough to observe breakthrough and do not sample so often that costs are excessive. Table 9.5 provides some suggested sampling intervals. If tracer breakthrough is expected to occur shortly after tracer injection begins, a short sampling interval should be implemented. As the expected tracer breakthrough time increases, the sampling interval can increase as well. The sampling interval depends on expected tracer breakthrough times, which can be estimated as described on the following page.

Table 9.4 Solubilities of Selected Tracers

Tracer	Solubility in Distilled Water (lb/bbl)	Recommended Injection Concentration (lb/bbl)
Ammonium Nitrate	1,280	200
Ammonium Thiocyanate	420	200
Potassium Bromide	187	100
Sodium Bromide	278	100
Potassium Iodide	446	200
Sodium Iodide	556	100
Sodium Chloride	125	50

Table 9.5 Suggested Sampling Intervals for Detecting Tracer Breakthrough

Expected Tracer Breakthrough	Sampling Interval
1 day	1–2 hours
2 days	2–3 hours
3 days	4–8 hours
4–7 days	8–16 hours
1–2 weeks	Once a day
2–4 weeks	Every other day
1 or more months	Once a week

An estimate of frontal advance rate, v_r, can be obtained by assuming uniform radial flow:

$$v_r = 5.6146 \frac{q_w}{2\pi r h \phi} \qquad (9.7.8)$$

where

v_r = radial frontal advance rate (ft/day)
q_w = injection rate of tracer injector (bbl/day)
h = net pay thickness (ft)
r = radial distance from injector (ft)
ϕ = porosity (fraction)

The term $2\pi r h$ in the denominator is the cross-sectional area that is perpendicular to the direction of fluid flow. If we assume uniform, radial flow, the time t_b for the tracer to reach the producer is estimated as the ratio of radial distance divided by radial frontal advance rate:

$$t_b = \frac{r}{v_r} \qquad (9.7.9)$$

where t_b is tracer breakthrough time (days). The effect of sweep efficiency can be included in the estimates of radial frontal advance rate and tracer breakthrough times by modifying the cross-sectional area transverse to fluid flow in the denominator of Eq. (9.7.8).

Data obtained from the sampling program should be promptly evaluated to determine when enough tracer appears at a well to be significant. Observed tracer production should exceed background levels of tracer. For each production well, plot either tracer concentration versus time or tracer concentration versus pore volume of produced water. Tracer breakthrough will appear as a significant increase in tracer production followed by a decline to background levels of tracer as the tracer slug moves through the reservoir. An estimate of the total amount of tracer reaching each well can be obtained from a variety of sources, including analytical calculations and computer flow models.

CS.9 Valley Fill Case Study: Production

Six vertical wells are productive in the channel. Their locations are shown in Figure CS.1A, and production data for a year are shown in Table CS.9A. The wells maintained their initial rates through the first year of history. The wells are perforated in the upper 72 feet of reservoir. Each well was brought on-line at an interval of 45 days. Inner radius of tubing is 0.25 feet, and there is no known skin. The bottomhole flowing pressure P_{wf} is estimated for an oil gradient of 0.32 psia per foot and a depth of 8,400 feet.

Production data for the Valley Fill reservoir is presented in Table CS.9B. Historical production data cover 365 days. Note that GOR is constant, and water production is very small. The oil rate in Table CS.9B shows the oil rate at the time reported. For example, the oil rate in the field is 200 STB/D at 90 days and increases to 300 STB/D when Well 3 begins to produce on the 91st day.

Table CS.9A Production Well Information

Well	Well On-Line	Wellbore Radius		P_{wf}	Initial Oil Rate	GOR
No.	Days	Feet	Skin	psia	STB/D	SCF/STB
1	0	0.25	0	2,700	100	392
2	46	0.25	0	2,700	100	392
3	91	0.25	0	2,700	100	392
4	136	0.25	0	2,700	100	392
5	181	0.25	0	2,700	100	392
6	226	0.25	0	2,700	100	392

Table CS.9B Field Production Data

Time	P_{avg}	Oil Rate	Water Rate	WOR	Gas Rate	GOR
Days	psia	STB/D	STB/D	STB/STB	MCF/D	SCF/STB
5	3,989	100.0	0.0	0.0	39.0	392.0
45	3,961	100.0	0.0	0.0	39.2	392.0
90	3,897	200.0	0.0	0.0	78.4	392.0
135	3,801	300.0	0.3	0.0	117.6	392.0
180	3,673	600.0	0.6	0.0	235.0	392.0
225	3,511	600.0	1.2	0.0	235.0	392.0
270	3,315	600.0	1.9	0.0	235.0	392.0
365	2,898	600.0	3.4	0.0	235.0	392.0

Fieldwide average pressure (P_{avg}) in Table CS.9B ranges from an initial pressure of approximately 4,000 psia to 2,898 psia at 365 days. Since the bubble point pressure is approximately 2,015 psia, the field is continuing to produce as an undersaturated oil reservoir.

Exercises

9-1. Show that $q^{-1} = at + q_i^{-1}$ is a solution of $dq/dt = -aq^2$, where a, q_i are constants. What is the value of q at $t = 0$?

9-2A. Show that $q = q_i e^{-at}$ is a solution of the decline curve equation $dq/dt = -aq^{n+1}$ for the exponential decline case.

9-2B. Plot oil flow rate as a function of time for a well that produces 10,000 barrels per day with a decline factor $a = 0.06$ per year. Time should be expressed in years and should range from 0 to 50 years.

9-2C. When does flow rate drop below 1,000 barrels per day?

9-3. The initial oil rate of an oil well is 1,000 barrels per day. The rate declines to 800 barrels per day after two years of continuous production. Assume the decline is exponential. When will the oil rate be 100 barrels per day? Express your answer in years after the beginning of production.

9-4. The water cut of an oil well that produces 1,000 STB oil per day is 20 percent. What is the water production rate for the well? Express your answer in STB water per day. What is the WOR?

9-5. *Tracer Test Design Problem:* Suppose we have the following data: $r = 200$ ft, $h = 15$ ft, $\phi = 0.20$, $S_w = 0.55$, $\rho_T = 62.4$ lbm/ft^3, $c_T = 10$ ppm, and $q_w = 500$ bbl/day. Calculate the amount of tracer that is needed for the test and estimate when tracer breakthrough occurs.

9-6. Assuming that the initial well rates in Table CS.9A are maintained for 365 days, estimate cumulative oil recovery for the field after 365 days of production.

9-7. Use the volumetric estimate of oil in place from Exercise 5-7 and the cumulative oil recovery calculated in Exercise 9-6 to determine the recovery factor for the Valley Fill reservoir after 365 days.

10 Rock–Fluid Interactions

The distribution and flow of fluids in the reservoir have a significant impact on reservoir management. Laboratory measurements of multi-phase fluid flow in porous media show that fluid behavior depends on the properties of the solid material and the fluids that are present. The focus of this chapter is on the representation of rock–fluid interactions using physical parameters such as relative permeability and capillary pressure (see Ahmed, 2000; Carlson, 2003; Tiab and Donaldson, 2003; Honarpour et al., 2006; and Christiansen, 2006).

10.1 Interfacial Tension

A surface free energy that results from electrical forces is present on all interfaces between solids and fluids and between immiscible fluids. These forces cause the surface of a liquid to occupy the smallest possible area and act like a membrane. Interfacial tension (IFT) refers to the tension between liquids at a liquid–liquid interface. Surface tension refers to the tension between fluids at a gas–liquid interface.

Interfacial tension is energy per unit of surface area, or force per unit length. The units of IFT are typically expressed in milli-Newtons per meter or the equivalent dynes per centimeter. The value of IFT depends on the composition of the two fluids at the interface between phases. Table 10.1 lists a few examples.

Interfacial tension can be estimated using the Macleod-Sugden correlation. The Weinaug-Katz variation of the Macleod-Sugden correlation is

$$\sigma^{1/4} = \sum_{i=1}^{N_c} P_{chi} \left(x_i \frac{\rho_L}{M_L} - y_i \frac{\rho_V}{M_V} \right) \qquad (10.1.1)$$

where

σ = interfacial tension (dyne/cm)
P_{chi} = parachor of component i [(dyne/cm)$^{1/4}$/(g/cm^3)]
M_L = molecular weight of liquid phase
M_V = molecular weight of vapor phase
ρ_L = liquid phase density (g/cm^3)
ρ_V = vapor phase density (g/cm^3)
x_i = mole fraction of component i in liquid phase
y_i = mole fraction of component i in vapor phase

Integrated Reservoir Asset Management. DOI: 10.1016/B978-0-12-382088-4.00010-4

Table 10.1 Examples of Interfacial Tension

Fluid Pair	IFT Range (mN/m or dyne/cm)
Air–Brine	72–100
Oil–Brine	15–40
Gas–Oil	35–65

Parachors are empirical parameters. The parachor of component i can be estimated using the molecular weight M_i of component i and the empirical regression equation

$$P_{chi} = 10.0 + 2.92M_i \tag{10.1.2}$$

This procedure works reasonably well for molecular weights ranging from 100 to 500. Fanchi (1990) provides a more accurate procedure for a wider range of molecular weights.

10.2 Wettability

Wettability is the ability of a fluid phase to wet a solid surface preferentially in the presence of a second immiscible phase. The wetting, or wettability, condition in a rock–fluid system depends on IFT. Changing the type of rock or fluid can change IFT and hence the wettability of the system. Adding a chemical such as surfactant, polymer, corrosion inhibitor, or scale inhibitor can alter wettability.

Wettability is measured by contact angle, which is always measured through the denser phase and is related to interfacial energies by

$$\sigma_{os} - \sigma_{ws} = \sigma_{ow} \cos\theta \tag{10.2.1}$$

where

σ_{os} = interfacial energy between oil and solid (dyne/cm)
σ_{ws} = interfacial energy between water and solid (dyne/cm)
σ_{ow} = interfacial energy, or IFT, between oil and water (dyne/cm)
θ = contact angle at oil–water–solid interface measured through the water phase (degrees)

Contact angles for oil-wet and water-wet examples are illustrated in Figure 10.1.

Wettability is usually measured in the laboratory. Table 10.2 presents examples of contact angles for different wetting conditions. Several factors can affect laboratory measurements of wettability. Wettability can be changed by contact of the core during coring with drilling fluids or fluids on the rig floor, and by contact of the core during core handling with oxygen or water from the atmosphere. Laboratory fluids should also be at reservoir conditions to obtain the most reliable measurements of wettability.

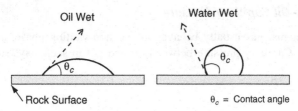

Figure 10.1 Contact angle and wettability.

Table 10.2 Examples of Contact Angle

Wetting Condition	Contact Angle (in degrees)
Strongly water wet	0–30
Moderately water wet	30–75
Neutrally wet	75–105
Moderately oil-wet	105–150
Strongly oil-wet	150–180

10.3   Capillary Pressure

Capillary pressure is the pressure difference across the curved interface formed by two immiscible fluids in a small capillary tube. The pressure difference is

$$P_c = P_{nw} - P_w \qquad (10.3.1)$$

where

P_c = capillary pressure (psi)
P_{nw} = pressure in non-wetting phase (psi)
P_w = pressure in wetting phase (psi)

10.3.1   Oil–Water Capillary Pressure

Oil is the non-wetting phase in a water-wet oil–water reservoir. Capillary pressure for an oil–water system is

$$P_{cow} = P_o - P_w \qquad (10.3.2)$$

where

P_o = pressure in the oil phase (psia)
P_w = pressure in the water phase (psia)

Capillary pressure increases with height above the oil–water contact (OWC) as water saturation decreases.

10.3.2   Gas–Oil Capillary Pressure

In gas–oil systems, gas usually behaves as the non-wetting phase, and oil is the wetting phase. Capillary pressure between oil and gas in such a system is

$$P_{cgo} = P_g - P_o \qquad\qquad (10.3.3)$$

where

P_g = pressure in the gas phase (psia)
P_o = pressure in the oil phase (psia)

Capillary pressure increases with height above the gas–oil contact (GOC) as the wetting phase saturation decreases.

10.3.3   Capillary Pressure Measurement

Capillary pressure is usually determined in the laboratory by centrifuge experiments that provide a relationship between capillary pressure P_c and water saturation S_w. A typical P_c versus S_w curve has the following features (see Figure 10.2):

- The drainage P_c curve starts at $S_w = 100$ percent.
- Water saturation S_w decreases as oil is forced into the rock. The pressure required to force the first droplet of oil into the rock is called entry pressure (or threshold pressure).
- As pressure increases above entry pressure, more oil enters the rock. Eventually no further reduction in S_w occurs. The minimum S_w is called irreducible water saturation, S_{wirr}.
- P_c can be converted to an equivalent height. This height is referenced to the free water level, which is the level at which the OWC would occur in the absence of a porous medium—that is, at porosity $\phi = 100$ percent. At the free water level, $P_c = 0$.
- The zone of rapidly changing S_w above the entry pressure and below S_{wirr} defines the transition zone between the oil reservoir and the water column.
- The OWC is the elevation at which oil saturation $S_o > 0$ first appears. In practice, other definitions of OWC might be used, such as the deepest occurrence of water-free oil production.

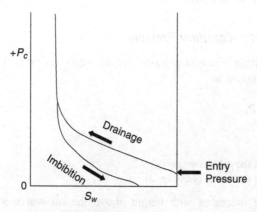

Figure 10.2 Capillary pressure and hysteresis.

10.3.4 Capillary Pressure and Pore Radius

Equilibrium between fluid phases in a capillary tube is satisfied by the relationship *force up = force down*. These forces are expressed in terms of the radius r of the capillary tube, the contact angle θ, and the interfacial tension σ. The forces are given by

force up = IFT acting around perimeter of capillary tube
$$= \sigma \cos\theta \times 2\pi r$$

and

force down = density gradient difference × cross-sectional area
× height h of capillary rise in tube

The density gradient Γ is the weight of the fluid per unit length per unit cross-sectional area. For example, the density gradient of water Γ_w is approximately 0.433 psia/ft at standard conditions. If we assume an air-water system, the force down is

force down $= (\Gamma_w - \Gamma_{air})\pi r^2 h$

where the cross-sectional area of the capillary tube is πr^2. Capillary pressure P_c is defined as the force per unit area; thus

$$P_c = \text{force up}/\pi r^2 = \text{force down}/\pi r^2$$

Expressing capillary pressure in terms of force up per unit area gives

$$P_c = \frac{2\pi r \sigma \cos\theta}{\pi r^2} = \frac{2\sigma \cos\theta}{r} \tag{10.3.4}$$

where

r = pore radius (cm)
σ = interfacial (or surface) tension (mN/m or dyne/cm)
θ = contact angle (degrees)

Equation (10.3.4) shows that an increase in pore radius will cause a reduction in capillary pressure, while a decrease in IFT will cause a decrease in capillary pressure. This effect is illustrated in Figure 10.3, which shows the behavior of two immiscible fluids, one a wetting phase and the other a non-wetting phase. Reducing capillary tube diameter increases capillary pressure and allows the wetting phase to move higher in the small-diameter capillary tube than it can in the larger-diameter capillary tubes.

Capillary pressure in reservoirs depends on the interfacial tension between two immiscible fluids, the contact angle between rock and fluid, and the pore radius

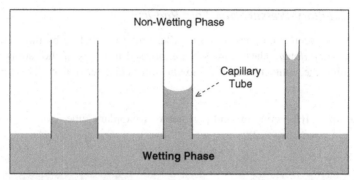

Figure 10.3 Capillary tubes and capillary pressure.

of the rock. The contact angle is a function of wettability, and pore radius is a microscopic rock property. Equation (10.3.4) shows that P_c decreases as pore radius increases. Rocks with large-pore radii usually have a larger permeability than rock with smaller-pore radii. Thus, high-permeability rocks have lower P_c than lower-permeability rocks that contain the same fluids.

Capillary forces explain why water is retained in oil and gas zones. In water-wet reservoirs, water coats rock surfaces and is preferentially held in smaller pores. Non-wetting hydrocarbon phases occupy the central parts of larger pores. Rocks with large pores are associated with low P_c and low S_{wirr}. Examples of rocks with large pores are coarse-grained sand, coarse-grained oolitic carbonates, and vuggy carbonates. High P_c and high S_{wirr} are associated with fine-grained reservoir rocks.

10.3.5 Capillary Pressure Correction

The capillary pressure measured in the laboratory is traditionally corrected to reservoir conditions by applying the correction

$$P_{c(res)} = P_{c(lab)}\eta_{corr}, \quad \eta_{corr} \equiv \frac{(\sigma|\cos\theta|)_{res}}{(\sigma|\cos\theta|)_{lab}} \tag{10.3.5}$$

where σ is interfacial tension (IFT), and θ is wettability angle (Amyx et al., 1960; Christiansen, 2006). The subscripts *lab* and *res* refer to the laboratory conditions and the reservoir conditions, respectively. If laboratory measurements of IFT are not available, IFT can be estimated from the Macleod-Sugden correlation for pure compounds or the Weinaug-Katz correlation for mixtures (Fanchi, 1990).

A problem with the capillary correction in Eq. (10.3.5) is that it requires data that are often not well known—namely, interfacial tension and wettability contact angle at reservoir conditions. Rao and Girard (1997) described a laboratory technique for measuring wettability using live fluids at reservoir temperature and pressure. Alternative methods for correcting capillary pressure include assuming the contact angle factors out or calibrating capillary pressure curves to be consistent with well log estimates of transition zone thickness.

10.3.6 Capillary Pressure Hysteresis

Capillary pressure depends on the history of the rock–fluid system. Capillary pressure is a function of saturation that depends on the direction of saturation change. The direction of saturation change depends on the historical events that led to the present distribution of fluids in the formation. There are two cases to consider:

- *Drainage curve*: wetting phase saturation decreases as the non-wetting phase saturation increases.
- *Imbibition curve*: wetting phase saturation increases as the non-wetting phase saturation decreases.

The difference in paths between the drainage and imbibition curves is called hysteresis. Oil trapped in an immobile state at the end of imbibition is residual oil saturation, S_{or}. Figure 10.1 illustrates the capillary pressure hysteresis effect.

10.4 Correlation of Capillary Pressure to Rock Properties

Rock samples with different pore-size distribution, permeability, and porosity will yield different capillary pressure curves. Relationships between capillary pressure and rock properties do exist, and we will look at some of them here.

10.4.1 Leverett's J-Function

Leverett's *J*-function is a technique for correlating capillary pressure to water saturation and rock properties:

$$J(S_w) = \frac{P_{c(lab)}}{\sigma_{lab}|(\cos\theta)_{lab}|}\left(\sqrt{\frac{K}{\phi}}\right)_{lab} \qquad (10.4.1)$$

where

$P_{c(lab)}$ = laboratory measured capillary pressure (psia)
$J(S_w)$ = Leverett's *J*-function
K = core sample permeability (md)
ϕ = porosity (fraction)
σ_{lab} = laboratory value of IFT (dyne/cm)
θ_{lab} = laboratory value of contact angle

Given $J(S_w)$, we can estimate capillary pressure at reservoir conditions as

$$P_{c(res)} = \frac{\sigma_{res}|(\cos\theta)_{res}|}{\left(\sqrt{\frac{K}{\phi}}\right)_{res}}J(S_w) \qquad (10.4.2)$$

where the value of $J(S_w)$ is obtained from the smooth curve constructed by the procedure in Table 10.3.

Table 10.3 Leverett's J-Function Procedure

Step	Task
A	Calculate $J(S_w)$ for each capillary pressure point
B	Plot $J(S_w)$ versus water saturation for all points
C	Draw a smooth curve through the points

10.4.2 P_c versus log K

The following procedure can be used to seek a correlation between capillary pressure and the logarithm of permeability. Begin by plotting water saturation S_w against log K for constant values of P_c:

1. At a given value of P_c, obtain the value of S_w for each core sample and plot log K versus S_w.
2. Repeat step 1 for enough values of P_c to cover the expected P_c range. Plot log K versus S_w for all values of P_c.
3. Enter the plot obtained in step 2 at any permeability and read S_w for each P_c value.

This method is reliable if approximately straight lines are obtained in a plot of P_c versus log K. If the data are scattered, it may be possible to improve the correlation by introducing another correlating parameter, such as porosity ϕ.

10.4.3 P_c versus log K and ϕ

The following procedure evaluates ϕ and log K as correlating parameters:

1. For a given value of P_c, plot log K versus S_w.
2. On the plot created in step 1, find two samples with similar ϕ and draw a straight line of constant ϕ. This line honors core data.
3. Draw additional constant-ϕ lines parallel to the line picked in step 2.
4. Enter the plot at a desired K, read to the intersection with the constant-ϕ line, and then read down to the S_w corresponding to P_c for the given ϕ, K.

Repeat steps 1 through 4 to find S_w for the P_c range expected in the reservoir. A plot of P_c versus S_w gives the P_c curve at the desired ϕ, K.

10.5 Equivalent Height and Transition Zone

Capillary pressure in a reservoir is responsible for the formation of a transition zone, or a zone in which multi-phase flow will occur. Transition zones may vary in thickness from a few feet to a few thousand feet. The size of the transition zone affects estimates of original hydrocarbons in place and the distribution of recoverable reserves. It is important to accurately characterize transition zones because of their potentially large impact on reservoir economics. Heymans (1997) provides an example in which a 90-foot-thick transition zone, in an edge-water drive reservoir with nearly a 1,000-foot oil column, contributed more than 30 percent of the estimated original oil in place. Rigorous volumetrics and material balance calculations were within 2 percent agreement. The transition zone was less than 10 percent of the total thickness of the reservoir, yet the volume of hydrocarbons in a relatively thin transition zone that formed a ring around the example reservoir had a significant impact on original oil in place estimates. In another example, Byrnes and Bhattacharya (2006) discussed factors that influenced recovery from transition zone reservoirs in shallow shelf carbonates.

Expressing capillary pressure P_c in terms of forcedown leads to the expression

$$P_c = \frac{\pi r^2 h(\Gamma_w - \Gamma_{air})}{\pi r^2} = h(\Gamma_w - \Gamma_{air}) \qquad (10.5.1)$$

where

h = height of capillary rise (ft)
P_c = capillary pressure (psia)
Γ_w = water, or wetting phase, density gradient (psia/ft)
Γ_{air} = air, or non-wetting phase, density gradient (psia/ft)

Solving for h yields a relationship between capillary pressure P_c and equivalent height h—namely,

$$h = \frac{P_c}{(\Gamma_w - \Gamma_{air})} \qquad (10.5.2)$$

Equivalent height is inversely proportional to the difference in density gradients between two immiscible phases. The equivalent height is the height above the free fluid level of the wetting phase, where the free fluid level is the elevation of the wetting phase at $P_c = 0$. For example, Ahmed (2000) defines free water level as the elevation where capillary pressure equals zero at 100 percent water saturation and the water–oil contact is the uppermost depth in the reservoir where water saturation is 100 percent. Figure 10.4 illustrates these definitions for an oil–water transition zone.

The oil–water transition zone is the zone between water-only flow and oil-only flow. A well producing from an oil–water transition zone will produce both oil and water. The transition zone is the part of the reservoir where 100 percent water saturation grades into oil saturation with irreducible water saturation. The top of a transition zone in an oil–water reservoir is the elevation at which water-free oil can be produced. It corresponds to the depth at which mobile water saturation first appears. The bottom of an oil–water transition zone is the shallowest depth at which oil-free water is produced.

Similar transition zones exist at the interface between any pair of immiscible phases. The thickness of the transition zone between the wetting phase and the non-wetting phase is the difference in equivalent height between the wetting phase contact (the uppermost depth in the reservoir where wetting phase saturation is 100 percent)

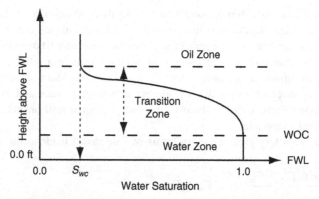

Figure 10.4 An oil–water transition zone.
Note: WOC is water–oil contact, FWL is free water level, and S_{wc} is irreducible or connate water saturation.

and the height where the wetting phase saturation is irreducible. The relatively large difference in density between gas and liquid phases results in a smaller transition zone thickness than the relatively small difference between two liquid phase densities.

10.6   Effective Permeability and Relative Permeability

The change in saturation as a function of time affects the movement of different fluid phases through the interconnected pore space. The measure of interconnectedness of pore space is permeability. When more than one phase is moving, the permeability needs to be modified to account for the relative movement of two or more phases in the pore space. Multi-phase flow is represented using relative permeability curves. We present the concepts of effective permeability and relative permeability in this section.

Absolute permeability is a measure of the capacity of the rock to transmit a single fluid. Permeability to each phase in the interconnected pore space is called effective permeability. Interfacial tension associated with the mixing of immiscible fluids tends to increase resistance to flow of fluids through the interconnected pore space. Consequently, the sum of the effective permeabilities for each phase ℓ is less than the absolute permeability; thus

$$\sum_{\ell=1}^{N} (k_r)_{eff} \leq k_{abs} \tag{10.6.1}$$

where N is the number of phases present. The effective permeability of multi-phase flow is represented by defining relative permeability.

Suppose we make the following assignments:

- k_o = effective permeability to oil
- k_w = effective permeability to water
- k_g = effective permeability to gas

Given the value of effective permeability, we define relative permeability as the ratio of effective permeability to reference permeability; thus:

$$k_r = \frac{k_{eff}}{k_{ref}} \tag{10.6.2}$$

where

k_{eff} = effective permeability of fluid (md)
k_{ref} = reference permeability (md)

The reference permeability is typically the absolute permeability of air. If we designate the reference or base permeability as k, we obtain the following often used effective permeabilities:

$$\text{Oil:} \quad k_{ro} = \frac{k_o}{k}$$

$$\text{Water:} \quad k_{rw} = \frac{k_w}{k} \tag{10.6.3}$$

$$\text{Gas:} \quad k_{rg} = \frac{k_g}{k}$$

It can be seen from these definitions that relative permeability varies between 0 and 1 because $k_{eff} \leq k$. The sum of the relative permeabilities over all phases ℓ at the same time and place satisfies the inequality

$$\sum_{\ell=1}^{N} k_{r\ell} \leq 1 \tag{10.6.4}$$

10.6.1   Oil–Water Two-Phase Relative Permeability

Consider a water-wet system containing oil and water phases, as illustrated in Figure 10.5. The relative permeability k_r curves are a function of water saturation S_w. Typically, the shape of a k_r curve is concave upward. The relative permeability of phase ℓ ($k_{r\ell}$) goes to 0 before the corresponding phase saturation S_ℓ reaches 0. The last observation shows that one fluid phase cannot be completely displaced from a porous medium by injecting another phase.

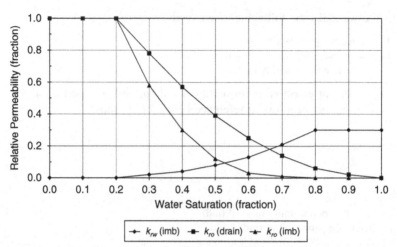

Figure 10.5 Typical water–oil relative permeability curves.

The saturation where k_r goes to 0 is the end point saturation of the relative permeability curve. For water, the end point saturation is called the irreducible water saturation S_{wirr} or the connate water saturation S_{wc}. For oil, the end point saturation is called residual oil saturation S_{or}. If the relative permeability curves were obtained for water displacing oil, the end point saturation for the oil relative permeability curve is residual oil saturation to waterflood S_{orw}.

10.6.2 Hysteresis: Imbibition versus Drainage

The shape of a relative permeability curve depends on its history. A drainage k_{ro} curve results when the wetting phase saturation decreases as the non-wetting phase saturation increases. An example of a drainage system is the displacement of water by oil during oil migration and accumulation. An imbibition k_{ro} curve, on the other hand, occurs when the wetting phase saturation increases as the non-wetting phase saturation decreases. It is exemplified by a waterflood in a water-wet reservoir. Drainage, imbibition, and wettability can often be surmised from the shape of relative permeability curves. Figure 10.5 illustrates drainage and imbibition k_r curves in a water-wet oil–water system. The k_{ro} imbibition curve is noticeably different from the k_{ro} drainage curve.

10.6.3 Relative Permeability and Wettability

Wettability can be inferred by examining relative permeability curves. Several indicators are described in this section. The wettability indicators are not a substitute for laboratory measurements, but they can aid in the characterization of the fluid flow environment. One wettability indicator is the point of intersection on the saturation axis of two relative permeability curves. The point of intersection implies the following wettability:

- Water-wet system if the curves cross at $S_w > 0.5$
- Oil-wet system if the curves cross at $S_w < 0.5$

Another wettability indicator is the relative permeability to water k_{rw} at S_{or}:

- Typically $k_{rw}(S_{or}) \leq 0.3$ for water-wet system
- Typically $k_{rw}(S_{or}) \geq 0.5$ for oil-wet system

The value of relative permeabilities at end point saturations can indicate wettability as follows:

- For water-wet system, $k_{rw}(S_{or}) \ll k_{ro}(S_{wirr})$
- For oil-wet system, $k_{rw}(S_{or}) \approx k_{ro}(S_{wirr})$

An oil-wet system usually satisfies the inequality $S_{wirr} \leq 0.1$.

Changes in the wettability conditions of the core can significantly affect relative permeability. Ideally, relative permeability should be measured in the laboratory under the same conditions of wettability that exist in the reservoir. One way to approximate this ideal is to use preserved, native state core samples. In practice, most relative permeability data are obtained using restored state cores in the laboratory.

10.6.4 Gas–Liquid Two-Phase Relative Permeability

Gas is the non-wetting phase during two-phase flow of gas and oil, and two-phase flow of gas and water. Critical gas saturation, S_{gc}, is the gas saturation at which gas first begins to flow in a reservoir. The value of S_{gc} usually varies from 0 to 10 percent, and a reasonable value in the absence of measurements is a value between 2 and 5 percent. If the relative permeability curves were obtained for gas displacing oil, the end point saturation for the oil relative permeability curve is residual oil saturation to gas flood S_{org}.

Hysteresis occurs in gas relative permeability in a manner similar to that in oil–water systems. Residual or trapped gas saturation (S_{gr} or S_{gt}) is the maximum value of S_g, where the relative permeability to gas $k_{rg} = 0$. Typical values of S_{gr} or S_{gt} range from 15 to 40 percent.

10.6.5 Two-Phase Relative Permeability Correlations

Many correlations exist for two-phase relative permeabilities. The correlations have been developed from a variety of sources, such as capillary tube models, statistical models of capillary bundles, microscopic network models, and empirical correlations. Correlations are typically used in the absence of laboratory measurements or multiphase production data. Correlations are not a substitute for laboratory measurements; they should be used in conjunction with laboratory data or when laboratory data are not available.

The correlations for two-phase relative permeability of Honarpour and colleagues (1982) show that relative permeability values depend on rock type and wettability. The correlations are presented as a function of wettability and lithology.

10.6.6   Saturation Exponent Method

Another representation of relative permeability curves for multiphase flow is the saturation exponent method. The saturation exponent method is based on an empirical functional relationship between saturation and relative permeability. The following equations are used to fit smooth curves over any desired saturation range.

Water:

$$k_{rw} = k_{rw}^* \left(\frac{S_w - S_{wc}}{1 - S_{orw} - S_{wc}} \right)^{n_w} \tag{10.6.5}$$

Oil in an oil–water system:

$$k_{row} = k_{row}^* \left(\frac{1 - S_{orw} - S_w}{1 - S_{orw} - S_{wc}} \right)^{n_{ow}} \tag{10.6.6}$$

Gas:

$$k_{rg} = k_{rg}^* \left(\frac{S_g - S_{gc}}{1 - S_{lrg} - S_{gc}} \right)^{n_g} \tag{10.6.7}$$

Oil in a gas–oil system:

$$k_{rog} = k_{rog}^* \left(\frac{1 - S_{lrg} - S_g}{1 - S_{lrg}} \right)^{n_{og}} \tag{10.6.8}$$

The multiplier $(k_{r\ell}^*)$ is the end point relative permeability of phase ℓ at the corresponding end point saturation (S_{wc}, S_{orw}, or S_{lrg}). The saturation S_{lrg} is the total residual liquid saturation in the gas–oil system (including S_{wc}). The equations must satisfy $k_{rog} = k_{row}$ when $S_w = S_{wirr}$ and $S_g = 0$ for both water–oil and gas–oil systems to be physically consistent. To simplify data adjustments, these equations can be fit to laboratory data or relative permeability correlations.

The exponent (n_ℓ) for phase ℓ controls the curvature of the k_r curve. The geometric curves associated with the exponent's different values are summarized as follows:

Exponent n_ℓ	Function of Saturation
= 1	Straight line
> 1	Concave upward
< 1	Concave downward

10.6.7   Three-Phase Relative Permeability

Most multi-phase flow phenomena involve the movement of two phases through the same interconnected pore space at the same time. The most frequently occurring two-phase relative permeability systems are oil–water relative permeability, gas–oil relative permeability, and gas–water relative permeability.

Three-phase relative permeabilities can be measured in the laboratory using techniques such as steady state and dynamic displacement. Laboratory measurements can be costly, time-consuming, and inaccurate. Consequently, three-phase relative permeability curves are usually estimated from models of three-phase behavior using two-phase relative permeability curves. Three-phase relative permeability models usually estimate oil relative permeability from two sets of two-phase relative permeability data: water displacing oil and gas displacing oil at irreducible water saturation, S_{wc}. Stone (1973) provided two of the most commonly used three-phase relative permeability methods (Carlson, 2003; Christiansen, 2006).

A comparison of three-phase k_{ro} correlations with laboratory data shows that a reliable three-phase relative permeability algorithm that can be applied with confidence to any reservoir does not exist. If three-phase flow is expected to have a major impact on project performance, it is wise to perform a sensitivity study to determine the impact of several different three-phase k_{ro} correlations on model results.

10.6.8 Practical Notes

Relative permeability data should be obtained using experiments that best model the type of displacement that is thought to dominate reservoir flow performance. For example, water–oil imbibition curves are representative of waterflooding, while water–oil drainage curves describe the movement of oil into a water zone.

Permeability and relative permeability describe flow of a particular fluid in a particular rock type. If the fluid system changes or the rock type changes, the appropriate values of permeability and relative permeability must be measured. For example, if a waterflood is planned for an oil reservoir that is being depleted, laboratory measured permeabilities need to represent the injection of water into a core with reservoir oil and connate water. Permeability measurements for a gasflood would not be consistent with the waterflood system. The permeability distribution and relative permeability curves used in reservoir engineering calculations need to reflect the type of processes that are expected to occur in the reservoir.

Relative permeability data are often measured and reported for laboratory analysis of several core samples from one or more wells in a field. The set of relative permeability curves should be sorted by lithology and averaged to determine a representative set of curves for each rock type. Several procedures exist for normalizing or averaging relative permeability data (see Schneider, 1987; Mattax and Dalton, 1990; Fanchi, 2000; Carlson, 2003).

10.7 Mobility, Relative Mobility, and Flow Capacity

Three frequently encountered concepts in multi-phase flow are mobility, mobility ratio, and flow capacity. Mobility is the ratio of effective phase permeability to phase viscosity, or $\lambda_\ell = k_\ell / \mu_\ell$, where ℓ denotes the phase. Fluid phase mobility for oil, water, and gas phases are

$$\lambda_o = \frac{k_o}{\mu_o}, \ \lambda_w = \frac{k_w}{\mu_w}, \ \lambda_g = \frac{k_g}{\mu_g} \tag{10.7.1}$$

where μ_ℓ is the viscosity of phase ℓ. Fluids with large mobilities can move at higher velocities than fluids with small mobilities. Gas, for example, has a small viscosity compared to liquids and will generally have a larger mobility than liquids. Oil and water often have comparable mobilities because they have comparable viscosities.

Relative mobility $\lambda_{r\ell}$ of phase ℓ is the relative permeability of the phase divided by the phase viscosity

$$\lambda_{r\ell} = \frac{k_{r\ell}}{\mu_\hbar} \tag{10.7.2}$$

Mobility is equal to relative mobility times reference permeability.

Total mobility is the sum of the individual phase mobilities:

$$\lambda_T = \lambda_o + \lambda_w + \lambda_g \tag{10.7.3}$$

The saturation corresponding to a minimum in total mobility can be found by plotting total mobility versus saturation.

10.7.1 Mobility Ratio

Mobility ratio is defined as the mobility of the displacing fluid λ_D behind the front divided by the mobility of the displaced fluid λ_d ahead of the front; thus

$$M = \frac{\lambda_D}{\lambda_d} \tag{10.7.4}$$

In a displacement process, a mobility ratio that is greater than 1 is considered unfavorable, while a mobility ratio that is less than 1 is considered favorable. This is discussed in more detail in Chapter 14.

An example of mobility ratio is the mobility ratio of water to oil for a waterflood:

$$M_{w,o} = \frac{(\lambda_w)_{S_{or}}}{(\lambda_o)_{S_{wc}}} = \frac{k_{rw}(S_{or})/\mu_w}{k_{ro}(S_{wc})/\mu_o} \tag{10.7.5}$$

In this case, relative permeability to water is evaluated at residual oil saturation S_{or}, and relative permeability to oil is evaluated at connate water saturation S_{wc}. Notice that absolute permeability factors out of the expression for mobility ratio. Consequently, mobility ratio can be calculated using either mobilities or relative mobilities.

10.7.2 Flow Capacity

Flow capacity is a measure of the ability of a fluid to move in a specified interval. It is the product of permeability and thickness. The specific expression for flow capacity depends on definitions of the permeability and thickness terms. A straightforward expression for the flow capacity of an interval with N_L elements is the arithmetic average

$$KH = \sum_{i=1}^{N_L} k_i h_i \qquad (10.7.6)$$

where k_i, h_i are the permeability and thickness of the ith element.

CS.10 Valley Fill Case Study: Rock–Fluid Interaction Data

Two-phase relative permeability curves are shown in Tables CS.10A and CS.10B for an oil-water system and a gas–oil system, respectively. These curves were calculated using the correlations of Honarpour and colleagues (1982) for water-wet sandstone with the following end point saturations: initial oil saturation is 70 percent, residual oil saturation is 25 percent, irreducible water saturation is 30 percent, and critical gas saturation is 3 percent. The analytical correlations for relative permeability curves was used because no explicit measurements exist for irreducible water saturation, critical gas saturation, relative permeability curves, or capillary pressure curves. These measurements could be made on core from the productive interval if core is available. Values of irreducible water saturation and critical gas saturation are estimated based on analogous fields. Capillary pressure effects are neglected for now. In a more complete study, the sensitivity of the dynamic flow model to uncertainty in rock–fluid interaction effects should be studied.

Table CS.10A Oil–Water Relative Permeability

Water Saturation	K_{row}	K_{rw}
0.00	1.0	0.0
0.30	1.0	0.0
0.35	0.59	0.01
0.40	0.32	0.02
0.45	0.18	0.034
0.50	0.08	0.046
0.55	0.03	0.068
0.60	0.01	0.09
0.65	0.001	0.128
0.70	0.0001	0.166
0.75	0.0	0.2
0.80	0.0	0.24
1.00	0.0	0.24

Table CS.10B Gas–Oil Relative Permeability

Gas Saturation	K_{rog}	K_{rg}
0.00	1.0	0.0
0.03	0.75	0.0
0.05	0.59	0.02
0.10	0.32	0.09
0.15	0.18	0.16
0.20	0.08	0.24
0.25	0.03	0.33
0.30	0.01	0.43
0.35	0.001	0.55
0.40	0.0	0.67
0.45	0.0	0.81
0.50	0.0	1.0
1.00	0	1.0

Exercises

10-1. Plot the oil–water relative permeability curves in Table CS.10A as a function of water saturation.

10-2. Use fluid property data at 4,014.7 psia in Table CS.2A and relative permeability in Table CS.10A to prepare plots of oil and water relative mobilities as functions of water saturation.

10-3. Plot the gas–oil relative permeability curves in Table CS.10B as a function of gas saturation.

10-4. Use fluid property data in Table CS.2A and relative permeability in Table CS.10B to determine oil and gas relative mobilities.

10-5. Suppose the capillary pressure of an oil–water system is 250 psi and the pressure of the water phase is 3,500 psi. What is the pressure of the oil phase?

10-6. Consider the following oil–water relative permeability table:

S_w	k_{rw}	k_{row}
0.30	0.000	1.000
0.35	0.005	0.590
0.40	0.010	0.320
0.45	0.017	0.180
0.50	0.023	0.080
0.55	0.034	0.030
0.60	0.045	0.010
0.65	0.064	0.001
0.70	0.083	0.000
0.80	0.120	0.000

     (a) What is the residual oil saturation?

     (b) What is the connate water saturation?

     (c) What is the relative permeability to oil at connate water saturation for the oil–water relative permeability curves?

     (d) What is the relative permeability to water at residual oil saturation?

10-7. Assume oil viscosity is 2.0 cp and water viscosity is 0.50 cp. Calculate the mobility ratio of a waterflood for water displacing oil using the oil–water relative permeability curves given in Exercise 10-6. Is the mobility ratio favorable or unfavorable?

10-8. Will the height of a transition zone be greater for a reservoir with small pore throats or large pore throats? *Hint:* Combine Eqs. (10.3.4) and (10.5.2).

11 Reservoir Characterization

Reservoir characterization is the process of preparing a quantitative representation of a reservoir using data from a variety of sources and disciplines. Kelkar defined "reservoir characterization" as "the process of integrating various qualities and quantities of data in a consistent manner to describe reservoir properties of interest at inter-well locations" (2000, p. 25). All of the information collected at various scales in the reservoir characterization process must be integrated into a single, comprehensive, and consistent representation of the reservoir.

The techniques described in this chapter illustrate technology that can be used to generate geologic maps and to characterize the reservoir. This chapter describes two fundamental topics in reservoir characterization: flow unit characterization and reservoir mapping.

11.1 Flow Units

The correct relationship between data measured at two different scales may be difficult to determine. For example, permeability is often obtained from both pressure transient testing and routine core analysis. The respective permeabilities, however, may appear to be uncorrelated because they represent two different measurement scales. The integration of data obtained at different scales is often referred to as the scale-up problem.

An important task of the scale-up problem is to develop a detailed understanding of how measured parameters vary with scale. The focus on detail in one or more aspects of the reservoir modeling process can obscure the fundamental character of the reservoir in a flow model study. One way to integrate available data within the context of a "big picture" is to apply the concept of flow unit.

Ebanks defined the flow unit as "a volume of rock subdivided according to geological and petrophysical properties that influence the flow of fluids through it" (1987, abstract). This definition was later modified to state that a flow unit is "a mappable portion of the total reservoir within which geological and petrophysical properties that affect the flow of fluids are consistent and predictably different from the properties of other reservoir rock volumes" (Ebanks et al., 1993, p. 282). Typical geologic and petrophysical properties are listed in Table 11.1.

Static reservoir models are digital representations of the reservoir. They are constructed by subdividing the reservoir into an ensemble, or collection, of representative elementary volumes (REV), as shown in Figure 11.1. The REV is the volume

Integrated Reservoir Asset Management. DOI: 10.1016/B978-0-12-382088-4.00011-6

Table 11.1 Properties Typically Needed to Define a Flow Unit

Geologic	Petrophysical
Texture	Porosity
Mineralogy	Permeability
Sedimentary Structure	Compressibility
Bedding Contacts	Fluid Saturations
Permeability Barriers	

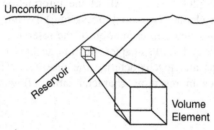

Figure 11.1 Representative elementary volume.

element that is large enough to provide statistically significant average values of parameters describing flow in the contained volume but is small enough to provide a meaningful numerical approximation of the fundamental flow equations. Traditional models represent an REV as a Cartesian cell, illustrated in Figure 11.1, while many modern models allow a variety of shapes to be used to represent REVs. Each REV contains a set of rock properties, such as porosity, permeability, and bulk modulus. Fayers and Hewett (1992) cautioned that it is "somewhat an act of faith that reservoirs can be described by relatively few REV types at each scale with stationary average properties." Today, the number of cells in a static reservoir model is increasing as modelers seek to improve the quality of the digital representation of a reservoir.

The REV concept is not the same as the flow unit concept. A flow unit is a contiguous part of the reservoir that has similar flow properties as characterized by geological and petrophysical data. Flow units usually contain one or more REVs. Ebanks and colleagues (1993) identified the following characteristics of a flow unit:

1. A specific volume of reservoir composed of one or more reservoir quality lithologies; adjacent non-reservoir quality rock types; and associated fluids
2. Correlative and mappable at the interwell scale
3. Zonation can be recognized, such as on well logs
4. May be in communication with other flow units

11.1.1 The Modified Lorenz Plot

Several flow unit identification techniques are proposed in the literature, such as the modified Lorenz plot used by Gunter and colleagues (1997). A simplified variation of the modified Lorenz plot technique is to identify a flow unit by plotting cumulative flow capacity as a function of depth. Cumulative flow capacity F_m is calculated as

$$F_m = \frac{\sum\limits_{i=1}^{m} k_i h_i}{\sum\limits_{i=1}^{n} k_i h_i}; \; m = 1, \ldots, n \qquad (11.1.1)$$

where n is the total number of reservoir layers. The layers are numbered in order from the shallowest layer at index $i = 1$ to the deepest layer at index $i = m$. The cumulative flow capacity F_m is the value of Eq. (11.1.1) at depth

$$Z_m = Z_0 + \sum\limits_{i=1}^{m} h_i \qquad (11.1.2)$$

where Z_0 is the depth to the top of layer 1 from a specified datum. A flow unit will appear on the plot as a line with constant slope. A change in slope is interpreted as a change from one flow unit to another, as illustrated in Figure 11.2.

Another plot that can be used to identify flow units is a plot of cumulative flow capacity F_m in Eq. (11.1.1) versus a cumulative storage capacity Φ_m defined by

$$\Phi_m = \frac{\sum\limits_{i=1}^{m} \phi_i h_i}{\sum\limits_{i=1}^{n} \phi_i h_i}; \; m = 1, \ldots, n \qquad (11.1.3)$$

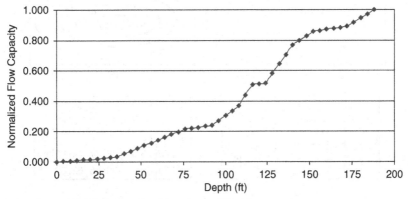

Figure 11.2 Identifying flow units using the modified Lorenz plot.

Again, n is the total number of reservoir layers, and the layers are numbered in order from the shallowest layer $i = 1$ to the deepest layer $i = m$. The analyst again looks for changes in slope in the plot of F_m versus Φ_m.

11.2 Traditional Mapping

The mapping process is the point where geological and geophysical interpretations have their greatest impact on the reservoir's representation in dynamic flow models. A sampling of authors who have discussed mapping and reservoir characterization includes Harris (1975), Harpole (1985), Tearpock and Bischke (1991), Haldorsen and Damsleth (1993), Uland et al. (1997), Tearpock et al. (2002), and Carlson (2003).

A geologic or static reservoir model must include a structure of the reservoir and the distribution of rock properties within that structure. The properties that must be digitized for use in a flow model include: elevations or structure tops; permeability in three orthogonal directions; porosity; gross thickness; net-to-gross thickness; saturations; and, where appropriate, descriptions of special geologic features (e.g., faults, fractures, vugs, and karsts). Additional maps are needed for flow models that integrate petrophysics and traditional flow models. In particular, distributions of moduli and bulk density can be used to calculate seismic velocities. The resulting maps are digitized by overlaying a grid on them and reading a value for each cell. The digitizing process is shown in Figure 11.3.

The contouring step is a point where geological interpretation is included in the flow model. The following technical contouring guidelines apply:

1. Contour lines do not branch.
2. Contour lines do not cross.
3. Contour lines either close or run off the map.
4. Steep slopes have close contour lines.
5. Gentle or flat slopes have contour lines that are far apart.

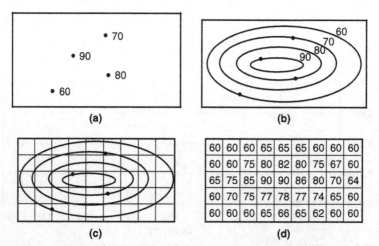

Figure 11.3 The digitizing process: (a) gather data, (b) contour data, (c) overlay grid, and (d) digitize data.

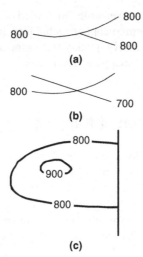

Figure 11.4 Examples of contour tips.

The first three of the preceding guidelines are illustrated in Figure 11.4. Discontinuities in contour lines are possible, but they must be justified by the inferred existence of geologic discontinuities, such as faults and unconformities. While tedious, traditional hand contouring can let a geologist imprint a vision on the data that many computer algorithms will miss.

If a project is very large, several geologists, geophysicists, and engineers may have contributed to the static reservoir model. Different geoscientists or engineers may provide different interpretations of the same data set. For example, two or more petrophysicists may differ on defining well log cutoffs, or geologists may disagree on picking the depth of formation tops. These types of differences introduce variations in the model that may prove significant.

Another type of problem may appear when incorrect data are entered into a digital model. Errors may range from transposed digits to unit conversion errors. Some of these errors can be spotted by contouring and visualizing the data. Gross errors will be evident. The resolution of the model depends on the resolution of the grid. A fine grid divides the reservoir into many small cells. It gives the most accurate numerical representation but has the greatest computational expense. A coarse grid has fewer cells, but the coarse cells must be larger than the fine cells to cover the same model volume. As a result, the coarse grid is less expensive to run than a fine grid, but it is also less accurate numerically.

The loss of accuracy is evident when a coarse grid is used to model small geologic features or fluid flow at the interface between phases, such as fluid contacts and displacement fronts. Thus, fine-grid modeling is often the preferred choice to achieve maximum numerical accuracy. It is important to recognize, however, that a fine grid that is covering an area defined by sparse data can give the illusion of accuracy. Uncertainty analysis can help quantify the uncertainty associated with a model study.

The gridding process is most versatile when used with an integrated 3-D reservoir mapping package. Modern mapping techniques include computer-generated maps that can be changed relatively quickly once properly set up. The next section introduces computer-generated mapping techniques.

11.3   Computer-Generated Mapping

An important function of geologic maps is to present values for a spatially distributed property at any point in a volume from a set of control point values. Control point values correspond to property values measured at wells or determined by seismic methods that apply to volume of interest. Control points can also be imposed by a modeler using soft data such as seismic indications of structure boundaries. Maps of spatially distributed properties can be generated by computer using a variety of techniques.

After an algorithm has computed a surface, modelers may want to edit the surface. An easy method is to add data points to force a contour to move to a certain location. More complex computer programs allow the imposition of trends onto the data. The character of the reservoir conceptualized by the modeler should be adequately represented in the final computer-generated map.

Computer-generated maps may not include all of the detailed interpretations a geologist might wish to include in the model, particularly with regard to faults, but the maps generated by computer in a 3-D mapping program do not have the problems often associated with the stacking of 2-D plan view maps—namely, physically unrealistic layer overlaps. Layer overlaps need to be corrected before the history match process begins. Dahlberg (1975) presented one of the first analyses of the relative merits of hand-drawn and computer-generated maps.

Another problem with computer-generated maps is the amount of detail that can be obtained. Computer-generated maps can describe a reservoir with a much finer grid than the resolution typically used in a flow model. For example, a computer mapping program may use a grid with several million cells to represent the reservoir, yet reservoir simulation grids are typically less than one million cells. This means that the reservoir representation in the computer mapping program must be upscaled, or coarsened, for use in a reservoir simulator.

Many attempts have been made to find the most realistic process for upscaling data, but no widely accepted scale-up method is in use today. Christie and Blunt (2001) present a comparison of upscaling techniques in the tenth Society of Petroleum Engineers (SPE) comparative solution project. For additional discussion of scale-up, see Slatt and Hopkins (1990), Christie (1996), King and Mansfield (1999), Chawathé and Taggart (2004), Lasseter and Jackson (2004), and Stern (2005). Gorell and Bassett (2001) discuss the relative merits of different workflows. Dogru (2000) and Dogru et al. (2009) noted that scale-up is not needed if the dynamic flow model uses the same scale as the static reservoir model.

11.3.1   Inverse Distance Weighting

One of the simplest algorithms that can be coded into a computer program to generate a map is to distribute property values over a surface or within a layer by using inverse distance weighting of all applicable control point values. The formula for inverse distance weighting is

$$V_x = \frac{\sum_{i=1}^{N}(V_i/d_i)}{\sum_{i=1}^{N}(1/d_i)} \tag{11.3.1}$$

where V_x is the value of the property at x calculated from N known values $\{V_i\}$ of the property at distances $\{d_i\}$ from x. Inverse distance weighting assigns more weight to control points close to location x and less weight to control points that are farther away. The weighting factor is the inverse of control point distance from x. For example, the value at point x that is at distances $\{d_A, d_B\}$ from two known values $\{V_A, V_B\}$ is

$$V_x = \frac{\dfrac{V_A}{d_A} + \dfrac{V_B}{d_B}}{\dfrac{1}{d_A} + \dfrac{1}{d_B}} \tag{11.3.2}$$

Figure 11.5 illustrates the inverse distance weighting example in Eq. (11.3.2) with two control points. If only one value V_C is known ($N = 1$), then $V_x = V_C$ for all values of x.

11.3.2   Weighted Averaging

Inverse distance weighting is an example of a technique that uses control points in the neighborhood of an unknown point to estimate the property value at the point. A more general expression for distributing an attribute using a weighted average is

$$Z_{avg} = \frac{\sum_{i=1}^{N}W(r_i,R) \times Z_i}{\sum_{i=1}^{N}W(r_i,R)} \tag{11.3.3}$$

where

Z_{avg} = weighted average value of attribute Z
Z_i = value of attribute Z at control point i
W = weighting function
r_i = distance from the interpolated point to control point i
R = user specified search radius
N = number of control points

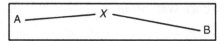

Figure 11.5 Inverse distance weighting with two control points {A, B}.

The search radius R constrains the number of control points N that are used to determine the weighted average of the attribute. An example of a weighting function is shown in Eq. (11.3.1)—namely, $1/d_i$. Another example of a weighting function with a search radius is

$$W(r,R) = \left(1 - \frac{r}{R}\right)^2 \times \left(\frac{R}{r}\right)^e \qquad (11.3.4)$$

where the value of exponent e is specified by the modeler.

11.3.3   Trend Surface Analysis

A technique for determining the spatial distribution of a property by computer is to fit a surface through a set of control points. This technique is referred to as trend surface analysis, and it uses regression to fit a line through all of the control point values in a given direction. The regression model for linear trend surface analysis is

$$Z_{obs} = a_0 + a_1 x_{loc} + a_2 y_{loc} \qquad (11.3.5)$$

where

$\qquad\qquad Z_{obs} = $ the observed value of attribute Z at the control point
$\qquad \{x_{loc}, y_{loc}\} = $ the {x-axis, y-axis} locations of the control point
$\qquad \{a_0, a_1, a_2\} = $ regression coefficients

Equation (11.3.5) can be extended to be a quadratic function of control point location. Quadratic trend surface analysis can fit a curved surface to data and is therefore useful for representing geologic structures such as anticlines or synclines.

11.4   Geostatistics and Kriging

A fundamental objective of the reservoir characterization process is to distribute rock properties throughout the volume of interest. Two modern methods for spatially distributing rock properties are reservoir geophysics and geostatistics. Information obtained from reservoir geophysics is improving our ability to "see" between wells in a deterministic sense. By contrast, geostatistics provides a reservoir characterization that is statistical. Many modelers view geostatistics as the method of choice for sophisticated reservoir flow modeling. Are these methods competing or complementary?

The discussion in this section presents several points about geostatistics that can help answer this question.

Geostatistics, also known as spatial statistics, is a branch of "applied statistics" that attempts to describe the distribution of a property in space. It assumes that a spatially distributed property exhibits some degree of continuity. Porosity and permeability are examples of spatially dependent properties that are suitable for geostatistical description. Much of our discussion of geostatistics here is based on publications by Isaaks and Srivastava (1989), Hirsche et al. (1997), Deutsch and Journel (1998), Chambers et al. (2000), Clark and Harper (2000), and Davis (2002).

Geostatistics consists of a set of mathematical tools that employ the assumptions that properties are correlated in space and are not randomly distributed. The geological context of the data must be considered in addition to spatial relationships between data. Geostatistical algorithms provide formalized methods for integrating data of diverse type, quality, and quantity.

A geostatistical analysis has several goals, including the following:

1. Acquiring an understanding of the spatial relationships and correlations between reservoir properties
2. Modeling those relationships with mathematical expressions
3. Developing an understanding of the uncertainty associated with the reservoir properties and the conceptual geologic model
4. Determining if a deterministic or stochastic approach is appropriate for the creation of a reservoir model

A deterministic model is a single realization, or representation, of reservoir geology. The uncertainty associated with a deterministic model can be studied by estimating the sensitivity of the model to uncertainties in available data. A stochastic model is a set of realizations obtained from the probability distributions developed during the geostatistical analysis of data. The shape of a probability distribution is defined by the proximity and quality of local data within the context of a spatial correlation model. By its nature, stochastic modeling propagates the uncertainty of the input parameters.

Stochastic modeling has two goals: to preserve the heterogeneity inherent in a geological system as a means of creating more realistic and useful simulation models and to quantify the uncertainty in the geologic model by generating many possible realizations. The stochastic model should incorporate multiple data types with varying degrees of quality and quantity. The data should represent different measurement scales.

The process of preparing a geologic model requires the development of a structural and stratigraphic framework using available seismic and well data. Multiple realizations may be generated and used to quantify uncertainty in the geologic model. The process of translating point observations to a conceptual geologic model is a sequential process. It is also an iterative process if a match of time-dependent (dynamic) data is included in the preparation of the final reservoir model. Once the framework exists, both a lithofacies model and petrophysical properties can be incorporated into the flow model.

11.4.1 Semivariograms

Spatial correlation of an attribute is quantified by the semivariogram, which is a plot of semivariance versus range. Davis says that the semivariance "is used to express the rate of change of a regionalized variable along a specific orientation" (2002, p. 254). Semivariance is a measure of the degree of spatial dependence between values of attribute Z at two different locations or points in space. The separation distance between the spatial points is the lag h. The semivariance $\gamma(h)$ is a function of lag h between two observations $Z(x)$ and $Z(x + h)$ of attribute Z; thus

$$\gamma(h) = \frac{1}{2N(h)} \sum_{i=1}^{N(h)} [Z(x_i) - Z(x_i + h)]^2 \qquad (11.4.1)$$

where $N(h)$ is the number of data pairs that are approximately separated by lag h.

Figure 11.6 illustrates three important features of the semivariogram. The sill is the maximum value of the semivariogram for attribute Z and is also the variance σ^2 of the measured data, where σ is the standard deviation. The nugget in Figure 11.6 is the value of the semivariance at zero lag. A nonzero value of the nugget is due to factors such as sampling error and short-range variability of the parameter. In fact, the term "nugget" refers to the observation that the lag for a finite-size gold nugget can never equal zero.

The range in Figure 11.6 is an estimate of the maximum correlation length between two points at separation distance h. A spatial correlation between values of attribute Z exists at values of the lag less than the range.

Several types of semivariogram models exist. For example, the exponential model is

$$\gamma(h) = C_0 + C_1 \left\{ 1 - \exp\left(-\frac{h}{a}\right) \right\} \qquad (11.4.2)$$

and the Gaussian model is

$$\gamma(h) = C_0 + C_1 \left\{ 1 - \exp\left(-\frac{h^2}{a^2}\right) \right\} \qquad (11.4.3)$$

where

$h \geq 0 = $ lag
$C_0 = $ the nugget
$C_1 = $ the sill
$a = $ the range of influence

Semivariogram modeling is performed by fitting a semivariogram model to experimental data. The resulting semivariogram is a measure of the spatial dependence of reservoir attributes such as porosity, permeability, and net thickness. The semivariogram model is used to predict values of the modeled attribute at unsampled locations. Figure 11.7 illustrates a fit to data by a semivariogram model.

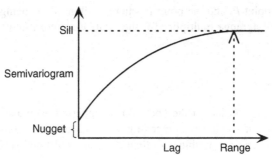

Figure 11.6 Characterizing a semivariogram.

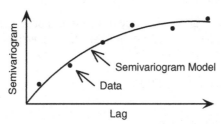

Figure 11.7 A semivariogram.

11.4.2 Kriging

One widely used technique for estimating attribute values is kriging. Kriging is named after South African mining engineer D. G. Krige, who helped pioneer the development of geostatistical methods in the 1950s. Kriging is a linear weighted average method. The weights used in kriging are based on the semivariogram model of spatial correlation. It is instructive to make these points explicit.

The kriging equation for estimating the value of attribute Z_P at point P from a set of n control points with attribute values $\{Z_i : i = 1, 2, \ldots, n\}$ is

$$Z_P = \sum_{i=1}^{n} w_i Z_i \tag{11.4.4}$$

The attribute Z_P may be a rock property such as porosity or permeability. The weights $\{w_i : i = 1, 2, \ldots, n\}$ are calculated from the set of n equations

$$
\begin{aligned}
&w_1\gamma(h_{11}) + w_2\gamma(h_{12}) + \ldots + w_n\gamma(h_{1n}) + \lambda = \gamma(h_{1P}) \\
&w_1\gamma(h_{21}) + w_2\gamma(h_{22}) + \ldots + w_n\gamma(h_{2n}) + \lambda = \gamma(h_{2P}) \\
&\vdots \\
&w_1\gamma(h_{n1}) + w_2\gamma(h_{n2}) + \ldots + w_n\gamma(h_{nn}) + \lambda = \gamma(h_{nP})
\end{aligned}
\tag{11.4.5}
$$

The semivariogram $\gamma(h_{ij})$ is the semivariogram at lag distance h_{ij} between two points (P_i, P_j). The semivariogram $\gamma(h_{iP})$ is the semivariogram at lag distance h_{iP}

between control point P_i and the point P where attribute Z_P is being estimated. The constant λ is the Lagrange multiplier for the "unbiased" constraint

$$\sum_{i=1}^{n} w_i = 1 \qquad (11.4.6)$$

Equation (11.4.4) is considered the "best linear unbiased estimate" (BLUE) of Z_P, and the procedure for solving the preceding set of equations is considered ordinary kriging. Universal kriging combines ordinary kriging and trend surface analysis.

11.4.3 Kriging Accuracy

One method of determining the accuracy of the values obtained by ordinary kriging is to calculate the variance σ_{OK}^2 of the ordinary kriging estimate. The variance is

$$\sigma_{OK}^2 = \sum_{i=1}^{n} w_i \gamma(h_{iP}) + \lambda - \gamma(h_{PP}) \qquad (11.4.7)$$

where the Lagrange multiplier is

$$\lambda = \sum_{i=1}^{n} w_i \gamma(h_{iP}) - \sum_{i=1}^{n}\sum_{j=1}^{n} w_i w_j \gamma(h_{ij}) \qquad (11.4.8)$$

Equations (11.4.7) and (11.4.8) can be solved once the weights have been calculated. Equation (11.4.8) can be used to check the value of the Lagrange multiplier obtained by solving Eq. (11.4.6).

Another method of determining the accuracy of the values obtained by an estimation technique is to treat a sampled (known) data point as an unknown point at the test location. The estimation technique is used to calculate the parameter at the test location, and the resulting value is then compared with the known data point. The accuracy of the estimation process can be quantified by calculating the semivariance of actual values relative to the estimated values. The resulting semivariance provides a cross-validation of the original semivariogram model and provides information about the quality of the estimation technique. Notice that this model cross-validation procedure could be applied to any computer-based estimation technique.

11.5   Geostatistical Modeling

Geostatistical modeling refers to the procedure for determining a set of reservoir realizations. The realizations depend on both the spatial relationships between the data points and their statistical correlation as a function of separation in space. Chambers and colleagues (2000) identified seven steps in a geostatistical study that are provided in Table 11.2.

Table 11.2 Steps in a Geostatistical Study

Step	Task
1	Data mining
2	Spatial continuity analysis and modeling
3	Search ellipse design
4	Model cross-validation
5	Kriging
6	Conditional simulation
7	Uncertainty assessment

11.5.1   Step 1: Data Mining

Data mining is the study of available data. Data analysis helps the modeler locate anomalies and errors in the data and gain familiarity with the data. Familiarity with the data set helps the modeler optimize the usefulness of the data because geostatistics is subject to interpretation and relies on experience. The modeler can use a variety of techniques, many graphical, to analyze the data. The analysis is designed to detect and model patterns of spatial variability and interdependence.

11.5.2   Step 2: Spatial Continuity Analysis and Modeling

The spatial relationship(s) associated with data are computed and then modeled. This process is analogous to (1) plotting data on a cross-plot (computing) and then (2) fitting a line to the data with linear regression (modeling). The plotted points make up the experimental semivariogram, and the line that is fit to the data points is the model semivariogram.

Semivariogram modeling is achieved by calculating the semivariance of data and then fitting a semivariogram model to the plot of semivariance versus lag. The semivariogram contains information about the correlation range between spatially separated values of an attribute. The semivariogram model is used to predict the value of the modeled attribute at locations away from control points.

11.5.3   Step 3: Search Ellipse Design

The separation between data points is specified by the lag distance. Semivariogram calculations can compare data points in all directions or in specified directions. It is possible to study the anisotropy of a reservoir attribute by investigating the correlation of data points in a specified direction. In this case, the lag is a vector with both magnitude and orientation. The search for the number $N(h)$ of data points at a fixed lag h is done by specifying a neighborhood around h. The neighborhood is shown in Figure 11.8, and the orientation of the lag vector with magnitude h is specified by the angle θ. Orientation of lag may be included in the analysis by dividing the areal distribution of properties into a finite number of sectors, such as four quadrants or eight octants.

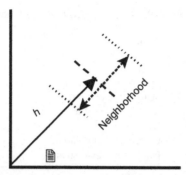

Figure 11.8 A lag neighborhood.

The neighborhood of the lag vector is sometimes referred to as a search bin, and the size of the neighborhood is the bin size. Data points within the bin are included in the calculation of the semivariance of a point at the origin with a point in the bin at lag h. If the search depends on the magnitude of h, but not the orientation, the search is circular with a radius h, and the spatial distribution of the parameter will be isotropic. If the search is allowed to depend on the magnitude and orientation of h, the search is elliptical and the spatial distribution of the parameter can be anisotropic.

11.5.4 Step 4: Model Cross-Validation

The variograms developed previously are used to estimate the value of a parameter at an unsampled location. One widely used estimation technique is kriging, which is a linear weighted average method that is similar to inverse distance weighting. The weights used in kriging are based on the semivariogram model of spatial correlation. The result of applying an estimation technique such as kriging is the determination of a spatial distribution of a parameter at all points of interest. The next question is, How accurate are the estimated values?

One method of determining the accuracy of the values obtained by an estimation technique is to treat a sampled (known) data point as an unknown point at the test location. The estimation technique is used to calculate the parameter at the test location, and the resulting value is then compared with the known data point. The accuracy of the estimation process can be quantified by calculating the semivariance of actual values relative to the estimated values. The resulting semivariance provides a cross-validation of the original semivariogram model and provides information about the quality of the estimation technique. Notice that this model cross-validation procedure could be applied to any computer-based estimation technique, including techniques like inverse distance weighting, which does not require geostatistics.

11.5.5 Step 5: Kriging

The kriging technique was introduced in Step 4. In this step, kriging or some other geostatistical mapping technique is used to complete the computer mapping process.

Mapping by hand or by computer requires some type of estimation procedure that can interpolate values between control points and extrapolate values in the region beyond control points to provide estimates of reservoir parameters that are needed in the entire region of interest. The kriging method can be computationally intensive when used with large data sets, but it can yield maps that exhibit the isotropy associated with the original data set. Kriging is a method that recognizes that data close together are more likely to be similar than data that are farther apart.

11.5.6 Step 6: Conditional Simulation

The maps that are created in geostatistics are geologic realizations. Each realization is also known as a stochastic image. The generation of stochastic images is called stochastic modeling. Two widely used approaches in stochastic modeling are the pixel-based approach and the object-based approach. The pixel-based approach uses semivariograms and estimation procedures such as kriging to generate a realization of a spatially distributed variable such as permeability. The object-based or Boolean approach populates 3-D space with distinct geologic bodies, such as sand lenses. In practice, both approaches may be combined.

A stochastic model is considered conditional when it honors both measured data and associated spatial correlations. The spatial correlations are represented mathematically by semivariograms. If a stochastic model is conditional, we can say that the model is constrained by the measured data or spatial models. Examples of conditioning or constraining data include seismic data, well log data, facies distributions, dimensions of geologic bodies, well test data, and production history. A properly conditioned stochastic model must honor all of the available constraints. The generation of constrained stochastic images is the goal of conditional simulation.

Stochastic modeling algorithm types include simulated annealing, sequential simulation, Boolean or object-based modeling, and others. For further discussion of these algorithms, consult Deutsch and Journel (1998) and Clark and Harper (2000).

11.5.7 Step 7: Uncertainty Assessment

One of the advantages of geostatistical modeling relative to deterministic modeling is the estimation of uncertainty during the stochastic modeling process. An assessment of uncertainty can be provided as part of a deterministic study by analyzing the sensitivity of the deterministic model to changes in reservoir parameters within physically meaningful ranges. Many of the techniques involved in stochastic modeling include a measure of uncertainty as a by-product of the calculation procedure.

A geostatistical study is expected to generate many equally probable, or equiprobable, realizations of a reservoir. Each realization should honor all of the available data. The suite of realizations can be used to provide probabilistic estimates of important reservoir management information such as original fluids in place and performance forecasts. The assessment of uncertainty should facilitate the decision-making process.

11.5.8 The Use and Abuse of Geostatistics

Hirsche and colleagues pointed out that "geostatistical reservoir characterization should not be done in isolation" (1997, p. 259). Geostatistics is like other reservoir characterization techniques: the technique is most successful when all of the available data are incorporated into the reservoir characterization process.

The violation of basic geostatistical assumptions, such as statistical continuity of the attribute throughout the space of interest, can lead to the creation of an inaccurate reservoir model. Inaccuracies in the model appear as errors in associated maps. Limited well control and biased sampling of well information are examples of real-world constraints that can violate the underlying assumptions of geostatistics. Abrupt changes in reservoir features, such as faults and high-permeability channels, are difficult to identify using geostatistics.

Geostatistics and stochastic modeling can be used to integrate data, provide a realistic representation of reservoir heterogeneity, and quantify uncertainty. On the other hand, the existence of multiple realizations can be confusing and more expensive than the construction of a single deterministic representation of the reservoir. In addition, the stochastic images may look realistic but actually do a poor job of representing flow in the actual reservoir if an important flow mechanism is missed. The process of validating the reservoir model is made more complicated by the existence of multiple realizations. Researchers are presently trying to simplify the stochastic modeling process and improve the reliability of forecasts made using stochastic models.

11.6 Visualization Technology

Mapping technology has evolved rapidly in the past decade because of the acceptance of advances in computer mapping and visualization technology. The last ten years have seen a conversion from 2-D mapping to 3-D and 4-D mapping. In 2-D mapping, planar maps were prepared and then stacked to form a composite 3-D map. It was very common to find nonphysical overlaps between surfaces because each 2-D map was prepared separately. For example, a gross thickness map might exceed the gross thickness defined by the top and base maps of a formation. The excessive thickness would incorrectly extend into contiguous formations. One of the advantages of 3-D mapping is that the mapping takes place in three dimensions, so overlaps are avoided and a more realistic representation of the reservoir can be constructed using data displayed in three dimensions.

The use of 3-D mapping became feasible with the advent of computer systems that could manipulate three-dimensional data sets and project a 3-D image onto a two-dimensional surface. Improved computer processing speed and algorithms allow rapid calculation of 3-D views. A viewer can rotate the reservoir to any desired angle. Slices of the reservoir can be taken to see features that are influencing fluid flow. Reservoir compartments are easily seen.

Reservoir views can be projected in a manner that lets an audience see a 3-D image. One 3-D imaging technology requires the audience to wear colored glasses. Each eye receives a slightly different view, which makes the image appear to have depth as well as width and breadth. If a sequence of 3-D images is set in motion, it gives the effect of animation, or 4-D mapping.

Animation can be used to track a variety of physical processes, such as fluid flow or geologic deposition. Seismic data, for example, can be viewed as an animated sequence of time slices to watch the formation of geologic features such as fluvial systems. Arrays of fluid saturations from fluid flow simulators can be placed into an animated sequence to show the movement of fluid fronts in a reservoir. These images can be compared with time-lapse seismic data to evaluate the quality of the flow model.

CS.11 Valley Fill Case Study: Reservoir Structure

The Valley Fill reservoir is a meandering structure with a regional dip from north to south. The shape of the meandering structure is indicated by the compressional to shear velocity ratio distribution shown in Figure CS.5A in Chapter 5. Figure CS.11A shows a plan view of the wells in the reservoir.

A Cartesian grid has been overlain on the area of interest. The grid has 30 columns and 15 rows. A comparison of Figures CS.5A and CS.11A shows that the productive wells are in the channel and the dry holes are outside the channel. The depths to the top of the Valley Fill formation for the six producing wells are given in Table CS.11A.

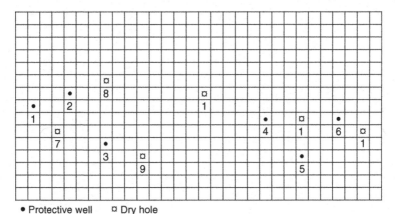

• Protective well   □ Dry hole

Figure CS.11A Well locations in 30×15 grid. The sides of each square gridblock are 200 feet long.

Table CS.11A Depths to Top of Structure

Well	Depth (ft)	Column	Row
1	8,435	2	8
2	8,430	5	7
3	8,450	8	11
4	8,440	21	9
5	8,455	24	12
6	8,440	27	9

Exercises

11-1. Sketch a contour map of the top of the structure using the data shown in Table CS.11A, the well locations shown in Figure CS.11A, and the seismic data shown in Figure CS.5A. The contour map should cover the entire area shown in Figure CS.11A. Remember that the Valley Fill reservoir has a regional dip.

11-2. Digitize the contour map prepared in Exercise 11-1. Present the values in array form using the array shown in Figure CS.11A.

11-3. Use the data in Table C.11A, the distance scale in Figure C.11A, and inverse distance weighting to estimate the formation top at the midpoint of the following grid cell locations: (a) column 6, row 9; and (b) column 15, row 11.

11-4. Compare the values obtained in Exercises 11-2 and 11-3. Explain the differences.

11-5A. Plot the exponential semivariogram as a function of lag distance in the range $0 \leq h \leq 2{,}000$ for nugget $= 0$, sill $= 500$, and range of influence $= 200$.

11-5B. Plot the Gaussian semivariogram as a function of lag distance in the range $0 \leq h \leq 2{,}000$ for nugget $= 0$, sill $= 500$, and range of influence $= 200$.

11-5C. Compare the plots in parts A and B of this exercise.

11-6. List the significant changes in slope in Figure 11.2. Identify the zone where the permeability is highest.

12 Fluid Displacement

Many reservoir management options include the injection of one fluid to help displace another. Fluid displacement processes require interaction between the displacing fluid and the displaced fluid. The interface between displacing and displaced fluids is the fluid front. Movement of the fluid front and the breakthrough time associated with the production of injected fluids at producing wells are indicators of sweep efficiency. This chapter discusses immiscible fluid displacement, frontal advance, and well patterns.

12.1 Fractional Flow

Suppose we have two phases, A and B, flowing through a porous medium. The fractional flow f_A of phase A is the flow rate q_A of phase A divided by the total flow rate q_T of the two phases A and B; thus

$$f_A = \frac{q_A}{q_T} = \frac{q_A}{q_A + q_B} \qquad (12.1.1)$$

where q_B is the flow rate of phase B. Fractional flow may also be expressed in terms of mobilities. If we assume Darcy's law applies and neglect gravity and capillary pressure, then the fractional flow of phase A is

$$f_A = \frac{\lambda_A}{\lambda_A + \lambda_B} = \frac{1}{1 + \dfrac{\lambda_B}{\lambda_A}} \qquad (12.1.2)$$

where λ_A and λ_B are the mobilities of phases A and B, respectively. If only one phase (q_A) is flowing, then $q_B = 0$ and $f_A = 1$, as expected. A similar definition of fractional flow applies for phase B.

As an example, the fractional flow of water is the ratio of water production rate to total production rate. In the case of an oil–water system, the fractional flow of water is given by

$$f_w = \frac{q_w}{q_t} = \frac{q_w}{q_w + q_o} \qquad (12.1.3)$$

Integrated Reservoir Asset Management. DOI: 10.1016/B978-0-12-382088-4.00012-8

where

f_w = fractional flow of water
q_w = water volumetric flow rate (RB)
q_o = oil volumetric flow rate (RB)
q_t = total volumetric flow rate (RB)

Notice that flow rates are expressed in terms of reservoir volumes. The fractional flow of oil f_o and the fractional flow of water are related by $f_w = 1 - f_o$ for an oil–water system. Fractional flow should have a value between 0 and 1.

12.1.1 The Simplified Fractional Flow Equation

A simplified fractional flow equation is obtained by replacing flow rates with Darcy's law in the definition of fractional flow. If we neglect gravity, Darcy's law in one spatial dimension is

$$q_\ell = -\frac{kk_{r\ell}A}{\mu_\ell}\frac{\partial P_\ell}{\partial x}$$

(12.1.4)

where

A = cross-sectional area
P_ℓ = the pressure of phase ℓ

Darcy's law says that the flow rate is proportional to the pressure gradient. The minus sign shows that the direction of flow is opposite to that of the direction of increasing pressure.

If we neglect capillary pressure, we have the equality of phase pressures $P_w = P_o$. Substituting Eq. (12.1.4) into Eq. (12.1.3) and neglecting capillary pressure gives

$$f_w = \frac{\dfrac{k_{rw}}{\mu_w}}{\dfrac{k_{rw}}{\mu_w} + \dfrac{k_{ro}}{\mu_o}}$$

(12.1.5)

Equation (12.1.5) can be expressed in terms of mobilities as

$$f_w = \frac{1}{1 + \dfrac{k_{ro}}{k_{rw}}\dfrac{\mu_w}{\mu_o}} = \frac{1}{1 + \dfrac{\lambda_o}{\lambda_w}}$$

(12.1.6)

The construction of Eq. (12.1.6) is based on the following simplifying assumptions: Darcy's law adequately describes flow rate, and capillary pressure and gravity are negligible. Given these assumptions, we can calculate f_w at reservoir conditions.

12.1.2 The Fractional Flow Equation with Gravity

Gravity can be included in the fractional flow equation as follows. First, consider a two-phase flow of oil and water in a tilted linear system. Darcy's law, including capillary pressure and gravity effects for linear flow, is

$$q_w = -\frac{kk_{rw}A}{\mu_w}\left(\frac{\partial P_w}{\partial x} + \rho_w g \sin\alpha\right)$$

$$\tag{12.1.7}$$

$$q_o = -\frac{kk_{ro}A}{\mu_o}\left(\frac{\partial P_o}{\partial x} + \rho_o g \sin\alpha\right)$$

where

α = dip angle of formation
g = gravitational constant

If we differentiate capillary pressure for a water–oil system with respect to position x along the dipping bed, we find

$$\frac{\partial P_{cow}}{\partial x} = \frac{\partial P_o}{\partial x} - \frac{\partial P_w}{\partial x}\tag{12.1.8}$$

Combining Eqs. (12.1.7) and (12.1.8) gives

$$\frac{\partial P_{cow}}{\partial x} = -\frac{(q_t - q_w)\mu_o}{Akk_{ro}} - \rho_o g \sin\alpha + \frac{q_w\mu_w}{Akk_w} + \rho_w g \sin\alpha\tag{12.1.9}$$

where we have used $q_t = q_o + q_w$. Use the density difference

$$\Delta\rho = \rho_w - \rho_o\tag{12.1.10}$$

in Eq. (12.1.9), rearrange terms and simplify to obtain

$$\frac{q_w}{Ak}\left(\frac{\mu_o}{k_{ro}} + \frac{\mu_w}{k_{rw}}\right) = \frac{q_t\mu_o}{Akk_{ro}} + \frac{\partial P_{cow}}{\partial x} - g\Delta\rho\sin\alpha\tag{12.1.11}$$

Collecting terms gives the fractional flow to water f_w in conventional oil field units:

$$f_w = \frac{q_w}{q_t} = \frac{1 + 0.001127\dfrac{Akk_{ro}}{\mu_o q_t}\left[\dfrac{\partial P_{cow}}{\partial x} - 0.433(\gamma_w - \gamma_o)\sin\alpha\right]}{1 + \dfrac{k_{ro}}{k_{rw}}\dfrac{\mu_w}{\mu_o}}\tag{12.1.12}$$

where

A = cross-sectional area of flow system (ft^2)
k = absolute permeability (md)
k_{ro} = relative permeability to oil
k_{rw} = relative permeability to water
μ_o = oil viscosity (cp)
μ_w = water viscosity (cp)
P_{cow} = oil–water capillary pressure (psi) = $P_o - P_w$
x = direction of linear flow (ft)
α = dip angle of formation (degrees)
γ_o = oil specific gravity (pure water = 1)
γ_w = water specific gravity (pure water = 1)

The general expression for f_w includes all three of the terms governing immiscible displacement: the mobility term $(k_{ro}/k_{rw})(\mu_w/\mu_o)$, the capillary pressure term $\partial P_{cow}/\partial x$, and the gravity term $(\gamma_w - \gamma_o)\sin\alpha$. The capillary pressure and gravity terms are multiplied by $1/q_t$ in Eq. (12.12). If total flow rate for a waterflood is large, the impact of capillary pressure and gravity effects can be neglected.

12.1.3 Gas Fractional Flow

A similar analysis can be performed to determine the fractional flow of gas f_g. The result for a gas–oil system is

$$f_g = \frac{1 + 0.001127 \dfrac{Akk_{ro}}{\mu_o q_t'}\left(\dfrac{\partial P_{cgo}}{\partial x} - 0.433(\gamma_g - \gamma_o)\sin\alpha\right)}{1 + \dfrac{k_{ro}}{k_{rg}}\dfrac{\mu_g}{\mu_o}} \tag{12.1.13}$$

where oil phase properties are defined after Eq. (12.1.12) and the remaining variables are

k_{rg} = relative permeability to gas
μ_g = gas viscosity (cp)
P_{cgo} = gas–oil capillary pressure = $P_g - P_o$ (psi)
γ_g = gas specific gravity (pure water = 1)
q_g = gas volumetric flow rate (RB/day)
q_t' = total volumetric flow rate = $q_o + q_g$ (RB/day)

Immiscible displacement of oil by gas is analogous to water displacing oil with the water terms replaced by gas terms. In general, the gravity term in f_g should not be neglected unless q_t' is very high because of the specific gravity difference between gas and oil.

12.2   The Buckley-Leverett Theory

One of the simplest and most widely used methods of estimating the advance of a fluid displacement front in an immiscible displacement process is the Buckley-Leverett method (Buckley and Leverett, 1942). Their theory determines the rate at which an injected water bank moves through a porous medium. The approach uses fractional flow theory and is based on the following assumptions:

- Flow is linear and horizontal.
- Water is injected into an oil reservoir.
- Oil and water are both incompressible.
- Oil and water are immiscible.
- Gravity and capillary pressure effects are negligible.

The following analysis can be found in Collins (1961), Dake (1978), Wilhite (1986), Craft et al. (1991), Towler (2002), and Warner (2007).

Frontal advance theory is an application of the law of conservation of mass. Flow through a small volume element, as shown in Figure 12.1, with length Δx and cross-sectional area A can be expressed in terms of total flow rate q_t as

$$q_t = q_o + q_w \tag{12.2.1}$$

where q denotes volumetric flow rate at reservoir conditions, and the subscripts $\{o, w, t\}$ refer to oil, water, and total rate, respectively. The rate of water entering the element on the left-hand side (LHS) is

$$q_t f_w = \text{entering LHS} \tag{12.2.2}$$

for a fractional flow to water f_w. The rate of water leaving the element on the right-hand side (RHS) is

$$q_t(f_w + \Delta f_w) = \text{leaving RHS} \tag{12.2.3}$$

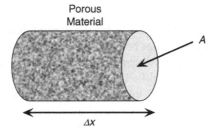

Porous
Material

A

Δx

Figure 12.1 Flow geometry.

The change in water flow rate across the element is found by performing a mass balance. The movement of mass for an immiscible, incompressible system gives

$$
\begin{aligned}
\text{water rate} &= \text{water entering} - \text{water leaving} \\
&= q_t f_w - q_t(f_w + \Delta q_t f_w) \\
&= -q_t \Delta f_w
\end{aligned} \tag{12.2.4}
$$

The change in water saturation per unit time is the water rate in Eq. (12.2.4) divided by the pore volume of the element; thus

$$
\frac{\Delta S_w}{\Delta t} = -\frac{q_t}{A\phi}\frac{\Delta f_w}{\Delta x} \tag{12.2.5}
$$

In the limit as $\Delta t \to 0$ and $\Delta x \to 0$, we pass to the differential form of Eq. (12.2.5) for the water phase:

$$
\frac{\partial S_w}{\partial t} = -\frac{q_t}{A\phi}\frac{\partial f_w}{\partial x} \tag{12.2.6}
$$

A similar equation applies to the oil phase:

$$
\frac{\partial S_o}{\partial t} = -\frac{q_t}{A\phi}\frac{\partial f_o}{\partial x} \tag{12.2.7}
$$

Since f_w depends only on S_w, we can write the derivative of fractional flow as

$$
\frac{\partial f_w}{\partial x} = \frac{df_w}{dS_w}\frac{\partial S_w}{\partial x} \tag{12.2.8}
$$

Substituting $\partial f_w/\partial x$ into $\partial S_w/\partial x$ yields

$$
\frac{\partial S_w}{\partial t} = -\frac{q_t}{A\phi}\frac{df_w}{dS_w}\frac{\partial S_w}{\partial x} \tag{12.2.9}
$$

It is not possible to solve Eq. (12.2.9) for the general distribution of water saturation $S_w(x, t)$ in most realistic cases because the problem is nonlinear. For example, water fractional flow is usually a nonlinear function of water saturation. It is therefore necessary to consider a simplified approach to solving Eq. (12.2.9).

We begin by considering the total differential of $S_w\,(x, t)$:

$$
\frac{dS_w}{dt} = \frac{\partial S_w}{\partial x}\frac{dx}{dt} + \frac{\partial S_w}{\partial t} \tag{12.2.10}
$$

Equation (12.2.10) is simplified by choosing x to coincide with a surface of fixed S_w so $dS_w/dt = 0$ and

$$\left(\frac{dx}{dt}\right)_{S_w} = -\frac{\left(\frac{\partial S_w}{\partial t}\right)}{\left(\frac{\partial S_w}{\partial x}\right)} \tag{12.2.11}$$

Substituting Eqs. (12.2.8) and (12.2.9) into Eq. (12.2.11) gives the Buckley-Leverett frontal advance equation:

$$\left(\frac{dx}{dt}\right)_{S_w} = -\frac{q_t}{A\phi}\left(\frac{df_w}{dS_w}\right)_{S_w} \tag{12.2.12}$$

The derivative $(dx/dt)_{S_w}$ is the velocity of the moving plane with water saturation S_w, and the derivative $(df_w/dS_w)_{S_w}$ is the slope of the fractional flow curve. The integral of the frontal advance equation gives

$$x_{S_w} = \frac{W_i}{A\phi}\left(\frac{df_w}{dS_w}\right)_{S_w} \tag{12.2.13}$$

where

x_{S_w} = distance traveled by a particular S_w contour (ft)
W_i = cumulative water injected (cu ft)
$(df_w/dS_w)_{S_w}$ = slope of fractional flow curve

12.2.1 The Water Saturation Profile

A plot of S_w versus distance using Eq. (12.2.13) and typical fractional flow curves leads to the physically impossible situation of multiple values of S_w at a given location. A discontinuity in S_w at a cutoff location x_c is needed to make the water saturation distribution single valued and to provide a material balance for wetting fluids. The procedure is summarized in the following section.

12.3 Welge's Method

Welge (1952) published an approach that is widely used to perform the Buckley-Leverett frontal advance calculation. Welge's approach is best demonstrated using a plot of f_w versus S_w, as shown in Figure 12.2.

A line is drawn with its intercept at irreducible water saturation S_{wirr}—the water saturation S_w in front of the waterflood—and tangent to a point on the f_w curve. The resulting tangent line is called the breakthrough tangent, or slope, which appears in Figure 12.2. Water saturation at the flood front S_{wf} is the point of tangency on the f_w curve. The water-oil flood front is sometimes called a shock front because of the abrupt change from irreducible water saturation in front of the waterflood to S_{wf}.

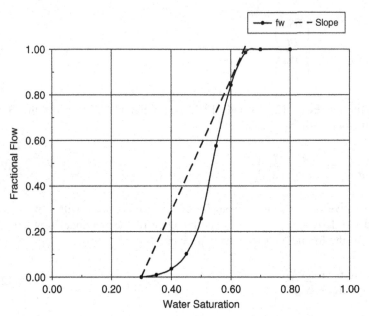

Figure 12.2 Welge's method.

Fractional flow of water at the flood front is f_{wf} and occurs at the point of tangency S_{wf} on the f_w curve. Average water saturation behind the flood front S_{wbt} is the intercept of the main tangent line with the upper limiting line, where $f_w = 1.0$. In summary, when injected water reaches the producer, Welge's approach gives these results:

- The water saturation at the producing well is S_{wf}.
- The average water saturation behind the front is S_{wbt}.
- The producing water cut at reservoir conditions is f_{wf}.

Welge's approach can be used to obtain other useful information about the waterflood. The time to water breakthrough at the producer is

$$t_{bt} = \frac{LA\phi}{q_i(df_w/dS_w)_{S_w}} \tag{12.3.1}$$

where

$$
\begin{aligned}
t_{bt} &= \text{breakthrough time (days)} \\
q_i &= \text{injection rate (RCF/day)} \\
(df_w/dS_w)_{S_w} &= \text{slope of main tangent line} \\
L &= \text{linear distance from injection well to production well (ft)} \\
A &= \text{cross-sectional area perpendicular to flow (ft}^2) \\
\phi &= \text{Porosity (fraction)}
\end{aligned}
$$

Cumulative water injected is given by

$$Q_i = \frac{1}{(df_w/dS_w)_{S_{wf}}} \qquad (12.3.2)$$

where Q_i is the cumulative pore volume of injected water. The slope of the water fractional flow curve with respect to water saturation df_w/dS_w evaluated at the water saturation at breakthrough S_{wbt} gives cumulative water injection Q_i at breakthrough.

12.3.1 The Effects of Capillary Pressure and Gravity

In the absence of capillary pressure and gravity effects, the flood front propagates as a relatively "sharp" step function, or pistonlike displacement. The characterization of the front as sharp or pistonlike is only approximate. In an ideal piston displacement, only one phase would flow on either side of the front.

The presence of capillary pressure leads to the imbibition of water ahead of the front. This causes a change in the behavior of produced fluid ratios. Rather than an abrupt increase in the water–oil ratio (WOR) associated with pistonlike displacement, the WOR will increase gradually as the leading edge of the mobile water reaches the well and is produced. In addition, the WOR will begin to increase sooner than it would have in the absence of capillary pressure. By contrast, gravity causes high S_w values to lag behind the front. The result is a smeared or "dispersed" flood front.

12.4 Frontal Advance

The stability of a flood front can influence the efficiency of fluid displacement. A front is stable if it retains the shape of the interface between displaced and displacing fluids as the front moves through the medium. An analysis of frontal stability is presented here in terms of a specific example: the advance of a water–oil displacement front. Front stability is then studied using linear stability analysis.

The displacement of one phase by another may be analytically studied if we simplify the problem to displacement in a linear, homogenous porous medium. Let us first consider the displacement of oil by water in a porous medium of length L inclined at an angle Θ relative to the horizontal plane, as shown in Figure 12.3. Water is downstructure of the front located at x_f and oil is upstructure. Gravity is

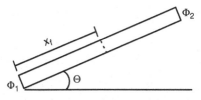

Figure 12.3 Geometry of frontal advance.

included in the analysis of frontal advance in a dipping reservoir by expressing the equations in terms of the potential of phase ℓ:

$$\Phi_\ell = P_\ell - \rho_\ell gx \sin \Theta \qquad (12.4.1)$$

Application of Darcy's law and the continuity equation leads to a pressure distribution described by Poisson's equation. The absence of sources or sinks in the medium reduces Poisson's equation to the Laplace equation for the water phase potential:

$$\frac{\partial^2 \Phi_w}{\partial x^2} = 0, \quad 0 < x < x_f \qquad (12.4.2)$$

The corresponding equation for oil phase potential is

$$\frac{\partial^2 \Phi_o}{\partial x^2} = 0, \quad x_f < x < L \qquad (12.4.3)$$

Equations (12.4.2) and (12.4.3) apply to those parts of the medium containing water and oil, respectively. They assume that the fluids are incompressible and that the oil–water interface is a pistonlike displacement in the x-direction. The piston like displacement assumption implies a discontinuous change from mobile oil to mobile water at the displacement front. This concept differs from the Buckley-Leverett analysis presented previously.

Buckley-Leverett theory with Welge's method shows the existence of a transition zone as saturations grade from mobile oil to mobile water. The saturation profile at the interface between the immiscible phases depends on the fractional flow characteristics of the system. The present method of analysis has less structure in the saturation profile but is more readily suited for analyzing the stability of the displacement front.

The phase potentials at the flood front are related by

$$\Phi_o = \Phi_w + (\rho_o - \rho_w)gx_f \sin \Theta \qquad (12.4.4)$$

with continuity of phase velocities

$$\lambda_w \frac{\partial \Phi_w}{\partial x} = \lambda_o \frac{\partial \Phi_o}{\partial x} \qquad (12.4.5)$$

The boundary conditions for the phase potentials are

$$\Phi_w = \Phi_1 \quad \text{at} \quad x = 0 \qquad (12.4.6)$$

and

$$\Phi_o = \Phi_2 \quad \text{at} \quad x = L \qquad (12.4.7)$$

Capillary pressure is neglected in this formulation.

The linear relationships

$$\Phi_w = A'_w x + B'_w \tag{12.4.8}$$

and

$$\Phi_o = A'_o x + B'_o \tag{12.4.9}$$

are solutions of the second-order ordinary differential equations given by Eqs. (12.4.2) and (12.4.3). The coefficients are evaluated by substituting Eqs. (12.4.8) and (12.4.9) into Eqs. (12.4.2) and (12.4.3) and applying the boundary conditions. The coefficients are

$$A'_w = -\left[\frac{(\Phi_1 - \Phi_2) + x_f(\rho_o - \rho_w)g \sin \Theta}{ML + (1 - M)x_f}\right] \tag{12.4.10}$$

$$A'_o = MA'_w \tag{12.4.11}$$

$$B'_w = \Phi_1 \tag{12.4.12}$$

$$B'_o = \Phi_2 - A'_o L = \Phi_2 - MA'_w L \tag{12.4.13}$$

where M is mobility ratio

$$M = \frac{\lambda_w}{\lambda_o} \tag{12.4.14}$$

Frontal velocity v_f is given by

$$v_f \equiv \frac{dx_f}{dt} = \frac{v_w}{\phi(1 - S_{or} - S_{wc})} \tag{12.4.15}$$

where

S_{or} = residual oil saturation
S_{wc} = connate water saturation
v_w = the Darcy velocity of the water

The v_w phase is given by

$$v_w = -\lambda_w \frac{\partial \Phi_w}{\partial x} = -\lambda_w A'_w \tag{12.4.16}$$

The velocity of frontal advance in a dipping reservoir is found by substituting Eq. (12.4.16) into Eq. (12.4.15) to yield the result

$$\frac{dx_f}{dt} = \frac{\lambda_w}{\phi(1 - S_{or} - S_{wc})} \left\{ \frac{(\Phi_1 - \Phi_2) + [(\rho_o - \rho_w)g\sin\Theta]x_f}{ML + (1 - M)x_f} \right\} \tag{12.4.17}$$

The integral of Eq. (12.4.17) with respect to time gives frontal advance.

12.5   Linear Stability Analysis

The stability of frontal advance is determined by considering the rate of growth of a perturbation at the front. We first express the frontal advance velocity given by Eq. (12.4.17) in the general form

$$\frac{dx_f}{dt} = \frac{\alpha + \beta x_f}{\gamma + \delta x_f} \tag{12.5.1}$$

where the coefficients $\{\alpha, \beta, \gamma, \delta\}$ are independent of time and frontal location. Equation (12.5.1) is a nonlinear, first-order differential equation. Imposing a slight perturbation ε on the front location gives

$$\frac{d(x_f + \varepsilon)}{dt} = \frac{\alpha + \beta(x_f + \varepsilon)}{\gamma + \delta(x_f + \varepsilon)} \tag{12.5.2}$$

The velocity of propagation of the perturbation is given by the difference between Eqs. (12.5.2) and (12.5.1):

$$\frac{d\varepsilon}{dt} = \frac{\alpha + \beta x_f + \beta\varepsilon}{\gamma + \delta x_f + \delta\varepsilon} - \frac{\alpha + \beta x_f}{\gamma + \delta x_f} \tag{12.5.3}$$

Combining fractions and simplifying yields

$$\frac{d\varepsilon}{dt} = \left\{ \frac{\beta(\gamma + \delta x_f) - \delta(\alpha + \beta x_f)}{(\gamma + \delta x_f)^2 \left[1 + \dfrac{\delta\varepsilon}{\gamma + \delta x_f} \right]} \right\} \varepsilon \tag{12.5.4}$$

Further simplification is achieved by recognizing that the perturbation is slight, so that we have the approximation

$$\frac{1}{1 + \dfrac{\delta\varepsilon}{\gamma + \delta x_f}} \approx 1 - \frac{\delta\varepsilon}{\gamma + \delta x_f} \quad \text{for } \delta\varepsilon << \gamma + \delta x_f \tag{12.5.5}$$

Substituting Eq. (12.5.5) into Eq. (12.5.4) gives

$$\frac{d\varepsilon}{dt} = \left\{ \frac{\beta(\gamma + \delta x_f) - \delta(\alpha + \beta x_f)}{(\gamma + \delta x_f)^2} \right\} \left[1 - \frac{\delta\varepsilon}{\gamma + \delta x_f} \right] \tag{12.5.6}$$

Keeping only terms to first order in ε and simplifying gives

$$\frac{d\varepsilon}{dt} = \frac{(\beta\gamma - \delta\alpha)\varepsilon}{(\gamma + \delta x_f)^2} \tag{12.5.7}$$

Equation (12.5.7) has the solution

$$\varepsilon = \varepsilon_0 e^{t\tau} \tag{12.5.8}$$

where ε_0 is an integration constant, and

$$\tau = \frac{\beta\gamma - \delta\alpha}{(\gamma + \delta x_f)^2} \tag{12.5.9}$$

If τ is negative, the perturbation decays exponentially. If τ is greater than zero, the perturbation grows exponentially. Finally, if τ equals zero, the perturbation does not propagate because $d\varepsilon/dt = 0$ in Eq. (12.5.7).

We can now examine the stability of a displacement front. When comparing Eq. (12.5.1) with (12.4.17), we can make the identifications

$$\alpha = \frac{\lambda_w(\Phi_1 - \Phi_2)}{\phi(1 - S_{or} - S_{wc})} \tag{12.5.10}$$

$$\beta = \frac{\lambda_w(\rho_o - \rho_w)g \sin\Theta}{\phi(1 - S_{or} - S_{wc})} \tag{12.5.11}$$

$$\gamma = ML \tag{12.5.12}$$

$$\delta = (1 - M) \tag{12.5.13}$$

The resulting expression for the growth of a perturbation is

$$\begin{aligned} \frac{d\varepsilon}{dt} = &-\frac{\lambda_w\varepsilon}{\phi(1 - S_{or} - S_{wc})} \\ &\times \frac{(1 - M)(\Phi_1 - \Phi_2) + ML(\rho_o - \rho_w)g \sin\Theta}{[ML + (1 - M)x_f]^2} \end{aligned} \tag{12.5.14}$$

Equation (12.5.14) agrees with Eq. (7-104) in Collins (1961).

The zero growth rate of a perturbation is determined by setting the derivative to $d\varepsilon/dt = 0$ in Eq. (12.5.14). The resulting condition for zero growth rate is

$$(1 - M)(\Phi_1 - \Phi_2) + ML(\rho_o - \rho_w)g \sin \Theta = 0 \qquad (12.5.15)$$

If the medium is horizontal, the condition for a system without gravity is

$$(1 - M)\Delta P = 0, \quad \Delta P = P_1 - P_2 \qquad (12.5.16)$$

To see the effect of mobility ratio M on finger growth for the gravity-free case, we set $g = 0$ in Eq. (12.5.14) to get

$$\frac{d\varepsilon}{dt} = -\frac{\lambda_w \varepsilon}{\phi(1 - S_{or} - S_{wc})} \frac{(1 - M)\Delta P}{[ML + (1 - M)x_f]^2} \qquad (12.5.17)$$

The finger grows exponentially if $M > 1$, decays exponentially if $M < 1$, and does not propagate if $M = 1$. This type of stability analysis is the basis for the claim that a displacement is favorable if $M < 1$ and unfavorable if $M > 1$.

12.6 Well Patterns

The effectiveness of a displacement process depends on many factors, including some that are beyond our control, such as depth, structure, and fluid type. Other factors that influence displacement efficiency, however, can be controlled. These include the number and type of wells, well rates, and well locations. The distribution of wells is known as the well pattern. The selection of a development plan depends on a comparison of the economics of alternative development concepts. Reservoir flow models are especially useful tools for performing these studies.

The analytical techniques for describing displacement that we discussed previously apply to fluid displacement between one injection well and one production well. The alignment of the injector-producer pair represents a linear displacement process. It is the simplest pattern involving injection and production wells. A variety of other patterns may be defined, some of which are shown in Figure 12.4. A representative pattern element for the five-spot pattern is identified using shaded wells.

The ratio of the number of producing wells to the number of injection wells is shown in Table 12.1. The patterns depicted in Table 12.1 and Figure 12.4 are symmetric patterns that are especially effective for reservoirs with relatively small dip and large areal extent. The injectors and producers are generally interspersed. Other patterns in which injectors and producers are grouped together may be needed for reservoirs with significant dip. For example, a peripheral or flank injection pattern may be needed to effectively flood an anticlinal or monoclinal reservoir.

In addition to reservoir geometry and the displacement process, the well pattern depends on the distribution and orientation of existing production wells and the desired spacing of wells. Wells may be oriented vertically, horizontally, or at some

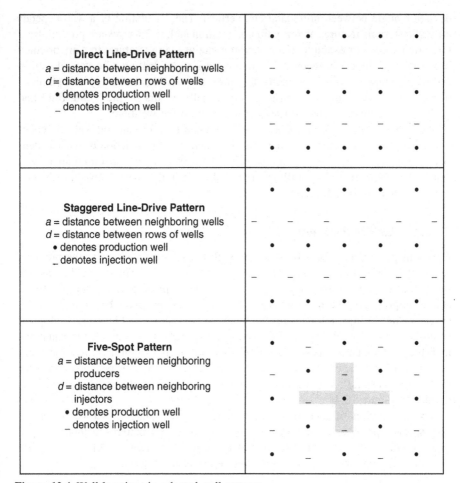

Figure 12.4 Well locations in selected well patterns.

Table 12.1 Producer-to-Injector Ratios for Common Well Patterns

Well Pattern	Producer–Injector Ratio
Four-Spot	2
Five-Spot	1
Direct Line-Drive	1
Staggered Line-Drive	1
Seven-Spot	1 : 2
Nine-Spot	1 : 3

deviation angle between horizontal and vertical. The orientation of a well depends on such reservoir features as formation orientation and, if fractures are present, fracture orientation. For example, if a reservoir contains many fractures that are oriented in a particular direction, recovery is often optimized by drilling a horizontal well in a direction that intersects as many fractures as possible. Recovery is optimized because recovery from fractured reservoirs usually occurs by producing fluid that flows from the matrix into the fractures and then to the wellbore.

Well spacing depends on the area being drained by a production well. A reduction in well spacing requires an increase in the density of production wells. The density of production wells is the number of production wells in a specified area. Well density can be increased by drilling additional wells in the space between wells in a process called infill drilling.

12.6.1 Pattern Recovery

Optimum performance may be achieved with the patterns defined in the previous section by controlling the rates of injectors and producers. These calculations can be performed analytically if we assume that the displacing and displaced fluids are incompressible, the mobility ratio is one, and the reservoir has uniform properties. Values of injection rates for the three patterns shown in Figure 12.4 are presented in Table 12.2 (Wilhite, 1986). Units and nomenclature for the rate equations in Table 12.2 are barrels per day for rate q; darcies for permeability k; feet for thickness h; feet for distances a and d, and wellbore radius r_w; pounds per square inch for pressure change ΔP; and centipoise for viscosity μ. The distances a and d are defined in Figure 12.4.

The calculation of analytical injection rates, even under a set of restrictive assumptions, provides a methodology for designing well patterns without using a reservoir simulator. More accurate estimates of injection rates under a less restrictive set of assumptions are obtained using reservoir simulators.

Table 12.2 Analytical Injection Rates for Selected Well Patterns

Pattern	Rate
Direct Line-Drive	$q_D = \dfrac{3.541\, kh\, \Delta P}{\mu\left[\ln\left(\dfrac{a}{r_w}\right) + 1.571\dfrac{d}{a} - 1.838\right]},\ \dfrac{d}{a} \geq 1$
Staggered Line-Drive	$q_S = \dfrac{3.541\, kh\, \Delta P}{\mu\left[\ln\left(\dfrac{a}{r_w}\right) + 1.571\dfrac{d}{a} - 1.838\right]}$
Five-Spot	$q_F = \dfrac{3.541\, kh\, \Delta P}{\mu\left[\ln\left(\dfrac{a}{r_w}\right) - 0.619\right]}$

Correlations of volumetric sweep efficiency with mobility ratio and permeability variation show that volumetric sweep efficiency declines as reservoir heterogeneity increases or mobility ratio increases, particularly for mobility ratios greater than one. This makes sense physically if we recall that mobility ratio is the mobility of the displacing fluid behind the front divided by the mobility of the displaced fluid ahead of the front. If the mobility of the displacing fluid is greater than the mobility of the displaced fluid, then the mobility ratio is greater than one and is considered unfavorable. On the other hand, if the mobility of the displacing fluid is less than the mobility of the displaced fluid, then the mobility ratio is less than one and is considered favorable. Unfavorable mobility ratios often occur when gas displaces oil or when water displaces high-viscosity oil. An example of a flood with a favorable mobility ratio is the displacement of low-viscosity oil by water.

CS.12 Valley Fill Case Study: Conceptual Model

It is often worthwhile to study simplified models of the system of interest to learn more about how displacement mechanisms may behave in the more complex full-field model. For example, a conceptual, one-layer model of the Valley Fill reservoir is a conceptual areal model. It can provide insight into the volumetrics of the reservoir and the pressure behavior of the field. The conceptual model is considered in more detail in the exercises.

Exercises

12-1. Find the following values for the example shown in Figure 12.2: fractional flow of water at the flood front f_{wf}, water saturation at the flood front S_{wf}, and average water saturation behind the flood front S_{wbt}.

12-2. Consider the following oil–water relative permeability table.

S_w	k_{rw}	k_{row}
0.30	0.000	1.000
0.35	0.010	0.590
0.40	0.020	0.320
0.45	0.034	0.180
0.50	0.046	0.080
0.55	0.068	0.030
0.60	0.128	0.010
0.65	0.166	0.001
0.70	0.200	0.0001
0.80	0.240	0.0000

At 3014.7 psia, oil viscosity = 0.594 cp and water viscosity = 0.503. Calculate and plot the fractional flow of water.

12-3. Find the following values for the data in Exercise 12-2: fractional flow of water at the flood front f_{wf}, water saturation at the flood front S_{wf}, and average water saturation behind the flood front S_{wbt}.

12-4. Find the mobility ratio of water to oil for the data in Exercise 12-2. Is the mobility ratio favorable or unfavorable?

12-5. The slope of the main tangent line for the data in Exercise 12-2 is 3.145. Suppose the water injection rate is 400 STB/D, the separation between injector and producer is 300 feet, the cross-sectional area is 40,000 ft^2, and the porosity is 15 percent. Estimate the time to water breakthrough at the producer using Eq. (12.3.1). Assume the water formation volume factor is 1.01 RB/STB.

12-6. Show that Eq. (12.4.8) is a solution of Eq. (12.4.2).

12-7. Determine the rate of finger growth of a unit mobility flood in a horizontal medium using Eq. (12.4.14). *Hint:* Set $M = 1$ in Eq. (12.4.14) and simplify.

12-8. Explain why the mobility ratio condition $M < 1$ is considered "favorable" for a displacement flood using Eq. (12.5.45).

12-9. Copy the file VFILL1_HM.DAT to ITEMP.DAT and run IFLO by double clicking on the IFLO.EXE file on your hard drive. Select option "Y" to write the run output to files. When the program ends, it will print "STOP". Close the IFLO window. You do not need to save changes. Copy the file ITEMP.ROF to VFILL1_HM.ROF, and copy ITEMP.ARR to VFILL1_HM.ARR. Open VFILL1_HM.ROF using a text editor. Search the file for INITIAL FLUID VOLUMES.

(a) How much oil is initially in place?

(b) How much water is initially in place?

(c) How much gas is initially in place?

(d) How much of the gas exists in a free gas phase?

12-10. Run the visualization program 3DVIEW and load the file VFILL1_HM.ARR created in Exercise 12-9. To load the file after 3DVIEW is open, click on the "File" button and select "Open Array File." Select the file called "VFILL1_HM.ARR" and click on the "OK" button. Select the oil saturation attribute at the beginning of the run. To select this attribute, click on the "Model" button and select "Select Active Attribute." From the list of options select "SO" for oil saturation. You are looking at the side of the reservoir. To see the top of the reservoir, place the cursor in the black field near the reservoir display, hold the left mouse button down, and pull the mouse toward you. You should see the reservoir image rotate. Continue rotating until you see the top of the reservoir. Sketch the image and indicate which part of the image represents the reservoir.

13 Reservoir Simulation

A reservoir characterization study is needed to prepare a static reservoir model for use in a dynamic flow model. The dynamic flow model is a set of input data for use by a computer program, called a reservoir simulator, that simulates fluid flow based on a set of fluid flow equations. The set of fluid flow equations expresses the mathematical representation of the physics that is incorporated in the reservoir simulator. Previous chapters have presented the physical concepts needed to formulate the fluid flow equations. The objective of this chapter is to deepen our understanding of dynamic flow models by outlining the assumptions and limitations associated with several widely used fluid flow equations. More discussion of reservoir simulation is provided in Ertekin et al. (2001) and Batycky et al. (2007). Workflows for building dynamic flow models using reservoir simulators are discussed in the next chapter.

13.1 Continuity Equation

Hydrocarbon mixtures contain molecular components with a range of molecular weights. The following analysis applies to the flow of a fluid that may contain a single component or a mixture of chemical components. The flow of fluid into and out of a single reservoir block is illustrated in Figure 13.1 for flow in the x-direction. The symbol J denotes fluid flux. Flux is defined as the rate of flow of mass per unit cross-sectional area normal to the direction of flow. The term "mass" refers to the mass of a component in the fluid. In our case, we assume fluid flows into the block at x with fluid flux J_x and out of the block at $x + \Delta x$ with fluid flux $J_{x + \Delta x}$. Applying the principle of conservation of mass to the system depicted in Figure 13.1, we have

Mass entering the block − mass leaving the block
= accumulation of mass in the block

We now consider a block with length Δx, width Δy, and depth Δz. The block's bulk volume is $\Delta x \Delta y \Delta z$. The mass entering the block in a time interval Δt is

$$\left[(J_x)_x \Delta y \Delta z + (J_y)_y \Delta x \Delta z + (J_z)_z \Delta x \Delta y \right] \Delta t = \text{Mass in} \qquad (13.1.1)$$

where we have generalized the expression to allow flux in the y- and z-directions as well. In the first term on the left-hand side, the notation $(J_x)_x$ denotes the x direction flux at location x, and $\Delta y \Delta z$ is the cross-sectional area that is perpendicular to the

Integrated Reservoir Asset Management. DOI: 10.1016/B978-0-12-382088-4.00013-X

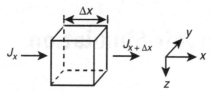

Figure 13.1 A coordinate convention.

direction of fluid flux. Analogous meanings apply to the remaining terms on the left-hand side of Eq. (13.1.1).

Corresponding to mass entering the block is a term for mass exiting the block. It has the form

$$[(J_x)_{x+\Delta x}\Delta y\Delta z + (J_y)_{y+\Delta y}\Delta x\Delta z + (J_z)_{z+\Delta z}\Delta x\Delta y]\Delta t$$
$$+ q\Delta x\Delta y\Delta z\Delta t = \text{mass out} \tag{13.1.2}$$

We added a source/sink term q to represent mass flow into (source) or out of (sink) the block. For example, q can represent well flow rate, aquifer influx rate, liquid flow from a matrix block into a fracture in a naturally fractured reservoir, or flow of gas from a coal matrix into a coal cleat. The source/sink term represents fluid entering or leaving the block through an object such as a well or the rock matrix of a naturally fractured reservoir. Production is represented by $q > 0$, and injection is represented by $q < 0$. In many cases, volumetric flow rate Q_ℓ for phase ℓ is used instead of mass flow rate q_ℓ, where $Q_\ell = q_\ell/\rho_{\ell sc}$ and $\rho_{\ell sc}$ is the density of phase ℓ at standard conditions of pressure and temperature.

Accumulation of mass in the block is the change in concentration of the mass of a component (C) in the block over the time interval Δt. If the concentration C is defined as the total mass of a component in the reservoir block divided by the block volume, then the accumulation term becomes

$$[(C)_{t+\Delta t} - (C)_t]\Delta x\Delta y\Delta z = \text{mass accumulation} \tag{13.1.3}$$

Using Eqs. (13.1.1) through (13.1.3) in the mass conservation equality, where $(C)_t$ denotes the concentration of a component at time t, and $(C)_{t+\Delta t}$ is the concentration at time $t + \Delta t$, we have

Mass in $-$ mass out $=$ mass accumulation

or

$$\left[(J_x)_x\Delta y\Delta z + (J_y)_y\Delta x\Delta z + (J_z)_z\Delta x\Delta y\right]\Delta t$$
$$- \left[(J_x)_{x+\Delta x}\Delta y\Delta z + (J_y)_{y+\Delta y}\Delta x\Delta z + (J_z)_{z+\Delta z}\Delta x\Delta y\right]\Delta t \tag{13.1.4}$$
$$- q\Delta x\Delta y\Delta z\Delta t = \left[(C)_{t+\Delta t} - (C)_t\right]\Delta x\Delta y\Delta z$$

Dividing Eq. (13.1.4) by $\Delta x \Delta y \Delta z \Delta t$ and rearranging gives

$$-\frac{(J_x)_{x+\Delta x} - (J_x)_x}{\Delta x} - \frac{(J_y)_{y+\Delta y} - (J_y)_y}{\Delta y} - \frac{(J_z)_{z+\Delta z} - (J_z)_z}{\Delta z}$$

$$-q = \frac{(C)_{t+\Delta t} - (C)_t}{\Delta t} \tag{13.1.5}$$

In the limit as Δx, Δy, Δz, and Δt go to zero, Eq. (13.5) becomes the continuity equation for a component in the fluid:

$$-\frac{\partial J_x}{\partial x} - \frac{\partial J_y}{\partial y} - \frac{\partial J_z}{\partial z} - q = \frac{\partial C}{\partial t} \tag{13.1.6}$$

Equation (13.1.6) is written in vector notation as

$$-\nabla \cdot \vec{J} - q = \frac{\partial C}{\partial t} \tag{13.1.7}$$

where $\nabla \cdot \vec{J}$ is divergence of flux. A continuity equation for mass is needed for each component in the block.

13.1.1 The Convection-Diffusion Equation

Suppose a solute with concentration C is mixing with a solvent. Then the flux of the solute has the form

$$\vec{J} = C \vec{v} - D \nabla C \tag{13.1.8}$$

where D is a scalar diffusion term. In general, the diffusion term can be a tensor. The form of Eq. (13.1.8) assumes diffusion is homogeneous and isotropic. If we assume there are no sources or sinks, then q is zero and the continuity equation, Eq. (13.1.7), becomes

$$-\nabla \cdot C \vec{v} + \nabla \cdot D \nabla C = \frac{\partial C}{\partial t} \tag{13.1.9}$$

If $\vec{v}$ and D are constant, we obtain the Convection-Diffusion (C-D) equation for the concentration of a solute mixing with a solvent:

$$D \nabla^2 C - \vec{v} \cdot \nabla C = \frac{\partial C}{\partial t} \tag{13.1.10}$$

The Laplacian operator in Cartesian coordinates has the form

$$\nabla^2 = \frac{\partial^2}{\partial x^2} + \frac{\partial^2}{\partial y^2} + \frac{\partial^2}{\partial z^2} \tag{13.1.11}$$

13.1.2 Incompressible Flow

The continuity equation for the flow of a fluid with density ρ, velocity $\vec{v}$, and no source or sink terms may be written as

$$\frac{\partial \rho}{\partial t} + \nabla \cdot \left(\rho \vec{v} \right) = 0 \qquad (13.1.12)$$

If we introduce the differential operator

$$\frac{D}{Dt} = \frac{\partial}{\partial t} + \vec{v} \cdot \nabla \qquad (13.1.13)$$

into Eq. (13.1.12), the continuity equation has the form

$$\frac{D\rho}{Dt} + \rho \nabla \cdot \vec{v} = 0 \qquad (13.1.14)$$

For an incompressible fluid, density is constant, and the continuity equation reduces to the following condition:

$$\nabla \cdot \vec{v} = 0 \qquad (13.1.15)$$

13.2 The Convection–Dispersion Equation

The one-dimensional C–D dupe equation with constant dispersion D, constant velocity v, and variable concentration C has the form

$$D\frac{\partial^2 C}{\partial x^2} - v\frac{\partial C}{\partial x} = \frac{\partial C}{\partial t} \qquad (13.2.1)$$

The C–D equation in Eq. (13.2.1) is referred to as a Convection-Diffusion equation when D is diffusion. The concentration $C(x, t)$ is a function of space and time. In this case, we assume that D and v are real, scalar constants. The dispersion term is $D(\partial^2 C/\partial x^2)$, and the convection term is $v(\partial C/\partial x)$. When the dispersion term is much larger than the convection term, the C–D equation behaves like the heat conduction equation, which is a parabolic partial differential equation (PDE). If the dispersion term is much smaller than the convection term, the C–D equation behaves like a first-order hyperbolic PDE.

An analytical solution for the C–D equation depends on boundary and initial conditions for concentration $C(x, t)$. In the case of linear fluid flow, we impose the boundary conditions $C(0, t) = 1$, $C(\infty, t) = 0$ for all times t greater than 0, and the initial condition $C(x, 0) = 0$ for all values of x greater than 0. The initial condition says that concentration $C(x, t)$ is 0 everywhere at the beginning of injection. The boundary condition at $x = 0$ says that concentration is injected at $x = 0$, and the boundary condition at $x = \infty$ says the injected concentration never reaches $x = \infty$. The corresponding solution (Peaceman, 1977) is

$$C(x,t) = \frac{1}{2}\left\{ \text{erfc}\left[\frac{x-vt}{2\sqrt{Dt}}\right] + e^{(vx/D)}\text{erfc}\left[\frac{x+vt}{2\sqrt{Dt}}\right]\right\} \qquad (13.2.2)$$

where the complementary error function erfc(y) is

$$\text{erfc}(y) = 1 - \frac{2}{\sqrt{\pi}}\int_0^y e^{-z^2}dz \qquad (13.2.3)$$

In many cases, the partial differential equations that describe the physics of fluid flow in porous media cannot be solved analytically. Two traditional numerical formulations for solving fluid flow equations are Implicit Pressure–Explicit Saturation (IMPES) and Newton-Raphson. The terms in the finite difference form of the flow equations are expanded in the Newton-Raphson procedure as the sum of each term at the current iteration level, plus a contribution due to a change of each term with respect to the primary unknown variables over the iteration.

To calculate the changes in primary variables, it is necessary to calculate derivatives, either numerically or analytically, of the flow equation terms. The derivatives are stored in a matrix called the acceleration matrix or the Jacobian. The Newton-Raphson technique leads to a matrix equation that is solved for the unknown change in primary variables. These changes are added to the value of the primary variables at the beginning of the iteration. If the changes are less than a specified tolerance, the iterative Newton-Raphson technique is considered complete, and the simulator proceeds to the next timestep. The Newton-Raphson technique is considered a fully implicit technique because all of the primary variables are calculated at the new time level.

The IMPES procedure solves for pressure at the new time level using saturations at the old time level and then uses the pressures at the new time level to explicitly calculate saturations at the new time level. One implementation of the IMPES procedure is similar to the Newton-Raphson technique except that flow coefficients are not updated in an iterative process.

Timestep size in a fully implicit model can be much larger than in an IMPES model. Arbitrarily large timestep sizes can lead to smeared spatial gradients of saturation or concentration (Lantz, 1971) and grid orientation effects (Fanchi, 1983). The smearing and grid orientation effects are referred to as numerical dispersion effects and result from the discretization of derivatives in time and space.

Numerical dispersion D^{num} in one spatial dimension may be written in the form

$$D^{num} = \frac{v}{2}\left(\Delta x \pm \frac{v\Delta t}{\phi}\right) \qquad (13.2.4)$$

with gridblock size Δx, timestep size Δt, frontal advance velocity v, and porosity ϕ. The "+" sign in front of the $v\Delta t/\phi$ term applies to the fully implicit formulation, and the "$-$" sign in front of the $v \Delta t / \phi$ term applies to IMPES. An increase in Δt in the fully implicit formulation increases D^{num}, while the same increase in Δt decreases D^{num} in an IMPES model. Although it appears that a judicious choice of Δx and Δt could eliminate D^{num} entirely in an IMPES model, the combination of Δx and Δt that yields $D^{num} = 0$ violates a numerical stability criterion. Numerical dispersion in an IMPES model is not as large as that associated with fully implicit techniques.

13.3 The Navier-Stokes Equation

Suppose a fluid with constant density ρ and constant viscosity μ is flowing with velocity $\vec{v}$ in the presence of a pressure distribution P. The equation of motion for the fluid is derived by applying the principle of conservation of momentum. The resulting flow equation is the Navier-Stokes equation

$$\rho \frac{D\vec{v}}{Dt} = \rho \left(\frac{\partial \vec{v}}{\partial t} + \vec{v} \cdot \nabla \vec{v} \right) = \mu \nabla^2 \vec{v} - \nabla P + \rho \vec{g} \tag{13.3.1}$$

The vector $\vec{g}$ is the acceleration of gravity and the differential operator D/Dt defined earlier in this chapter.

13.3.1 Euler's Equation

If viscous effects are negligible, the term involving viscosity on the right-hand side of Eq. (13.3.1) is set to zero, and the Navier-Stokes equation becomes Euler's equation:

$$\rho \frac{D\vec{v}}{Dt} = -\nabla P + \rho \vec{g} \tag{13.3.2}$$

13.3.2 Stokes's Equation

If inertial effects are negligible, the term $\rho D \vec{v} / Dt$ is set to zero, and the Navier-Stokes equation becomes the Stokes's equation

$$\mu \nabla^2 \vec{v} - \nabla P + \rho \vec{g} = 0 \tag{13.3.3}$$

13.3.3 Darcy's Law

We can use Eq. (13.3.3) to construct an expression for Darcy's law. First, we assume that the Laplacian $\nabla^2 \vec{v}$ in the viscous effects term is proportional to velocity $\vec{v}$ but opposite in direction, so

$$-\kappa \nabla^2 \vec{v} = \vec{v} \tag{13.3.4}$$

where κ is the proportionality constant. Substituting Eq. (13.3.4) into Eq. (13.3.3) gives

$$-\frac{\mu}{\kappa}\vec{v} - \nabla P + \rho\vec{g} = 0 \qquad (13.3.5)$$

Solving for velocity $\vec{v}$ gives the following expression for Darcy's law:

$$\vec{v} = -\frac{\kappa}{\mu}\left(\nabla P - \rho\vec{g}\right) \qquad (13.3.6)$$

13.4 Black Oil Model Equations

The equations of a black oil simulator are derived from mass conservation and the continuity equation

$$-\nabla \cdot \vec{J} - q = \frac{\partial C}{\partial t} \qquad (13.4.1)$$

The flow equations for an oil, water, and gas system are determined by specifying the fluxes and concentrations of the conservation equations for each of the three phases. A flux in a given direction can be written as the density of the fluid times its velocity in the given direction. Letting the subscripts o, w, and g denote oil, water, and gas, respectively, the fluxes become

$$\left(\vec{J}\right)_o = \frac{\rho_{osc}}{B_o}\vec{v}_o \qquad (13.4.2)$$

$$\left(\vec{J}\right)_w = \frac{\rho_{wsc}}{B_w}\vec{v}_w \qquad (13.4.3)$$

$$\left(\vec{J}\right)_g = \frac{\rho_{gsc}}{B_g}\vec{v}_g + \frac{R_{so}\rho_{gsc}}{B_o}\vec{v}_o + \frac{R_{sw}\rho_{gsc}}{B_w}\vec{v}_w \qquad (13.4.4)$$

where

$$R_{so} \text{ and } R_{sw} = \text{gas solubilities}$$
$$B_o, B_w, \text{ and } B_g = \text{formation volume factors}$$
$$sc = \text{standard conditions (usually } 60°F \text{ and } 14.7 \text{ psia in oil field units)}$$
$$\rho = \text{fluid densities}$$

Liquid hydrocarbon in the form of condensate is not included in this formulation. The velocities are assumed to be Darcy velocities, and their x components are

$$v_{xo} = -K_x\lambda_o\frac{\partial}{\partial x}\left[P_o - \frac{\rho_o g z}{144 g_c}\right] \qquad (13.4.5)$$

$$v_{xw} = -K_x \lambda_w \frac{\partial}{\partial x} \left[P_w - \frac{\rho_w g z}{144 g_c} \right]$$
(13.4.6)

$$v_{xg} = -K_x \lambda_g \frac{\partial}{\partial x} \left[P_g - \frac{\rho_g g z}{144 g_c} \right]$$
(13.4.7)

where

g = the acceleration of gravity in ft/sec^2
g_c = 32.174 ft/sec^2

Similar expressions can be written for the y and z components.

The relative phase mobility λ_ℓ is defined as the ratio of the relative permeability to flow of the phase divided by its viscosity; thus

$$\lambda_\ell = k_{r\ell}/\mu_\ell$$
(13.4.8)

The phase densities are related to formation volume factors and gas solubilities by

$$\rho_o = \frac{1}{B_o} \left[\rho_{osc} + R_{so} \rho_{gsc} \right]$$
(13.4.9)

$$\rho_w = \frac{1}{B_w} \left[\rho_{wsc} + R_{sw} \rho_{gsc} \right]$$
(13.4.10)

$$\rho_g = \frac{\rho_{gsc}}{B_g}$$
(13.4.11)

Besides fluxes, we also need concentrations. These are given by

$$C_o = \frac{\phi \rho_{osc} S_o}{B_o}$$
(13.4.12)

$$C_w = \frac{\phi \rho_{wsc} S_w}{B_w}$$
(13.4.13)

$$C_o = \phi \rho_{gsc} \left[\frac{S_g}{B_g} + R_{so} \frac{S_o}{B_o} + R_{sw} \frac{S_w}{B_w} \right]$$
(13.4.14)

where

ϕ = the porosity
S_ℓ = the saturation of phase ℓ

The saturations satisfy the constraint

$$S_o + S_w + S_g = 1$$
(13.4.15)

Combining Eq. (13.4.1) with Eqs. (13.4.2) through (13.4.15) gives a mass conservation equation for each component in the appropriate phase.

Oil Component in the Oil Phase

$$
-\left[\frac{\partial}{\partial x}\left(\frac{\rho_{osc}}{B_o}v_{xo}\right)+\frac{\partial}{\partial y}\left(\frac{\rho_{osc}}{B_o}v_{yo}\right)+\frac{\partial}{\partial z}\left(\frac{\rho_{osc}}{B_o}v_{zo}\right)\right]
$$
$$
-q_o=\frac{\partial}{\partial t}\left(\frac{\phi\rho_{osc}S_o}{B_o}\right)
$$

(13.4.16)

Water Component in the Water Phase

$$
-\left[\frac{\partial}{\partial x}\left(\frac{\rho_{wsc}}{B_w}v_{xw}\right)+\frac{\partial}{\partial y}\left(\frac{\rho_{wsc}}{B_w}v_{yw}\right)+\frac{\partial}{\partial z}\left(\frac{\rho_{wsc}}{B_w}v_{zw}\right)\right]
$$
$$
-q_w=\frac{\partial}{\partial t}\left(\frac{\phi\rho_{wsc}S_w}{B_w}\right)
$$

(13.4.17)

Gas Component in the Gas, Oil, and Water Phases

$$
-\frac{\partial}{\partial x}\left(\frac{\rho_{gsc}}{B_g}v_{xg}+\frac{R_{so}\rho_{gsc}}{B_o}v_{xo}+\frac{R_{sw}\rho_{gsc}}{B_w}v_{xw}\right)
$$
$$
-\frac{\partial}{\partial y}\left(\frac{\rho_{gsc}}{B_g}v_{yg}+\frac{R_{so}\rho_{gsc}}{B_o}v_{yo}+\frac{R_{sw}\rho_{gsc}}{B_w}v_{yw}\right)
$$
$$
-\frac{\partial}{\partial z}\left(\frac{\rho_{gsc}}{B_g}v_{zg}+\frac{R_{so}\rho_{gsc}}{B_o}v_{zo}+\frac{R_{sw}\rho_{gsc}}{B_w}v_{zw}\right)-q_g
$$
$$
=\frac{\partial}{\partial t}\left(\phi\rho_{gsc}\left(\frac{S_g}{B_g}+\frac{R_{so}S_o}{B_o}+\frac{R_{sw}S_w}{B_w}\right)\right)
$$

(13.4.18)

The densities at standard conditions are constants and can be divided out of the preceding equations. This reduces the equations to the following form:

Oil Component in the Oil Phase

$$
-\left[\frac{\partial}{\partial x}\left(\frac{v_{xo}}{B_o}\right)+\frac{\partial}{\partial y}\left(\frac{v_{yo}}{B_o}\right)+\frac{\partial}{\partial z}\left(\frac{v_{zo}}{B_o}\right)\right]-\frac{q_o}{\rho_{osc}}=\frac{\partial}{\partial t}\left(\frac{\phi S_o}{B_o}\right)
$$

(13.4.19)

Water Component in the Water Phase

$$
-\left[\frac{\partial}{\partial x}\left(\frac{v_{xw}}{B_w}\right)+\frac{\partial}{\partial y}\left(\frac{v_{yw}}{B_w}\right)+\frac{\partial}{\partial z}\left(\frac{v_{zw}}{B_w}\right)\right]-\frac{q_w}{\rho_{wsc}}=\frac{\partial}{\partial t}\left(\frac{\phi S_w}{B_w}\right)
$$

(13.4.20)

Gas Component in the Gas, Oil, and Water Phases

$$-\frac{\partial}{\partial x}\left(\frac{v_{xg}}{B_g}+\frac{R_{so}}{B_o}v_{xo}+\frac{R_{sw}}{B_w}v_{xw}\right)$$

$$-\frac{\partial}{\partial y}\left(\frac{v_{yg}}{B_g}+\frac{R_{so}}{B_o}v_{yo}+\frac{R_{sw}}{B_w}v_{yw}\right)$$

$$-\frac{\partial}{\partial z}\left(\frac{v_{zg}}{B_g}+\frac{R_{so}}{B_o}v_{zo}+\frac{R_{sw}}{B_w}v_{zw}\right)-\frac{q_g}{\rho_{gsc}}$$

$$=\frac{\partial}{\partial t}\left(\phi\left(\frac{S_g}{B_g}+\frac{R_{so}S_o}{B_o}+\frac{R_{sw}S_w}{B_w}\right)\right)$$

$$(13.4.21)$$

13.4.1 Flow Equations in Vector Notation

Equations (13.4.19) through (13.4.21) are the traditional fluid flow equations that are numerically solved in a black oil simulator. They illustrate the computational complexity of a three-dimensional, three-phase black oil simulator. Equivalent but much simpler appearing forms of the flow equations are presented in terms of vector operators as

$$-\nabla\cdot\frac{\vec{v}_o}{B_o}-\frac{q_o}{\rho_{osc}}=\frac{\partial}{\partial t}\left(\frac{\phi S_o}{B_o}\right)$$

$$(13.4.22)$$

$$-\nabla\cdot\frac{\vec{v}_w}{B_w}-\frac{q_w}{\rho_{wsc}}=\frac{\partial}{\partial t}\left(\frac{\phi S_w}{B_w}\right)$$

$$(13.4.23)$$

$$-\nabla\cdot\left(\frac{\vec{v}_g}{B_g}+\frac{R_{so}}{B_o}\vec{v}_o+\frac{R_{sw}}{B_w}\vec{v}_w\right)-\frac{q_g}{\rho_{gsc}}$$

$$=\frac{\partial}{\partial t}\left(\phi\left(\frac{S_g}{B_g}+\frac{R_{so}S_o}{B_o}+\frac{R_{sw}S_w}{B_w}\right)\right)$$

$$(13.4.24)$$

The meanings of the symbols are summarized in Table 13.1.

13.5 Integrated Flow Model Equations

The fluid flow simulator that was used to prepare the files accompanying this text is an integrated flow model called IFLO (Fanchi, 2006a). IFLO is a multi-phase,

Table 13.1 Nomenclature for Flow Equations

Symbol	Meaning
B_ℓ	Formation volume factor of phase ℓ
K	Absolute permeability
$k_{r\ell}$	Relative permeability of phase ℓ
P_ℓ	Pressure of phase ℓ
q	Source/sink flow rate
$R_{i\ell}$	Solubility of soluble component i in phase ℓ
S_ℓ	Saturation of phase ℓ
v_i	Volume fraction of soluble component i
μ_ℓ	Viscosity of phase ℓ
ρ	Density
ϕ	Porosity

multidimensional, pseudocomponent simulator. It is a computer program written in FORTRAN 90/95. IFLO can be used to solve fluid flow problems in up to three phases. The flow equations for the pseudocomponents in IFLO are as follows.

Stock Tank Oil

$$\nabla \cdot \frac{Kk_{ro}^e}{\mu_o^e B_o} \nabla \Phi_o - \frac{q_o}{\rho_{osc}} = \frac{\partial}{\partial t}\left(\phi \frac{S_o}{B_o}\right) \tag{13.5.1}$$

Water Plus Surfactant

$$\nabla \cdot \frac{Kk_{rw}}{\mu_w B_w} \nabla \Phi_w - \frac{q_w}{\rho_{wsc}} = \frac{\partial}{\partial t}\left(\phi \frac{S_w}{B_w}\right) \tag{13.5.2}$$

Surfactant

$$\nabla \cdot x_s \frac{Kk_{rw}}{\mu_w B_w} \nabla \Phi_w - x_s \frac{q_w}{\rho_{wsc}} = \frac{\partial}{\partial t}\left(\phi x_s \frac{S_w}{B_w}\right) \tag{13.5.3}$$

Soluble Species

$$\nabla \cdot \left[v_i \frac{Kk_{rg}^e}{\mu_i^e B_i} \nabla \Phi_i + v_i R_{io} \frac{Kk_{ro}^e}{\mu_o^e B_o} \nabla \Phi_o + v_i R_{iw} \frac{Kk_{rw}}{\mu_w B_w} \nabla \Phi_w \right]$$

$$- \frac{q_i}{\rho_{isc}} = \frac{\partial}{\partial t}\left\{ \phi v_i \left[\frac{S_g}{B_i} + R_{io} \frac{S_o}{B_o} + R_{iw} \frac{S_w}{B_w} \right] \right\} \tag{13.5.4}$$

Table 13.2 Nomenclature for Flow Equations

Symbol	Meaning
B_ℓ	Formation volume factor of phase ℓ
K	Absolute permeability
$k_{r\ell}$	Relative permeability of phase ℓ
N_s	Number of soluble species
q	Source/sink flow rate
$R_{i\ell}$	Solubility of soluble component i in phase ℓ
S_ℓ	Saturation of phase ℓ
v_i	Volume fraction of soluble component i
x_s	Surfactant volume fraction
μ_ℓ	Viscosity of phase ℓ
μ_i	Gas phase viscosity including effects of soluble component i
ρ	Density
Φ_ℓ	Potential of phase ℓ
ϕ	Porosity

Note that in Eq. (13.5.4), $i = \{g, 1, \ldots, N_s\}$. Table 13.2 presents the nomenclature for the symbols in Eqs. (13.5.1) through (13.5.4). The superscript e indicates that an effective fluid property is being calculated, and the subscript sc refers to standard conditions.

The solution procedure for the flow equations is outlined in the subsections that follow. The formulation of fluid flow equations is presented in more detail by Ammer et al. (1991) and Fanchi (2000).

13.5.1   Volume Integration and Discretization

The fluid flow equations just presented are discretized using volume integration and finite difference techniques. The volume integration procedure is illustrated by integrating the oil flow equation over a block m with volume V_m; thus

$$\int_{V_m} \left[\nabla \cdot \frac{K k_{ro}^e}{\mu_o^e B_o} \nabla \Phi_o - \frac{q_o}{\rho_{osc}} \right] dV = \int_{V_m} \frac{\partial}{\partial t} \left(\phi \frac{S_o}{B_o} \right) dV \tag{13.5.5}$$

The divergence theorem is used to replace the volume integral over the convection term on the left-hand side of Eq. (13.5.5) with a surface integral. Applying the divergence theorem gives

$$\int_{S_{em}} \frac{K k_{ro}^e}{\mu_o^e B_o} \nabla \Phi_o \cdot \hat{n} dS - \int_{V_m} \left[\frac{q_o}{\rho_{osc}} \right] dV = \frac{\partial}{\partial t} \int_{V_m} \left(\phi \frac{S_o}{B_o} \right) dV \tag{13.5.6}$$

where block volume V_m corresponds to the volume V, and the surface S is the external surface S_{em} of the gridblock m. The surface integral represents fluid flow across the block boundaries.

The spatially discretized material balance equation for oil is

$$\frac{dM_o}{dt} + Q_o = \Delta A_o \Delta \Phi_o \tag{13.5.7}$$

where the volume integral over rate is

$$Q_o = \int\limits_{V_m} \frac{q_o}{\rho_{osc}} dV = \frac{q_o}{\rho_{osc}} V_m \tag{13.5.8}$$

The volume integral over the accumulation term is

$$M_o = \int\limits_{V_m} \left(\phi \frac{S_o}{B_o} \right) dV = \phi \frac{S_o}{B_o} V_m \tag{13.5.9}$$

and the surface integral is

$$\Delta A_o \Delta \Phi_o = \int\limits_{S_{em}} \frac{K k_{ro}^e}{\mu_o^e B_o} \nabla \Phi_o \cdot \hat{n} dS \tag{13.5.10}$$

The term A_o represents oil phase transmissibility, and oil phase potential is

$$\Delta \Phi_o^{n+1} = \Delta P^{n+1} - \Delta \gamma_o^n D \tag{13.5.11}$$

The variable P is oil phase pressure, D is depth to the center of the gridblock, and γ_o is the specific weight of the oil phase. The time derivative in Eq. (13.5.11) is replaced by

$$\frac{1}{\Delta t} \left[M_o^{n+1} - M_o^n \right] + Q_o^{n+1} = \Delta A_o^{n+1} \Delta \Phi_o^{n+1} \tag{13.5.12}$$

The superscript n denotes the present time level t^n, and the superscript $n+1$ denotes the future time level t^{n+1}. Time step size Δt equals $t^{n+1} - t^n$.

The preceding formulation is a fully implicit formulation because all of the variables are assessed at the future time level in Eq. (13.5.11). IMPES is invoked by approximating transmissibilities, capillary pressures, and densities at time level $n+1$ with their values at time level n. The resulting flow equation is

$$\frac{1}{\Delta t} \left[M_o^{n+1} - M_o^n \right] + Q_o^{n+1} = \Delta A_o^n \Delta \Phi_o^{n+1} \tag{13.5.13}$$

Similar equations apply to the other flow equations.

13.5.2   The Multivariable Newton-Raphson IMPES Procedure

The IMPES equations just developed are solved using an iterative technique that is illustrated using the oil flow equation. The residual form of Eq. (13.5.13) is

$$R_o^\ell = \frac{1}{\Delta t}\left[M_o^\ell - M_o^n\right] + Q_o^\ell - \Delta A_o^n \Delta \Phi_o^\ell \tag{13.5.14}$$

where the superscript ℓ denotes the iteration level for the variables that are desired at time level $n + 1$. The primary variables for a saturated block are ΔP, ΔS_w, ΔS_g, and $\{\Delta v_i : i = 1, \ldots, N_s\}$. Gas saturation is replaced by bubble point pressure P_b in the set of primary variables for a saturated block. The variable switching logic used to treat blocks undergoing phase transitions is described in Ammer et al. (1991). The solution process is designed to find the values of the primary variables that drive the residuals to zero in all gridblocks for all components. Ammer and colleagues (1991) refer to the solution procedure as the multivariable Newton-Raphson IMPES method.

13.6   The Well Model

In this section we show how to use porosity-permeability-pressure relationships to estimate the effect of pressure-dependent rock properties on fluid flow into a well. Consider the case of radial flow into a vertical well. Volumetric flow rate Q_ℓ for phase ℓ is proportional to pressure differential ΔP, so

$$Q_\ell = PI\Delta P \tag{13.6.1}$$

where the proportionality factor is the productivity index PI. The pressure differential is the difference between reservoir pressure and flowing wellbore pressure, or

$$\Delta P = P_{res} - P_{fwb} \tag{13.6.2}$$

Pressure differential is greater than zero ($\Delta P > 0$) for production wells and less than zero ($\Delta P < 0$) for injection wells.

By applying the radial form of Darcy's law, we can write the productivity index for radial flow into a vertical well as

$$PI = \frac{Q_\ell}{\Delta P} = \frac{0.00708\, K_e h_{net}}{\mu_\ell B_\ell [\ln(r_e/r_w) + S]} \tag{13.6.3}$$

where the meaning and unit of each term is

μ_o = viscosity of phase ℓ (cp)
B_ℓ = FVF of phase ℓ (RB/STB)
r_e = drainage radius (ft)

r_w = wellbore radius (ft)
S = dimensionless skin
K_e = effective permeability (md) = $k_{r\ell} K_{abs}$
$k_{r\ell}$ = relative permeability of phase ℓ
K_{abs} = absolute permeability (md)
h_{net} = net thickness (ft)
Q_ℓ = volumetric flow rate of phase ℓ (STB/D)

The productivity index (PI) is positive for both an injection well and a production well. We can verify our sign conventions for the pressure differential and the flow rate by noting that a production well has a positive pressure differential ($\Delta P > 0$) and a positive flow rate ($Q_\ell > 0$), so the productivity index in Eq. (13.6.1) is positive. Similarly, an injection well has a negative pressure differential ($\Delta P < 0$) and a negative flow rate ($Q_\ell < 0$), so, once again, the productivity index in Eq. (13.6.1) is positive.

A value of the effective drainage radius for a vertical well in the center of a rectangular block with cross-sectional area $\Delta x \Delta y$ can be estimated from Peaceman's formula (1978):

$$r_e \approx r_o = 0.14(\Delta x^2 + \Delta y^2)^{\frac{1}{2}} \tag{13.6.4}$$

Equation (13.6.4) applies to an isotropic system—that is, a system in which lateral permeability does not depend on direction. For a well in a square block with isotropic permeability, we have $\Delta x = \Delta y$ and $r_e \approx r_o = 0.2\Delta x$. For a well in a rectangular block and an anisotropic system, the effective permeability can be estimated as

$$K = \sqrt{K_x K_y} \tag{13.6.5}$$

In this case, lateral permeability depends on direction and the directional components of permeability are not equal—thus $K_x \neq K_y$. The equivalent well block radius for an anisotropic system must account for the dependence of permeability on direction. The effective drainage radius is approximated as the equivalent well block radius

$$r_e \approx r_o = 0.28 \frac{\left[(K_y/K_x)^{\frac{1}{2}}\Delta x^2 + (K_x/K_y)^{\frac{1}{2}}\Delta y^2\right]^{\frac{1}{2}}}{(K_y/K_x)^{\frac{1}{4}} + (K_x/K_y)^{\frac{1}{4}}} \tag{13.6.6}$$

Some of the terms in PI depend on time-varying pressure and saturation, while other factors change relatively slowly or are constant with respect to time. Slowly varying, or quasi-stationary, terms can be separated into the form

$$PI = \frac{k_{ro}}{\mu_o B_o} PID \tag{13.6.7}$$

where the quasi-stationary factors are collected in the PID term—that is,

$$PID = \frac{0.00708 \, K_{abs} h_{net}}{[\ln(r_e/r_w) + S]} \qquad (13.6.8)$$

In the following, we provide a procedure for calculating the changes in PID as pressure changes.

13.6.1 Overview of PI Calculation Procedure

The productivity index calculation begins with the estimation of seismic velocities from a petrophysical algorithm. The petrophysical algorithm is a set of correlations that depends on porosity, effective pressure, and clay content. The correlations are general functional forms that are used to calculate moduli for a given lithology. Seismic attributes include acoustic impedance, shear velocity, and compressional velocity. Geomechanical properties are calculated from the seismic attributes and include Poisson's ratio, Young's modulus, and uniaxial compaction.

The dynamic seismic velocities from the petrophysical algorithm are used to calculate the dynamic Young's modulus and the Poisson's ratio. Dynamic geomechanical properties are converted to static properties using a general algorithm for converting from dynamic to static conditions. Static geomechanical properties are used to estimate uniaxial compaction, which is used to estimate net thickness for the PI calculation. A general functional form for relating porosity and permeability is then presented. Pressure-dependent permeability is calculated from the porosity-permeability relationship and pressure-dependent porosity.

The well PI calculation procedure is summarized in Table 13.3. The permeability and net thickness calculated from this procedure can be used to calculate well PI for a range of pressures, as we illustrate in a soft rock example. The sensitivity of PI to pressure is studied for a range of input variables, and we show that the quasi-stationary part of well PI can vary significantly as pressure changes in the reservoir.

Table 13.3 Well PI Calculation Procedure

Step	Task
1	Calculate porosity
2	Calculate bulk and shear moduli
3	Calculate dynamic seismic velocities
4	Calculate dynamic Young's modulus and Poisson's ratio
5	Convert dynamic variables to static variables
6	Calculate uniaxial compaction and correct net thickness
7	Determine porosity-permeability model
8	Calculate permeability as function of pressure-dependent porosity
9	Calculate PI

13.6.2 Porosity as a Function of Pressure (Step 1)

Porosity ϕ is calculated as a pressure-dependent function using the first-order approximation

$$\phi = \phi_0\left(1 + c_\phi \Delta P\right) \tag{13.6.9}$$

where the change in pore pressure P is $\Delta P = P - P_0$ and P_0 is the reference pressure for the reference porosity ϕ_0. Porosity compressibility is defined as

$$c_\phi = \frac{1}{\phi}\frac{\partial \phi}{\partial P} \tag{13.6.10}$$

13.6.3 The Net Thickness Correction (Steps 2 through 6)

The net thickness correction follows steps 2 through 6 in Table 13.3. The petro-elastic model presented in Chapter 4 is used to calculate seismic compressional velocity, shear velocity, and moduli. Dynamic estimates of Young's modulus and Poisson's ratio are obtained by calculating Young's modulus and Poisson's ratio from seismic velocities. Static values of Young's modulus and Poisson's ratio are needed for geomechanical calculations and are obtained by a conversion calculation. Uniaxial compaction Δh is estimated using

$$\Delta h = \frac{1}{3}\left(\frac{1+v}{1-v}\right)\phi c_\phi h_{net}(P_0)\Delta P \tag{13.6.11}$$

where

$\quad h_{net}(P_0) = $ net thickness at P_0
$\qquad \phi = $ porosity
$\qquad c_\phi = $ porosity compressibility
$\qquad \Delta P = P - P_0 = $ change in pore pressure

The correction to net thickness used in the PI calculation is obtained from the compaction calculation:

$$h_{net} = h_{net}(P_0) + \Delta h \tag{13.6.12}$$

If pore pressure decreases so that $\Delta P < 0$, net thickness will decrease as a result of compaction.

13.6.4 The Porosity-Permeability Model (Steps 7 and 8)

Measurements of porosity and permeability distributions in fields around the world have shown that the statistical distribution of porosity is often the normal (or Gaussian) distribution, and the statistical distribution of permeability is often log normal. Such observations suggest that porosity ϕ and permeability K are correlated. Two empirical relationships between porosity and permeability have been observed: a semilogarithmic model and a log-log model. They can be represented in a single algorithmic form as

$$\frac{K}{K_0} = a_1 \left(\frac{\phi}{\phi_0}\right)^{b_1} + a_2 \exp[b_2\phi] \tag{13.6.13}$$

where K_0 is the permeability corresponding to ϕ_0, and the coefficients $\{a_1, a_2, b_1, b_2\}$ are specified by the user.

The dependence of permeability on fluid pressure is related to the dependence of porosity on fluid pressure. The relationship between porosity and fluid pressure is expressed in terms of porosity compressibility in Eq. (13.6.9). By calculating changes in porosity as a function of changes in fluid pressure, we can use Eq. (13.6.12) to estimate the corresponding change in permeability.

13.6.5 Calculate PI (Step 9)

The last step is to estimate the productivity index for radial flow into a vertical well, so

$$PI = \frac{0.00708 \, K_e h_{net}}{\mu_\ell B_\ell [\ln(r_e/r_w) + S]} \tag{13.6.14}$$

where the meaning and unit of each term is

μ_o = viscosity of phase ℓ (cp)
B_ℓ = FVF of phase ℓ (RB/STB)
r_e = drainage radius (ft)
r_w = wellbore radius (ft)
S = dimensionless skin
K_e = effective permeability (md) = $k_{r\ell} \, K_{abs}$
$k_{r\ell}$ = relative permeability of phase ℓ
K_{abs} = absolute permeability (md)
h_{net} = net thickness (ft)

CS.13 Valley Fill Case Study: Layering of Reservoir Flow Model

A reservoir flow model with a $30 \times 15 \times 5$ grid was used to model the Valley Fill reservoir. A plan view of the reservoir grid with well locations is shown in Figure CS.11A, and a plan view of the channel system is shown in Figure CS.5A.

The productive interval was subdivided into five layers to allow gravity segregation of fluids. Although gravity segregation is not significant in the case of single-phase flow, which is the historical situation in this case, it will become significant in forecasts of reservoir performance. For example, reservoir pressure in a base case pressure-depletion case will drop below the bubble point pressure and result in the formation of a gas cap.

The gas will segregate upstructure. This is best represented in a multilayer model. Similarly, waterflooding can result in downstructure water movement, which again is best represented in a multilayer model. The definition of a multilayer model is being done in anticipation of production performance forecasts and illustrates the need for planning in the performance of a reservoir management study.

Exercises

13-1. Use the expression for numerical dispersion in one spatial dimension presented in Section 13.2 to fill in the following table. We are assuming a displacement with a frontal advance of 0.5 ft/day, a gridblock size of 100 feet, and a porosity of 20 percent. The results should be expressed for both IMPES and fully implicit formulations.

Timestep Size Δt [days]	IMPES D^{num} [ft²/day]	Fully Implicit D^{num} [ft²/day]
1		
5		
10		
30		
45		

13-2. What is the equivalent well block radius of a gridblock with areal permeabilities $K_x = K_y = 100$ md, and gridblock sizes $\Delta x = \Delta y = 200$ ft?

13-3. What is the productivity index of an oil well that produces 1,000 STB/D with a drawdown of 50 psia?

13-4. Copy the file VFILL5_HM.DAT to ITEMP.DAT and run IFLO by double clicking on the IFLO.EXE file on your hard drive. Select option "Y" to write the run output to files. When the program ends, it will print "STOP." Close the IFLO window. You do not need to save changes. Copy ITEMP.ROF to VFILL5_HM.ROF, ITEMP.TSS to VFILL5_HM.TSS, and ITEMP.ARR to VFILL5_HM.ARR. Open VFILL5_HM.ROF using a text editor. Search the file for INITIAL FLUID VOLUMES.
(a) How much oil is initially in place?
(b) How much water is initially in place?
(c) How much gas is initially in place?
(d) How much of the gas exists in a free gas phase?

13-5. History match results are presented in file VFILL5_HM.TSS. Open VFILL5_HM.TSS using a text editor. Find model calculated oil and water production rates and plot them against the historical data shown in Table CS.9B.

13-6. Open VFILL5_HM.TSS using a text editor. Find model calculated GOR and plot it against the historical data shown in Table CS.9B.

13-7. Open VFILL5_HM.TSS using a text editor. The pressure reported in VFILL5_HM.TSS is the pore volume weighted average reservoir pressure. Plot it against the historical data shown in Table CS.9B.

13-8. Use the 3-D visualization program 3DView to look at file VFILL5_HM.ARR. Look at oil saturation (attribute SO) and describe what you see.

14 Data Management

Fields come in two categories: green and brown. A green field is an undeveloped field, while a brown field is a developed field. The main difference between green and brown fields is the availability of historical production and/or injection data for a brown field. Both green field and brown field flow modeling depend on data availability, and they differ in the amount of data that are available. This raises the topics of data acquisition and data management, which we briefly considered in previous chapters. In this chapter, we provide a more detailed discussion of sources of rock data, sources of fluid data, sources of production data, data management, and preparation of data for flow modeling.

14.1 Sources of Rock Data

Information about reservoir rock must be determined on multiple scales. The dependence of rock properties on scale was discussed in Chapter 5. Micro scale information depends on analysis of rock samples from the reservoir. Examples of rock samples include drill cuttings, sidewall core, and conventional core. Cuttings are bits of rock that are carried to the surface as fluids circulate through the borehole during the drilling process. Sidewall cores are obtained by using a small explosive to fire a cylinder into the formation adjacent to the wellbore, as shown in Figure 14.1, or by drilling into the sidewall. Conventional cores are obtained by adding a core barrel (essentially a hollow cylinder) to the drill string and collecting a core in the form of a cylinder as rotary drilling proceeds.

A conventional core provides the longest sample of rock from the reservoir and can be used for direct petrophysical measurements. The length of conventional core is useful for identifying the stratigraphy of the borehole. Cores are typically photographed and preserved. It is desirable to keep core mechanically intact, but in some cases a section of the core may be poorly consolidated or unconsolidated. Coring can also provide fluid samples.

Petrophysical tests can be applied to rock from core samples. Rock samples may be obtained in the form of small cylinders or plugs. The whole core can be tested, but the entire section of rock is used in the test. Some tests are destructive and do not allow use of the whole core section for other purposes.

Giga scale information about the reservoir is provided by seismic measurements and regional geologic studies. The acquisition, processing, and interpretation of

Integrated Reservoir Asset Management. DOI: 10.1016/B978-0-12-382088-4.00014-1

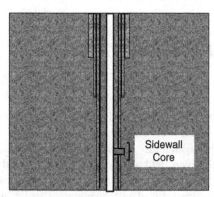

Figure 14.1 Acquiring sidewall core.

seismic data were discussed in Chapter 5. Measurements, such as well logging, pressure transient testing, tracer studies, and development geophysics, can provide information at scales between giga scale and micro scale.

Rock samples, especially core taken from the formation of interest, are a direct connection to reservoir rock. They can be used to obtain measurements of rock properties and rock–fluid interactions from reservoir rock. The benefits of acquiring core are offset by the cost of acquisition and the recognition that rock samples may not be representative of the flow unit or the reservoir environment. If possible, rock samples such as core from a well should be used to calibrate well log measurements. Seismic measurements can be calibrated by comparison with well logs and measurements of core properties.

14.2 Sources of Fluid Data

The best information about fluids is obtained from fluid samples. The goal of fluid sampling is to obtain a sample that is representative of the original *in situ* fluid. A well should be conditioned before the sample is taken. A well is conditioned by removing all fluids from within and around the wellbore that do not represent the *in situ* fluid. Nonrepresentative fluid, such as drilling mud, is removed by producing the well until the nonrepresentative fluid is replaced by original reservoir fluid flowing into the wellbore.

Surface sampling is easier and less expensive than subsurface sampling. Surface sampling can be accomplished by displacing one fluid by another in a sampling cylinder. If a surface sample is taken, the original *in situ* fluid must be reconstituted by combining separator gas and separator oil samples. The recombination step assumes accurate measurements of flow data at the surface, especially gas–oil ratio. Subsurface sampling from a properly conditioned well avoids the recombination step, but it is more difficult and costly than surface sampling, and it usually provides a smaller volume of sample fluid.

Subsurface sampling requires lowering a pressurized container to the production interval and subsequently trapping a fluid sample. Subsurface samples are often obtained using drill stem testing, especially when access to surface facilities is limited. Surface sampling allows the capture of a larger volume of fluid than subsurface sampling, but surface samples may not be as representative of *in situ* fluids.

The validity of fluid property data depends on the quality of the fluid sampling procedure. Several problems with sampling are possible: the displacement may be incomplete; mixing of *in situ* fluids with cylinder fluids may change the apparent composition of the *in situ* fluid; and corrosive gases in a gas sample may be absorbed by water or the cylinder walls. In addition, sample quality may be degraded during transport of the cylinder from the field to the laboratory, especially if the cylinder is leaking.

Gas reservoir fluid samples require the removal of liquid condensed in the tubing between the bottom of the well and the surface. The removal of condensed liquid in a wellbore can be achieved by producing the gas well at a sufficiently high gas rate. A momograph published by Turner and colleagues (1969) can be used to calculate the minimum flow rate needed for continuous removal of liquid from gas wells.

Once a sample has been acquired, it is necessary to verify the quality of the sample. This can be done by performing a compositional analysis and measuring such physical properties as density, molecular weight, viscosity, and interfacial tension. Compositional analysis of a fraction of the sample can include such tests as gas chromatography or low- and high-temperature distillation. The presence of cylinder leaks can be detected at the laboratory by measuring the cylinder pressure and verifying that it has not changed during transport.

After sample integrity is verified, several experiments may be performed to measure fluid properties that are suitable for reservoir engineering studies. The most common experiments include a combination of one or more of the following expansion tests: constant composition expansion (CCE); differential liberation (DL); and constant volume depletion (CVD). Other experimental tests include separator tests, swelling tests, and multiple contact tests. These tests are discussed in more detail by Pederson et al. (1989) and Whitson and Brulé (2000).

14.2.1 Constant Composition Expansion

A constant composition expansion (CCE) test provides information about pressure-volume behavior of a fluid without changes in fluid composition. The CCE test begins with a sample of reservoir fluid in a high-pressure cell at reservoir temperature and at a pressure in excess of the reservoir pressure. Figure 14.2 illustrates the steps in the process. The cell is labeled oil in Step A, and the volume of mercury is used to alter the pressure in the cell. The cell pressure is lowered in small increments, and the change in volume at each pressure is recorded. The procedure is repeated until the cell pressure is reduced to a pressure that is considerably lower than the saturation pressure. The original composition of the fluid in the cell does not change at any time during the test because no material is removed from the cell. The fluid may be either oil or gas with condensate. If the fluid is oil, the saturation pressure is the bubble point pressure. If the fluid is gas with condensate, the saturation pressure is the dew point pressure.

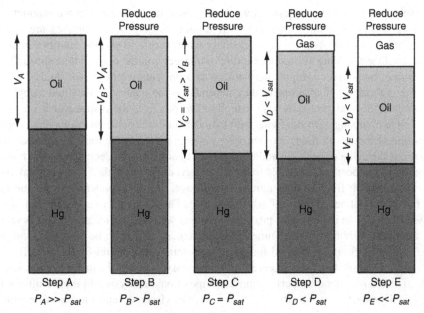

Figure 14.2 Constant composition expansion.

14.2.2 Constant Volume Depletion

The constant volume depletion (CVD) test provides information about a fluid system that is subjected to changes in pressure, composition, and phase volume. It is used to study the formation of a second phase, such as the evolution of gas from an oil sample or the formation of a condensate from a wet gas sample.

The test cell is charged with a reservoir fluid sample at reservoir temperature and the saturation pressure of the sample. If the fluid is a gas or gas condensate, the saturation pressure is the dew point pressure. If the fluid is oil, the saturation pressure is the bubble point pressure, as shown in Figure 14.3. The pressure is then lowered to a predetermined value by expanding the volume of the cell. The cell pressure is kept constant while the resulting liquid and gas phases equilibrate. Gas is withdrawn until the cell volume is restored to its initial volume at the new pressure. The composition of the produced vapor is determined, and the liquid volume is reported as a fraction of the initial cell volume. The process is repeated for a series of pressure increments.

14.2.3 Differential Liberation

The differential liberation (DL) test is used to determine the liberation of gas from live oil—that is, oil containing dissolved gas. A live oil sample is placed in a PVT cell at reservoir temperature and at a pressure above the bubble point pressure, as shown in Figure 14.4. The pressure is lowered in small increments, and the evolved gas is removed at each stage. The volume of the evolved gas and the volume of oil remaining in the cell are recorded.

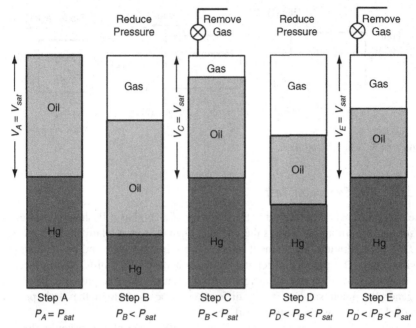

Figure 14.3 Constant volume depletion.

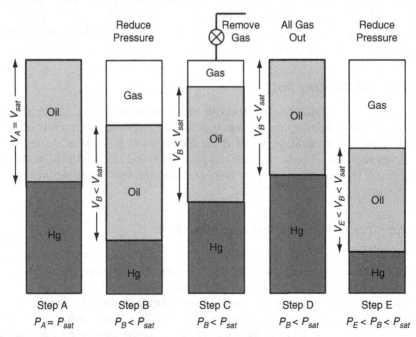

Figure 14.4 Differential liberation.

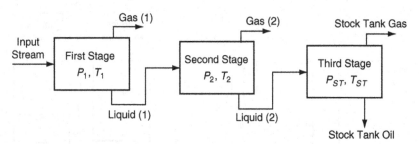

Figure 14.5 A multistage flash.

14.2.4 The Separator Test

The separator test is used to study the behavior of a fluid as it flashes from reservoir to surface conditions. A flash is the one-step change from a relatively high-pressure and high-temperature environment to a relatively low-pressure and low-temperature environment. The primary difference between a flash and a differential process is the magnitude of the pressure differential between stages. The pressure differential is generally much smaller in the differential process than in the flash process. Figure 14.5 illustrates a multistage flash.

The PVT cell in a separator test is charged with a carefully measured volume of reservoir fluid at reservoir temperature and saturation pressure. The cell pressure and temperature are then changed. Each change in pressure and temperature corresponds to a separator stage, and one or more stages may be used in the test. The volume of gas from each separation stage and the volume of the liquid that remains in the last stage are measured.

14.2.5 The Swelling Test

The swelling test simulates the behavior of reservoir oil during gas or solvent injection. Gas of known composition is added to a sample of reservoir oil in a series of steps. Gas addition begins at the bubble point pressure of the reservoir oil. After each gas addition step, the cell pressure is increased until the gas phase disappears. The pressure in the cell corresponds to the new saturation pressure of the mixture.

14.2.6 The Multiple Contact Miscibility Test

The multiple contact miscibility (MCM) test simulates the continuous multiple contact process associated with gas injection into oil. The MCM test measures vapor and liquid phase compositions after oil and gas are mixed in a PVT cell and allowed to reach equilibrium. If a condensation process is being studied, the vapor phase is purged, and the equilibrium liquid is again mixed with a fresh batch of gas. In the case of a vaporization process, the liquid phase is purged, and the equilibrium gas is mixed with a fresh batch of oil. The condensation process is analogous to phase behavior near the gas injection well. The vaporization process is analogous to phase changes occurring in the vicinity of the gas–oil interface. The procedure just described is repeated at several pressures and reservoir temperatures until only one phase remains in the cell after equilibration. This is the point when multiple contact miscibility is achieved and the corresponding pressure is considered the minimum miscibility pressure.

14.3   Sources of Field Performance Data

Field performance data should include information about surface and subsurface facilities. At a minimum, fluid injection and production rates and volumes should be acquired. In addition, measurements of well pressures and the performance of subsurface equipment such as electric submersible pumps (ESPs), flow lines, separators, compressors, and other facilities should be included in a field performance database. It may seem that important information could be lost in a sea of superfluous or less valuable information, but this typically assumes that operating facilities are functioning as expected. The value of additional information increases if something unexpected happens, such as a well leak, wellbore damage, pump failure, and so forth. A large inventory of information may contain clues to the problem and help the operator keep the field producing. One way to acquire fluid flow information is to implement a production performance monitoring system.

A production performance monitoring system should include procedures for well surveillance, quality control of gathered data, data storage, and periodic review and analysis of well performance data. Production from a well can be measured by separating oil, water, and gas phases in a test separator and then measuring the volume of each phase separately. This type of well test requires personnel to visit the site, connect the test separator, and perform the measurements. Tests are ordinarily conducted periodically, such as once a month.

Modern well surveillance techniques are providing continuous measurements. For example, well production can be continuously monitored using real-time sensors downhole or at the wellhead. Production monitoring systems can provide data continuously without requiring expensive special tests or surveys.

Production records can be combined with reservoir, petrophysical, and well completion data to create a well production model. A properly maintained production model should correctly represent well characteristics, identify wells with diminished deliverability, and generate reasonable production forecasts.

One example of a production monitoring system is Shell's Fieldware Flow Monitor (Cramer et al., 2000). Low-cost pressure transducers at the wellhead or manifold transmit signals to a supervisory control and data acquisition (SCADA) system or a distributed control system (DCS). SCADA or DCS software convert incoming pressure transducer signals into estimates of oil, water, and gas flow rates based on the results of previous well test measurements of GOR and BS&W (Basic Sediments and Water).

Effective surveillance provides information about a well—for example, when it is not producing or injecting as expected, is producing at abnormally high or low rates, or is exhibiting unstable behavior. Reservoir engineering and flow simulation calculations provide rate and pressure estimates that can be used for comparison with observations. Surveillance can inform the operator about how the well responds to changes in operating parameters such as gas-lift injection rate, pumping speed, or choke setting. In an instrumented oil field, permanently deployed instrumentation is used to monitor reservoir performance and modify production as needed.

14.4 Data Management

Reservoir management is most effective when as much relevant data as possible from all sources are collected and integrated into a reservoir management study. This requires the acquisition and management of data that can be expensive to acquire. As a consequence, it is important to estimate the value of the data by considering the cost of acquiring data in relation to the benefits that would result from its acquisition.

Decisions are needed to prioritize data needs based on project objectives, relevance, cost, and impact. Saleri pointed out that reservoir management is "essentially a data-to-knowledge-to-decision conversion process" (2005, p. 28). Stakeholders in the reservoir management process should be able to answer three questions:

1. Do you understand reservoir fundamentals?
2. Do you have mechanisms to independently verify your understanding?
3. Do you have the means to implement strategies that result from your understanding? Strategies include technologies and the capacity to apply them.

Project objectives depend on organizational goals. For example, Saleri (2005) stated that Saudi Aramco had four guiding principles: (1) sustainable performance, (2) life cycle economics, (3) low depletion rates, and (4) maximizing economic recovery. These principles are connected by a desire to continue operating for a long period of time and underlie rules for managing reservoirs. Examples of reservoir management rules are presented in Table 14.1.

Table 14.1 Examples of Reservoir Management Rules

Number	Rule
1	Do not allow migration of oil into aquifers or gas caps.
2	Do not operate below the bubble point.
3	Do not commingle production from different reservoirs.
4	Do not operate producers above their maximum allowable reservoir drawdown limits.
5	Do not operate injectors above their maximum allowable reservoir overburden.

Source: Adapted from Saleri (2005, p. 29).

Table 14.2 Data Planning Questions

Number	Question
1	Why are the data needed?
2	What type of data is it, and how much is needed?
3	When should the data be acquired?
4	When will it be used?
5	Who is responsible for acquiring the data?

Source: Adapted from Raza (1992, Figure 2).

Given a set of reservoir management guidelines, asset management team meetings can be used to prioritize data as long as all appropriate disciplines are adequately represented. Table 14.2 lists five questions that Raza (1992) proposed should be considered when planning to acquire data. The fifth question is designed to explicitly assign accountability.

If the data are to be used for a flow model study, it is important to gather all of the data that will be needed to run the simulator. Raza (1992) listed the issues in Table 14.3 that should be considered when data collection requires a significant commitment of resources. The last three issues explicitly assign responsibility and establish accountability.

An example of a data acquisition program was outlined by King and colleagues (2002) for the Takula oil field in Angola. Takula field data acquisition programs are listed in Table 14.4.

Table 14.3 Data Collection Issues

Number	Question
1	Where should the data be collected?
2	How often should the data be collected?
3	What is the correct procedure for collecting reliable and applicable data?
4	Who will audit the data for accuracy and pertinence?
5	Who will devise a database?
6	Who will maintain the database?

Source: Adapted from Raza (1992, p. 467).

Table 14.4 Takula Field Data Acquisition Programs

Number	Data Acquisition Program
1	Open-hole well log program
2	Wireline formation testing program
3	Core acquisition program
4	Cased-hole well log program
5	Well test program

Source: Adapted from King (2002).

The basic logging suite in the open-hole well log program consisted of a resistivity–gamma ray log and a neutron-density log. Formation pressure data were obtained in the wireline formation testing program. Seven wells were cored in the core acquisition program. Cores were used for conventional core analysis and special core analysis to provide rock and rock–fluid interaction data. Neutron logs were used initially in the cased-hole well log program to measure saturation data. A production logging tool was later run when water cut exceeded 30 percent in a well to determine which perforations were producing water. The well test program specified that every production well should be flow tested periodically (approximately every two months).

Time and cost constraints may rule out data acquisition by direct measurement. If information is needed to run the flow simulator, it must come from other sources. Typically, information from analogous reservoirs and correlations can be used to fill the gaps. In all cases, uncertainties associated with each data type must be identified as part of the data collection process.

Several types of uncertainties are possible. Error is associated with every measurement process. A measurement may be reasonably accurate, but the results of the measurement may only represent a small part of the system of interest. Measurements of important information may not be available or may not be made at all. Subjectivity may appear in the workflow when the processing and interpretation of measurements depend on the experience of and the judgments made by individuals involved in the workflow. The scale of a measurement may be inappropriate. Analysis of data may include approximations or assumptions that are inappropriate or inadequate, or that do not apply. It is possible that unknown processes, such as leaks and high-permeability flow paths, are significantly affecting the behavior of a reservoir but will not be included in a reservoir management study because no one knows the process that is involved. Measurements made by different organizations may have different degrees of quality.

Different companies have different philosophies about the quality of data. Everyone would like high-quality data, but the cost of acquiring it must be considered. For example, a company may decide that it is more important to be approximately correct than to spend the time and money needed to obtain a more accurate value of a datum that may have limited value. The philosophy of data acquisition may also vary from one project to another within a company. The data acquisition philosophy should be consistent throughout an organization to help ensure that data analogs taken from one reservoir are properly understood and applied to another.

Databases can be used to collect and organize data. This requires the expense of maintaining the database. A database should be updated on a timely basis and provide data in a raw form with information about factors that can affect data quality. It is also advisable to use only one database. Some companies use more than one database because different disciplines may prefer different databases for storing their data, and data may be stored at different locations. If data are collected in two or more databases, it increases the possibility that errors will occur in data entry or transfer, and corrections to one database may not appear in other databases.

Table 14.5 Data Analysis Questions

Number	Question
1	Are the data trustworthy?
2	What impact do the data have on achieving project objectives?
3	What is the uncertainty associated with the data?
4	Are the data from multiple sources? Are they consistent from one source to another?

Source: Adapted from Raza (1992, Figure 2).

Once the data have been collected, they must be analyzed. Table 14.5 summarizes Raza's list of questions that should be considered when analyzing data.

A review of available data may identify gaps or errors in the data. If it does, additional data should be obtained when possible. This may require special laboratory tests, depending on the objectives of the study. If additional field tests are needed, they should be requested and incorporated into the study schedule. Due to project costs and operating constraints, it may be difficult to justify the expense of acquiring more data or delaying the study while additional data are obtained. If measured data cannot be obtained during the scope of the study, then correlations or data from analogous fields will have to be used. Values and associated uncertainty ranges must be entered into the study, and it is necessary to select values that can be justified. The asset management team should take care to avoid underestimating the amount of work that may be needed to prepare the data. It can take just as long to collect and evaluate the data as it does to do the study.

One of the critical tasks of reservoir management is the acquisition and maintenance of an up-to-date database. A reservoir flow model study can help coordinate activities as an asset management team gathers the resources it needs to determine the optimum plan for operating a field. Collecting data for a model is a good way to ensure that every important technical variable is considered as the data are collected from the many disciplines that contribute to reservoir management. If model performance is especially sensitive to a particular parameter, then a plan should be made to reduce uncertainty in the parameter.

14.5 Data Preparation

The first step in preparing data for use in reservoir flow modeling is to set clear objectives for acquiring and evaluating the data with a focus on their quality and range of applicability. Honarpour and colleagues (2006) provided an example of a workflow for characterizing rock and fluid properties used in reservoir management. They also present a summary of fluid and rock property requirements for a variety of reservoirs and fluid types. Table 14.6 lists the types of data that are needed in a model study. A review of geophysical, geological, petrophysical, and engineering reports provides a background on how the project has been developed and what preconceived interpretations have been established.

Table 14.6 Data Required for a Simulation Study

Property	Sources
Permeability	Pressure transient testing, core analyses, correlations, well performance
Porosity, rock compressibility	Core analyses, well logs
Relative permeability and capillary pressure	Laboratory core flow tests
Saturations	Well logs, core analyses, pressure cores, single well tracer tests
Fluid property (PVT) data	Laboratory analyses of reservoir fluid samples
Faults, boundaries, fluid contacts	Seismic, pressure transient testing
Aquifers	Seismic, material balance calculations, regional exploration studies
Fracture spacing, orientation, connectivity	Core analyses, well logs, seismic, pressure transient tests, interference testing, wellbore performance
Rate and pressure data, completion and workover data	Field performance history

During the course of the study, it may be necessary not only to develop a new view of the reservoir but also to prepare an explanation of why the new view is superior to a previously approved interpretation. If significant gaps exist in the reports, particularly regarding the historical performance of the field, it is wise to update them.

The pressure range associated with fluid property data should cover the entire range of pressures expected to be encountered over the life of the field. The data should be smooth to enhance computational efficiency and to ensure data consistency. For example, the addition of the mole fractions of all components in a fluid property report should add up to one. Any inconsistencies or errors need to be identified and corrected or mitigated.

The range of uncertainty associated with data should be quantified. For example, porosity measurements may range from 17 to 21 percent and represent a normal distribution with a mean of 19 percent and a standard deviation of 0.5 percent. This information can be used to quantify uncertainty in static and dynamic model studies.

Flow units should be determined by reviewing geological and petrophysical data. It is possible to represent the behavior of a flow unit by defining a set of PVT and rock property tables for each flow unit. PVT property tables contain data that describe fluid properties, while rock property tables represent relative permeability and capillary pressure effects. Each set of PVT or rock property tables applies to a particular region of gridblocks; hence the collection of gridblocks to which a particular set of PVT or rock property tables applies is referred to as a PVT or rock region.

The parameters used to define a set of PVT and rock property tables typically contain uncertainty. For example, saturation end points for relative permeability curves and the shapes of relative permeability curves are not well known, even after measurements are made. Measurements can only be made on core samples obtained from wellbores, which sample a small fraction of the reservoir. Uncertainty can be due to measurement error and sampling limitations.

One of the essential tasks of the data acquisition stage is determining the history of field performance and select data that should be matched during the history-matching process. For example, if a gas–water reservoir is being modeled, the gas rate is usually specified and the water production is matched. By contrast, if an oil reservoir is being modeled, the oil rate is specified and both the water and gas production are matched.

CS.14 Valley Fill Case Study: Input Data Uncertainty

We have provided enough data in previous chapters to prepare an input data set for a reservoir flow simulator. Here we present uncertainties associated with some input data parameters that can be used in the flow modeling workflows discussed in the next chapter. To be specific, the three key parameters for the history match are porosity, permeability, and pressure at the water–oil contact (PWOC). Table CS.14A shows the range of uncertainties for the three parameters. More parameters could be added to the list, but this set is sufficient to illustrate flow modeling workflows.

Distributions for the Valley Fill parameters are not known. Typical distributions could be a normal distribution for porosity, a log normal distribution for permeability, and a uniform distribution or triangle distribution for PWOC. A triangle distribution for PWOC is shown in Figure CS.14A, which illustrates the probability density function versus PWOC. The maximum value of the probability density function is the mode. The value of the probability density function at the mode is

$$PDF(mode) = 2/(\max - \min) \qquad (14.CS.1)$$

Table CS.14A Key Input Parameters and Associated Uncertainties

Parameter	Values		
	Minimum	**Most Likely**	**Maximum**
Porosity	0.20	0.22	0.24
Permeability	37.5 md	150 md	600 md
PWOC	3,980 ft	4,000 ft	4,020 ft

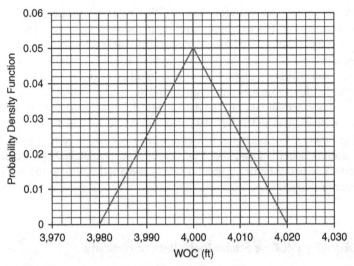

Figure CS.14A Illustrative triangle distribution for PWOC.

Exercises

14-1. The exercises in this section are designed to help you find your way around the flow simulator data file. Open file VFILL5_HM.DAT using a text editor. What is the length of a gridblock in the x-direction? *Hint:* See Section 2.1.1 in Appendix B.

14-2. Open file VFILL5_HM.DAT. What is the gross thickness of a gridblock? *Hint:* See Section 2.1.1 in Appendix B.

14-3. Open file VFILL5_HM.DAT. What is the range of depths to the top of the upper sand in layer 1? *Hint:* See Section 2.1.3 in Appendix B.

14-4. Open file VFILL5_HM.DAT. What is the porosity of each cell in the model? *Hint:* See Section 2.2.1 in Appendix B.

14-5. Open file VFILL5_HM.DAT. What is the permeability in the x-direction of each cell in the model? *Hint:* See Section 2.2.1 in Appendix B.

14-6. Open file VFILL5_HM.DAT. What is the permeability in the z-direction of each cell in the model? *Hint:* See Section 2.2.1 in Appendix B.

14-7. Open file VFILL5_HM.DAT. What is the residual oil saturation in the model? *Hint:* See Section 2.3.3 in Appendix B.

14-8. Open file VFILL5_HM.DAT. What is the bubble point pressure? *Hint:* See Section 2.6 in Appendix B.

14-9. Open file VFILL5_HM.DAT. What is the pressure at the WOC? *Hint:* See Section 2.8 in Appendix B.

14-10. Open file VFILL5_HM.DAT. What is the initial oil saturation in the model? *Hint:* See Section 2.8 in Appendix B.

14-11. Open file VFILL5_HM.DAT. What is the initial gas saturation in the model? *Hint:* See Section 2.8 in Appendix B.

14-12. Open file VFILL5_HM.DAT. How many oil wells are in the model? *Hint:* See Section 3.2 in Appendix B.

14-13. Open file VFILL5_HM.DAT. What is the flowing bottomhole pressure for each well in the model? *Hint:* See Section 3.2 in Appendix B.

14-14. Open file VFILL5_HM.DAT. What is the oil rate for each well in the model? *Hint:* See Section 3.2 in Appendix B.

15 Reservoir Flow Modeling

Different workflows exist for designing, implementing, and executing reservoir projects. A typical workflow needs to identify project opportunities; generate and evaluate alternatives; select and design the desired alternative; implement the alternative; operate the alternative over the life of the project, including abandonment; and then evaluate the success of the project so lessons can be learned and applied to future projects. Reservoir flow models, which are also known as dynamic models, can play a significant role in comparing alternatives, selecting the optimum reservoir management plan, and assessing the success of the project as it is being implemented and operated. Flow modeling workflows for green fields and brown fields are described next.

15.1 Green Field Modeling

Green fields include discovered, undeveloped fields and fields that have been discovered and delineated but are undeveloped. Well test information may be available, but green fields have little or no production information to constrain the selection of reservoir realizations or to calibrate uncertain model parameters. In this case, it is valuable to have a workflow that can quantify uncertainty. The workflow in Table 15.1 provides a probabilistic forecast that is useful for green fields. We briefly describe the steps and illustrate the workflow.

Step G1: Gather Data

Acquire and evaluate data as described in this chapter.

Step G2: Identify Key Parameters and Associated Uncertainties

A review of the data should yield information about key parameters and ranges of uncertainty that can be used for risk analysis. Murtha (1997) defines "risk analysis" as "any form of analysis that studies and hence attempts to quantify risks associated with an investment." Risk in this context refers to a potential "change in assets associated with some chance occurrences." Risk analysis generates probabilities associated with changes in model input parameters. Allowed parameter changes must be constrained within ranges that are typically determined by the range of available data, information from analogous fields, and the experience of the asset management team. Parameters that may be used in the study include any model

Integrated Reservoir Asset Management. DOI: 10.1016/B978-0-12-382088-4.00015-3

Table 15.1 Green Field Flow Modeling Workflow

Step	Task
G1	Gather data
G2	Identify key parameters and associated uncertainties
G3	Generate forecast of field performance results
G4	Generate distribution of field performance results

input information such as reservoir characterization parameters, geologic realizations or static models, rock–fluid interaction parameters, well properties, and well locations.

As an illustration, let us consider the Valley Fill reservoir discussed in previous chapters prior to the acquisition of historical production data. In this situation, the Valley Fill reservoir is a green field. To be specific, we identify three key parameters: porosity, permeability, and pressure at the water–oil contact (PWOC). Table CS.14A shows the range of uncertainties for three parameters: porosity, permeability, and PWOC.

Step G3: Generate Forecast of Field Performance Results

Every reservoir flow model run that uses a complete and unique set of model input parameters constitutes a trial. A set of trials is used to generate probability distributions. The number of trials depends on the number of parameters being considered and the desired level of analysis. Design of Experiments (DoEx) is a methodology for simultaneously and systematically varying a set of uncertain factors. The number of trials determined by experimental design depends on the number of factors and the number of values, or levels, used. A full factorial design uses all combinations of levels for all factors.

In our Valley Fill example, we have the three parameters shown in Table CS.14A and three different levels (minimum, most likely, and maximum). We design a set of trials that use the three parameters at each of the three values shown in the table. The number of trial runs is $3^3 = 27$, and they are shown in Table 15.2. Case 14 is the case in which the most likely value is used for each parameter.

The Valley Fill design is a three-level full factorial design. Figure 15.1 shows allowed cases for a two-level ($2^3 = 8$) and a three-level ($3^3 = 27$) factorial design. The two-level design uses minimum and maximum values, while the three-level design uses minimum, maximum, and most likely values. The three-level design covers more of the DoEx parameter space. The origin of the DoEx parameter space in Figure 15.1 is not used in the two-level design, but it is used in the three-level design, where it denotes the case (Case 14) that uses the most likely value of all three parameters. Other DoEx techniques have been developed to approximate a full factorial design by using fewer trials. Approximation techniques are necessary when the number of parameters increases, since the number of cases in a full factorial

Table 15.2 Valley Fill Dynamic Model Cases

Case	Porosity	Perm (md)	PWOC (psia)	Case	Porosity	Perm (md)	PWOC (psia)
1	0.2	37.5	3,980	15	0.22	150	4,020
2	0.2	37.5	4,000	16	0.22	600	3,980
3	0.2	37.5	4,020	17	0.22	600	4,000
4	0.2	150	3,980	18	0.22	600	4,020
5	0.2	150	4,000	19	0.24	37.5	3,980
6	0.2	150	4,020	20	0.24	37.5	4,000
7	0.2	600	3,980	21	0.24	37.5	4,020
8	0.2	600	4,000	22	0.24	150	3,980
9	0.2	600	4,020	23	0.24	150	4,000
10	0.22	37.5	3,980	24	0.24	150	4,020
11	0.22	37.5	4,000	25	0.24	600	3,980
12	0.22	37.5	4,020	26	0.24	600	4,000
13	0.22	150	3,980	27	0.24	600	4,020
14	0.22	150	4,000				

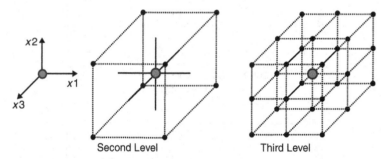

Figure 15.1 A DoEx parameter space.

design with m parameters with n levels is n^m. Yeten and colleagues (2005) presented a comparison of several different DoEx procedures.

All 27 trials were run using a reservoir flow model. Results for reservoir depletion are shown in Figures 15.2a–c for a period of three years. The models are comparable in the first few months but begin to diverge as the forecast continues. Since we do not have historical data to decide which forecast is correct, we need to consider all of them.

If we assume that the results of each model are normally distributed, we estimate the average cumulative oil recovery at the end of three years as 255 MSTB, with a standard deviation of 25 MSTB. Figure 15.3 shows the probability distribution for cumulative oil recovery for the green field.

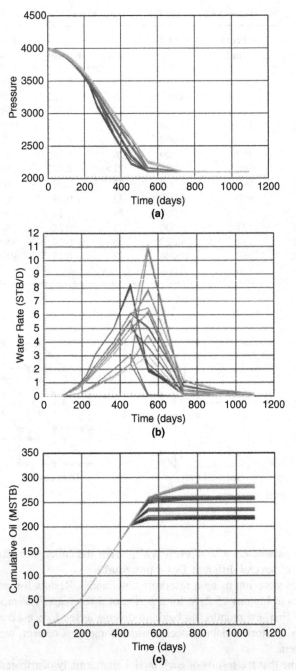

Figure 15.2 Dynamic model results for a green field: (a) pressure, (b) water rate, and (c) cumulative oil.

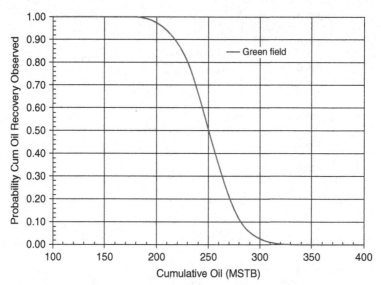

Figure 15.3 Green field cumulative oil recovery.

Step G4: Generate Distribution of Field Performance Results

The results in Step G3 can be used to identify key parameters. We can rank key parameters by developing a regression model of the reservoir flow model results. The regression model is called a proxy. Equation (15.2.1) is a linear regression model for our Valley Fill example:

$$y = a_0 + a_1 x_1 + a_2 x_2 + a_3 x_3 \qquad (15.1.1)$$

where the regression parameters $\{x_1, x_2, x_3\}$ represent porosity, permeability, and pressure at the water–oil contact (PWOC), respectively, and the regression variable y is cumulative oil recovery.

The coefficients $\{a_0, a_1, a_2, a_3\}$ are determined by regression analysis. Equation (15.2.1) is a model of a linear response surface that represents the relationship between input variables and output response (Kalla and White, 2005). A nonlinear response surface includes products of regression parameters, such as the following quadratic response surface:

$$y = a_0 + \sum_{j=1}^{3} a_j x_j + \sum_{i=1}^{3} \sum_{j=1}^{3} b_{ij} x_i x_j \qquad (15.1.2)$$

with additional regression coefficients $\{b_{ij}\}$. Figure 15.4 compares the linear regression model and a squared regression model without parameter interaction—that is, a quadratic regression model with $\{b_{ij} = 0$ if $i \neq j\}$ for the Valley Fill example. In this example, cumulative oil recovery is most affected by porosity and least affected by PWOC. Consequently, porosity is the heavy hitter.

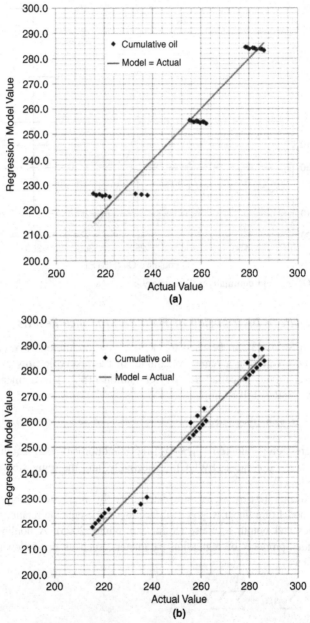

Figure 15.4 Proxy model results for green field cumulative oil recovery: (a) linear and (b) squared.

The proxy models can be used to perform a Monte Carlo analysis. The advantage of the Monte Carlo analysis is that hundreds or even thousands of proxy cases can be run to generate a probability distribution of results. The quality of the Monte Carlo analysis depends on the quality of the proxy and the distribution of regression parameters. Given the distributions and a proxy, a large number of trials can be run in a few minutes. The set of Monte Carlo runs can then be analyzed to determine the relative importance of regression parameters and the probability distribution associated with the regression variable.

The sophistication of the green field analysis workflow can imply a level of certainty that is deceiving. It is important to remember that reservoir flow model results depend on an assessment of uncertainties that are quantified based on limited sampling of data; selection of parameters; and associated uncertainty ranges, which may be incomplete or inadequate, calibrated to little or no field performance, and use parameter distributions that may be only approximate. Despite these limitations, the workflow for analyzing a green field is repeatable, can quantify uncertainty, and facilitates an objective appraisal of competing project options.

15.2   Brown Field Modeling

Brown fields are fields that have a significant development history. They may be modeled using deterministic or probabilistic reservoir forecasting workflows. History matching is used to calibrate the reservoir flow model. The history matching process may be considered an inverse problem because an answer already exists. We know how the reservoir performed; now we want to understand why. Our task is to find the set of reservoir parameters that minimizes the difference between the model performance and the historical performance of the field. This is not a unique problem, since the available data can usually be matched in more than one way.

Deterministic reservoir forecasting is the history matching workflow that was traditionally used in the twentieth century. In deterministic reservoir forecasting, a single reservoir realization is selected and matched to historical performance. Historical performance includes all data about the reservoir such as seismic data, well log data, well test data, core data, fluid property data, and production data. The uncertainty associated with a deterministic dynamic model can be estimated by evaluating the sensitivity of the model to uncertainties in available data. History match parameters are typically changed one at a time.

In probabilistic reservoir forecasting, a statistically significant collection, or ensemble, of reservoir realizations is prepared. Dynamic models are run for each possible realization, and the results of the dynamic model runs are then compared to historical performance of the reservoir. Design of experiments is used to simultaneously change multiple history match parameters. The dynamic models that match historical data within a specified set of tolerances are selected as the set of dynamic models that can be used for making predictions. A subset of the ensemble of realizations may be selected for matching to historical data to reduce the number of dynamic models that need to be studied.

In assisted history matching, a workflow process is designed to manage the dynamic model runs needed for history matching. The workflow process is typically controlled by a computer program that prepares and runs a set of dynamic models based on available static models and then collects output from the dynamic model runs for further analysis. Model output analysis usually includes a comparison of model results with historical data and identification of the subset of history matched runs. A statistical analysis of predictions from the subset of history matched runs provides probabilistic information about the quality of the history match.

The choice of history matching workflow depends on various factors, such as the objectives of the study, the complexity of reservoir performance, and the availability of history matching technology. Deterministic reservoir forecasting and probabilistic reservoir forecasting are discussed in the following sections.

15.3 Deterministic Reservoir Forecasting

Deterministic reservoir forecasting is an iterative process that makes it possible to integrate reservoir geoscience and engineering data. Starting with an initial reservoir description, the reservoir flow model is used to match and predict reservoir performance. If necessary, the input data are modified until an acceptable match is obtained. Figure 15.5 illustrates a history matching workflow. The arrows show the iterative character of the workflow.

There is no single, universally accepted strategy for altering input data to obtain a match of observed field performance with model results. Carlson (2003) pointed out that the guidelines suggested by two or more authors may actually contradict one another. Some guidelines are widely accepted. For example, authors have traditionally recommended matching pressure first and saturation second (for example, see Toronyi and Saleri, 1989; Williams et al., 1998). Table 15.3 presents one set of history matching guidelines.

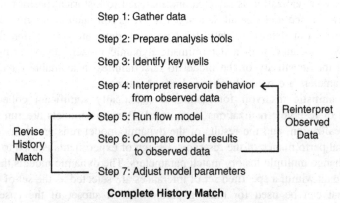

Step 1: Gather data

Step 2: Prepare analysis tools

Step 3: Identify key wells

Step 4: Interpret reservoir behavior
       from observed data

Step 5: Run flow model

Revise History Match

Step 6: Compare model results
       to observed data

Reinterpret Observed Data

Step 7: Adjust model parameters

Complete History Match

Figure 15.5 An illustrative history matching workflow.
Source: Adapted from Williams et al. (1998).

Table 15.3 Suggested History Matching Procedure

Step	Remarks
I	Match volumetrics with material balance and identify aquifer support.
II	Match reservoir pressure. Pressure may be matched both globally and locally. The match of average field pressure establishes the global quality of the model as an overall material balance. The pressure distribution obtained by plotting well test results at given points in time shows the spatial variation associated with local variability of field performance.
III	Match saturation-dependent variables. These variables include water–oil ratio (WOR) and gas–oil ratio (GOR).
IV	Match well flowing pressures.

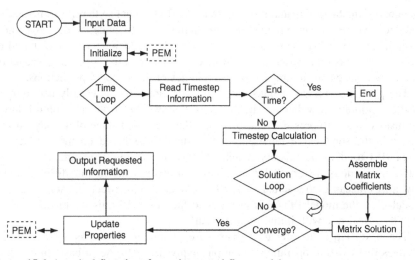

Figure 15.6 A typical flowchart for an integrated flow model.

Another history matching strategy is to combine time-lapse seismic reservoir monitoring with traditional flow modeling in a process called seismic history matching. Seismic history matching is an iterative process. Figure 15.6 shows a flowchart for integrating a petroelastic model (PEM) in a flow simulator (adapted from Fanchi, 2006a; Emerick et al., 2007). The resulting model can facilitate the comparison of model calculated seismic attributes with observed attributes. A seismic survey provides information among wells that can significantly constrain the input parameters entered into a satisfactory deterministic history match. Walker and Lane (2007) quantify improvements in the accuracy of forecasts and the extension of their useful life by the integration of time-lapse seismic data in the reservoir flow modeling process.

History matches are evaluated by comparing observed and calculated parameters. A clear understanding of study objectives should be the standard for deciding if a match is satisfactory. A model may be considered a match if it does not violate any known physical constraints. In other situations, not only must physical constraints be satisfied, but approved workflows for evaluating data must also be followed.

The result of a deterministic history match is a single flow model. The history matching workflow provided information about which parameters are the most important for matching historical field performance data. It is also possible that two or more flow models were found to yield acceptable history matches. Sensitivity analysis using the model can give insight into the quality of the match and its sensitivity to changes in important input parameters.

15.3.1 Predictions

The goal of the history match is to calibrate the flow model so that it can be used to make reliable performance predictions. Predictions help to guide the development of an optimized reservoir management plan. They should include a forecast based on the existing operating strategy, an evaluation of alternative operating scenarios, and an assessment of the impact of uncertainties on flow model predictions.

The prediction process begins with model calibration. It is usually necessary to ensure continuity in well rate when the modeler switches from rate control during the history match to pressure control during the prediction stage of a study. In Figure 15.7, the solid curve represents the predicted rate based on the productivity index (PI) used in the history match. A clear discontinuity in rate is observed between the end of history and the beginning of prediction. The rate difference usually arises because the actual well PI, especially skin effect, is not accurately modeled by the model PI. An adjustment to the model PI must be made to match final historical rate with initial predicted rate.

The next step is to prepare a base case prediction, which is a forecast that assumes that existing operating conditions apply. It establishes a basis from which to compare changes in field performance resulting from changes in existing operating conditions. In addition, a sensitivity analysis should be performed to provide insight into the uncertainty associated with model predictions.

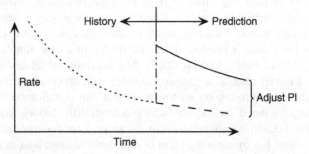

Figure 15.7 Model calibration.

Performance predictions can be used to better interpret and understand reservoir behavior. They also provide important commercial information for ranking prospective projects. Reservoir flow model results are combined with cost and price estimates to determine how much money will be available to pay for wells, compressors, pipelines, platforms, processing facilities, and any other items needed to implement the plan represented by the model. For this reason, the asset management team may be expected to generate flow predictions using a combination of reservoir parameters that yield better recoveries than would be expected if a less "optimistic" set of parameters had been used. Uncertainty analysis is a useful process for determining the likelihood that a set of parameters will be realized. Modern reserves classification systems are designed to present reserves estimates in terms of their probability of occurrence. The probabilistic representation of forecasts gives decision makers the information they need to make informed decisions.

15.4   Probabilistic Reservoir Forecasting

Brown fields are discovered fields that have been developed to the extent that historical performance data are available. The workflow in Table 15.4 provides a probabilistic forecast that is useful for brown fields. Each step is then described, and the workflow is illustrated.

Step B1: Gather Data

Acquire and evaluate data as described at the beginning of this chapter.

Step B2: Identify Key Parameters and Associated Uncertainties

A review of the data can yield information about key parameters and ranges of uncertainty that can be used for risk analysis. Allowed parameter changes must be contained within ranges that are typically determined by the range of available data, the information from analogous fields, and the experience of the asset management team.

Table 15.4 Brown Field Flow Modeling Workflow

Step	Task
B1	Gather data
B2	Identify key parameters and associated uncertainties
B3	Identify history match criteria and history match variables
B4	Generate forecast of field performance results
B5	Determine quality of history match (QoHM)
B6	Generate distribution of field performance results
B7	Verify workflow

Deterministic history matching is another source of information about which parameters are important for representing reservoir performance. One concern about deterministic history matching is that the workflow does not adequately capture uncertainty associated with different static model concepts. Given a particular static model, deterministic history matching can provide insight about some reservoir parameters. For example, effects such as casing leaks and high-permeability streaks can significantly impact reservoir flow yet be difficult to recognize or justify when preparing an uncertainty management plan.

In the Valley Fill reservoir example, we have 365 days of historical production data. To facilitate comparison with the green field study, we choose the same three parameters: porosity, permeability, and pressure at the water–oil contact (PWOC). Table CS.14A shows the range of uncertainties for the three parameters. Again, more parameters could be added to the list, but this set of three parameters is sufficient to illustrate the workflow.

Step B3: Identify History Match Criteria and History Match Variables

This step is the point where the brown field workflow begins to deviate significantly from the green field workflow. The availability of field history, such as pressure and production information, is used to decide which DoEx cases are acceptable. DoEx is used to establish a set of trials, and the results of each trial are then compared to criteria developed from historical field performance data DoEx trials. This procedure eliminates DoEx trials that are not considered acceptable representations of field performance based on the specified criteria.

In the Valley Fill example, we match data for reservoir pressure, cumulative oil recovery, and water production rate for the history match period. In our case, we use the oil rate to control production during the history match period, so we expect most, if not all, cases to match cumulative oil recovery during this period. The flow model is controlled by bottomhole flowing pressure in the second and third years, which can lead to cumulative oil recovery differences. Production GOR could be used, but it was not observed to change during the history match period and is not expected to be a sensitive indicator of model performance. This judgment can be changed if model GOR changes during the history match period. Notice that this is just one of many subjective decisions that are part of the brown field modeling workflow.

History match criteria may be defined in a number of ways. We illustrate the criteria using two objective functions and a set of constraints. The set of constraints is determined in terms of field performance data. In the Valley Fill example, we initially specify the following constraints:

- The model cumulative oil production at the end of the history match should be within 2 percent of the actual cumulative oil production.
- The model reservoir pressure should be within 10 percent of the actual reservoir pressure.
- The model water production rate should be within 10 percent of the actual water production rate.

The percentage changes can be based on fluctuations in data from this field or analogous fields. They can be modified if warranted by comparison of model results with actual data.

The objective functions used here are defined in terms of a normalized variable constructed from actual data (obs) and model results (calc). The normalized variable is

$$V = \eta(y_{obs} - y_{calc}) = \frac{y_{obs} - y_{calc}}{y_{obs,\,max} - y_{obs,\,max}} \tag{15.4.1}$$

where η is a normalization constant. The normalization constant is used to account for differences in units between history match variables that are combined in an objective function. In our case, we combine the reservoir pressure and the water production rate. Given the normalized variable, we define objective function 1 (OF1) as the sum of squares of the normalized variable and objective function 2 (OF2) as the sum of absolute values of the normalized variable, so

$$OF1 = \sum_{i=1}^{N} V_i^2$$

$$OF2 = \sum_{i=1}^{N} |V_i| \tag{15.4.2}$$

where N is the number of observations being evaluated. Objective functions with smaller values are considered to be better matches than objective functions with larger values.

In the Valley Fill example, objective functions need to be calculated for each of the 27 trials. The objective function for each trial includes contributions for two history match variables (reservoir pressure and water production rate) with eight observations per variable. The number of observations for each objective function is $N = 2 \times 8 = 16$.

Step B4: Generate Set of Field Performance Results

The trial cases identified by the DoEx study are run using the appropriate reservoir flow model to generate a set of dynamic model runs. Results (e.g., prod profiles) of reservoir depletion runs for reservoir pressure, water production rate, and cumulative oil recovery are shown in Figure 15.8. These results are the same as for the green field study, but now we have history match information to limit our set of acceptable runs. History match observations are shown in the figure as solid lines with diamond markers. The history match period extends from 0 to 365 days. All of the models have the same cumulative oil recovery during the history match period, so using the oil rate or the cumulative oil recovery as a history match variable would not help us to distinguish among cases using observations during the history match period. Substantial differences are present in the forecast period.

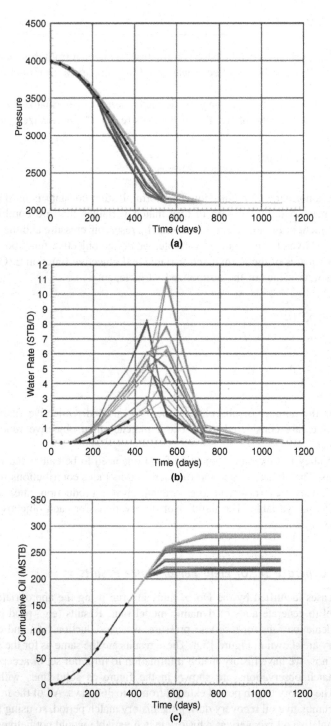

Figure 15.8 Dynamic model results for the Valley Fill example: (a) pressure, (b) water production rate, and (c) cumulative oil.

Step B5: Determine Quality of History Match

Figure 15.9 shows two objective functions for the Valley Fill runs, with each objective function presenting two groupings of cases. Cases with OF1 < 0.75 are part of the lower group, and cases with OF2 < 0.25 are also part of the lower group. The cases in the lower group are the cases that are used in the subset of trials that are considered history matched. The subset of history match cases based on objective function criteria contains 15 cases: 4 to 6, 10 to 15, and 19 to 24. The same subset applies to both objective functions in this particular case. A different definition of objective function could lead to a different subset of cases.

Figure 15.10 shows dynamic model results for the subset of 15 cases that satisfy the specified objective function criteria. Comparing Figure 15.10 with the results for the full set of 27 cases in Figure 15.8 shows the effects of constraining the dynamic model using historical production data.

The cases that can form the subset of history matched cases must satisfy the set of history match constraints specified in Step B3. In the Valley Fill example, we specified the following constraints:

- The model cumulative oil production at the end of the history match should be within 2 percent of the actual cumulative oil production.
- The model reservoir pressure should be within 10 percent of the actual reservoir pressure.
- The model water production rate should be within 10 percent of the actual water production rate.

There was no difference between actual cumulative oil recovery for the history match period and that calculated by the flow model, so the history match constraint based on cumulative oil recovery is satisfied but does not help us distinguish between cases.

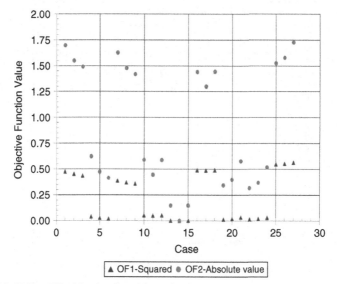

Figure 15.9 Valley Fill objective functions.

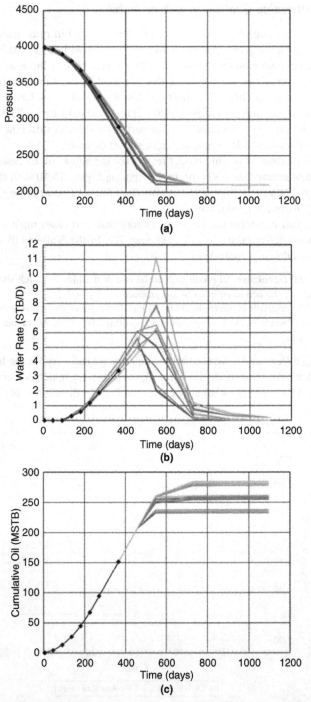

Figure 15.10 Valley Fill cases satisfying objective function criteria: (a) pressure, (b) water production rate, and (c) cumulative oil.

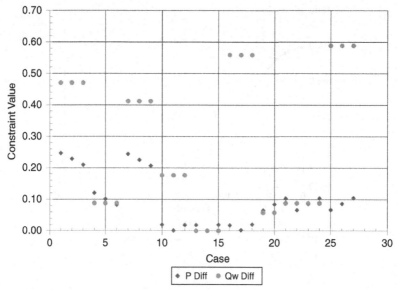

Figure 15.11 Valley Fill history match constraints.

The remaining two history match constraints based on pressure and water production rate are used to determine the cases that are considered satisfactory history matches. Figure 15.11 shows the difference between the model reservoir pressure and the actual reservoir pressure, and the difference between the model water production rate and the actual water production rate. Applying history match criteria, we find that the subset of history match cases based on the history match constraint criteria contains 8 cases.

Figure 15.12 shows the dynamic model results for the subset of 8 cases that satisfy the specified history match constraints. Comparing Figure 15.12 with the results for the full set of 27 cases in Figure 15.8 shows the effect of constraining the dynamic model using historical production data. Comparing Figure 15.12 with the results for the set of 15 cases in Figure 15.10 shows the effect of constraining the dynamic model using history match constraints rather than objective function criteria.

Step B6: Generate Distribution of Field Performance Results

The subset of history match runs has now been analyzed. Next we identify key parameters for preparing two proxies. The forecast proxy is used as it was in the green field analysis to predict the cumulative oil recovery for a set of history match parameters. A Monte Carlo analysis using the forecast proxy will use hundreds or even thousands of combinations of history match parameters. Not all of these trials satisfy history match constraints. A history match proxy for each history match parameter can be prepared. The resulting history match parameter proxy uses the

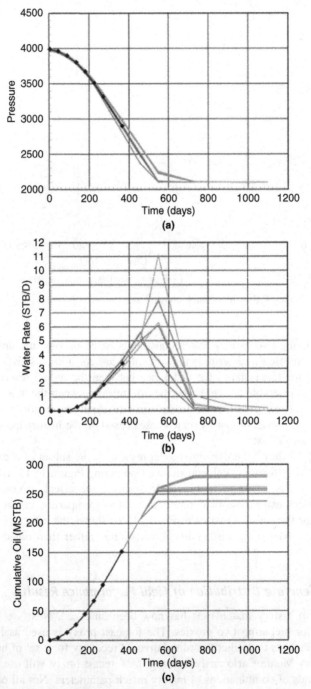

Figure 15.12 Valley Fill cases satisfying history match constraints: (a) pressure, (b) water production rate, and (c) cumulative oil.

set of history match parameters selected during the Monte Carlo analysis to determine if the Monte Carlo trial satisfies the applicable history match criteria. Those trials that do satisfy the history match criteria are included in the subset of history match cases. A probability distribution is fit to the cumulative oil recoveries for the subset of history match cases. The probability distribution provides information about reserves and model performance.

Step B7: Verify Workflow

Additional confidence in the brown field workflow can be obtained by verifying the workflow. One way to do this is to select a few cases to run with the reservoir flow model to verify that the history match proxy and forecast proxy are applicable. Three cases that are often the most important to verify are the P_{10}, P_{50}, and P_{90} cases obtained from the probability distribution for cumulative oil recovery obtained in Step B6 for the subset of the history match cases. Each of the reservoir flow model cases should satisfy the history match criteria and replicate the forecast proxy. Verification of the test cases using the dynamic model increases confidence in the workflow. By contrast, if one or more dynamic model test cases are not verified, confidence in workflow results is reduced, and it may be necessary to repeat the workflow with a modified set of input parameters.

15.5   Guidelines for Modern Flow Modeling

Modern flow modeling recognizes two types of workflows: green field workflow and brown field workflow. Both workflows are intended to be systematic procedures for quantifying uncertainty. The green field workflow applies to undeveloped fields, while the brown field workflow applies to developed fields—that is, fields with production-injection history. Two brown field workflows that are currently being used in industry are deterministic reservoir forecasting and probabilistic reservoir forecasting. The approach just presented treats deterministic reservoir forecasting and probabilistic reservoir forecasting as complementary workflows that help us to understand the reservoir and to improve reservoir management.

The workflow for deterministic reservoir forecasting is the traditional workflow. It can be used to gain familiarity with a field at a more sophisticated level than material balance analysis, and it provides insight into the impact of input parameters on dynamic model performance.

The workflow for probabilistic reservoir forecasting is still evolving (e.g., Kabir et al., 2004; Maschio et al., 2005; Nørgård, 2006 (MEPO); Passone and McRae, 2007; Yang et al., 2007). Workflows for quantifying uncertainty may include deterministic reservoir forecasting as a workflow for guiding work in Steps B1 through B3. If the analysis in Steps B4 through B6 does not yield a subset of history matched cases that contains enough cases to be statistically significant, it may be necessary to iterate between Steps B3 through B6 to improve the quality of the history match.

Probabilistic methods use DoEx to systematically generate a set of dynamic model runs and then develop response surfaces to generate probability distributions using Monte Carlo simulation. Amorim and Mocyzdlower recognize the value of the methodology in evaluating exploration prospects but caution the modeler to "take great care not to trust right away all the results generated by the proxy model" (2007, p. 3). Some Monte Carlo trials with a proxy may use a combination of input parameters that is unrealistic. Amudo and colleagues (2008) made a similar observation in their summary of lessons learned from applying probabilistic methods to a variety of reservoirs. One of the lessons they reported was that DoEx can reduce the number of simulations that are needed to assess uncertainty if the design provides a satisfactory representation of the system. If not, then repeating the workflow with a revised DoEx can increase the number of simulations that are ultimately used to assess uncertainty. The DoEx can be properly formulated and still be a poor design if it does not include parameters that represent an important, but unrecognized, mechanism that is influencing flow performance.

CS.15 Valley Fill Case Study: Deterministic History Match and Base Case Prediction

Case 14 is the deterministic history match model. The fluids originally in place in the history matched model are shown in Table CS.15A. A comparison of pressure performance of the history matched model with historical performance provides a test of the validity of the model. Similar results from the history matched model should show good agreement between historical and modeled oil and water production rates, and historical and model gas–oil ratios. The match of GOR is relatively simple because reservoir pressure is still above bubble point pressure at the end of history. Consequently, GOR is determined by a good representation of solution gas versus pressure for a single-phase liquid system. The match of pressure and produced fluids indicates that the channel observed using the seismic velocity ratio V_P/V_S has a significant impact on reservoir performance. Aquifer support was not needed to match either pressure decline or the relatively insignificant water production shown in Table CS.9B.

Two production forecasts were made based on the deterministic reservoir flow model (Case 14): a continuation of the primary depletion scenario and a pressure maintenance program based on waterflooding. The primary depletion scenario is the base case prediction, and waterflooding is an improved recovery process. Primary depletion is discussed here, and waterflooding is discussed in the next chapter.

The primary depletion forecast continues the current operating strategy using six vertical wells to produce the Valley Fill reservoir. The target oil rate is 100 STB/D per well, and the minimum flowing bottomhole pressure (BHP) is 2,100 psia for each well. The flowing BHP is greater than the bubble point pressure to prevent a transition from single-phase oil to two-phase gas and oil.

Table CS.15A Original Fluids in Place

Oil in Place	5,367 MMSTB
Water in Place	6,534 MMSTB
Gas in Place	2,104 MMSCF

Exercises

15-1. Estimate P_{10}, P_{50}, and P_{90} reserves for the green field distribution shown in Figure 15.3 using Eq. (1.2.1).

15-2. Suppose we have three factors with the two levels shown in the following table. Prepare a table showing the cases of a two-level, three-factor full factorial design. The value of each factor should be shown for each case.

Factor	Min	Max
A	−1	1
B	−1	1
C	−1	1

15-3. The proxy model with squared terms used for the results shown in Figure 15.4(b) is

$$y = a_0 + \sum_{j=1}^{3} a_j x_j + \sum_{j=1}^{3} b_j x_j^2$$

$$= 7645.244 x_1 + 0.077399 x_2 - 0.51211 x_3$$

$$- 14069.4 x_1^2 - 1.17289 \times 10^{-4} x_2^2 + 8.13 \times 10^{-5} x_3^2$$

Calculate the value of cumulative oil recovery using the preceding proxy model; the values for Case 14 shown in Table 15.2; and $x_1 = porosity$, $x_2 = permeability$, $x_3 = WOC$. How does the proxy model result compare to the flow model result of 258.7 MSTB?

15-4. Which cases in Figure 15.11 satisfy the history match constraint for pressure? Which cases satisfy the history match constraint for the water production rate? Which cases satisfy both constraints?

15-5. Copy the file VFILL5_DEPL.DAT to ITEMP.DAT and run IFLO by double clicking on the IFLO.EXE file on your hard drive. Select option "Y" to write the run output to files. When the program ends, it will print "STOP." Close the IFLO window. You do not need to save changes. Copy the files ITEMP.ROF to VFILL5_DEPL.ROF, ITEMP.TSS to VFILL5_DEPL.TSS, and ITEMP.ARR to VFILL5_DEPL.ARR. Open VFILL5_DEPL.ROF using a text editor. Search the file for INITIAL FLUID VOLUMES.
(a) How much oil is initially in place?
(b) How much water is initially in place?
(c) How much gas is initially in place?
(d) How much of the gas exists in a free gas phase?

15-6. Open VFILL5_DEPL.TSS using a text editor. Find model-calculated oil and water production rates and plot them through the end of the run. Compare your plot against the historical data shown in Table CS.9B. Do they agree?

15-7. Open VFILL5_DEPL.TSS using a text editor. The pressure in VFILL5_DEPL.TSS is the pore volume weighted average reservoir pressure. Plot it through the end of the run. Compare your plot against the historical data shown in Table CS.9B. Do they agree?

15-8. Open VFILL5_DEPL.TSS using a text editor.
   (a) What is the elapsed time of the model run (in days)?
   (b) What is the cumulative oil recovery at the end of the run?
   (c) What is the oil recovery factor at the end of the run?
15-9. Use the 3-D visualization program 3DView to look at file VFILL5_DEPL.ARR. Look at water saturation (attribute SW) and gas saturation (attribute SG). Describe what you see. Is there any free gas?

16 Modern Reservoir Management Applications

Chapter 1 introduced the concept of reservoir management, and it has been developed in the ensuing chapters. We discussed many topics that will help people with diverse backgrounds function more effectively on a reservoir asset management team. These teams were originally charged with the responsibility of managing oil and gas resources. Global demand for energy and societal demand for sustainable development have expanded the range of applications that come under the purview of reservoir asset management. We describe a few of these applications here.

16.1 Improved Oil Recovery

Improved recovery technology includes traditional secondary recovery processes such as waterflooding and immiscible gas injection, as well as enhanced oil recovery (EOR) processes. EOR processes are often classified as chemical, miscible, thermal, and microbial. A brief description of each of these processes is presented in this section. The literature on IOR is extensive, and for some more detailed discussions of EOR, including screening criteria and analyses of displacement mechanisms, see Taber and Martin (1983), Lake (1989), Martin (1992), Taber et al. (1997a,b), Green and Willhite (1998), and several articles in Holstein (2007) and Warner (2007). We first describe some IOR techniques before considering EOR processes.

16.1.1 Infill Drilling

Infill drilling is a means of improving sweep efficiency by increasing the number of wells in an area. Well spacing is reduced to provide access to unswept parts of a field. Modifications to well patterns and the increase in well density can change sweep patterns and increase sweep efficiency, particularly in heterogeneous reservoirs. Infill drilling can improve recovery efficiency, but it can also be more expensive than a fluid displacement process.

16.1.2 Immiscible Displacement

Waterflooding and gasflooding are two commonly used immiscible displacement processes. The concepts and techniques discussed in Chapter 12 apply to immiscible displacement processes.

Integrated Reservoir Asset Management. DOI: 10.1016/B978-0-12-382088-4.00016-5

Intelligent Wells and Intelligent Fields

It is often necessary in the management of a modern reservoir to alter the completion interval in a well. These adjustments are needed to modify producing well fluid ratios such as water–oil or gas–oil. One way to minimize the cost associated with completion interval adjustments is to design a well that can change the completion interval automatically. This is an example of an "intelligent well."

Intelligent wells are designed to give an operator remote control of subsurface well characteristics such as a completion interval. In addition, intelligent wells are being designed to provide information to the operator using downhole measurements of physical properties such as pressure, temperature, and seismic vibrations. One goal of intelligent well technology is to convey a stream of continuous and real-time information to the operator, who can monitor this information and make adjustments as needed to achieve reservoir management objectives.

Intelligent wells can be found in an intelligent field, or i-field, which is an integrated operation that uses improved information and computer technology to operate a field. The operation uses supervisory control and data acquisition (SCADA), which is computer technology designed to monitor the field. Capital investment in i-field technology can be larger than in fields without i-field technology, but a good design can result in a decrease in operating expenses by increasing automation of a field and allowing remote control of a field in a difficult environment. The desired result is less direct human intervention. For more discussion of recent developments in intelligent well completions, see Robinson (2007).

Chemical

Chemical flooding methods include polymer flooding, micellar-polymer flooding or surfactant-polymer flooding, and alkaline or caustic flooding. Polymer flooding is designed to improve the mobility ratio and fluid flow patterns of a displacement process by increasing the viscosity of the injected aqueous phase. In this case, high-molecular-weight polymer is added to injected water. Micellar-polymer flooding uses a detergent like solution to lower residual oil saturation to waterflooding. The polymer slug injected after the micellar slug is designed to improve displacement efficiency. Alkaline flooding uses alkaline chemicals that can react with certain types of *in situ* crude. The resulting chemical product is miscible with the oil and can reduce residual oil saturation to waterflooding.

Miscible

Miscible flooding methods include carbon dioxide injection, natural gas injection, and nitrogen injection. Miscible gas injection must be performed at a high enough pressure to ensure miscibility between the injected gas and the *in situ* oil. Miscibility is achieved when interfacial tension (IFT) between the aqueous and oleic phases is significantly reduced. The desired IFT reduction is typically from around 1 dyne/cm to 0.001 dyne/cm or less. Any reduction in IFT can improve displacement efficiency, and a near miscible process can yield much of the incremental oil that might be obtained from a miscible process. If reservoir pressure is not maintained above the minimum miscibility pressure (MMP) of the system, the gas flood will be an immiscible gas injection process.

Immiscible gas can be used as the principal injection fluid in a secondary displacement process, or it can be used as the injection fluid for a tertiary process. Two improved recovery processes based on immiscible gas injection are the double displacement process (DDP) and the second contact water displacement (SCWD) process. Both processes require the injection of immiscible gas into reservoirs that have been previously waterflooded. The processes require favorable gas–oil and oil–water interfacial tensions. Oil remaining after waterflooding can coalesce into a film when exposed to an immiscible gas. The oil film can be mobilized and produced by downdip gravity drainage (the DDP process) or by water influx from either an aquifer or water injection following the immiscible gas injection period. For more discussion of recent developments in miscible recovery processes, see Holstein and Stalkup (2007).

Thermal

Thermal flooding methods include hot water injection, steam drive, steam soak, and *in situ* combustion. The injection or generation of heat in a reservoir is designed to reduce the viscosity of *in situ* oil and to improve the mobility ratio of the displacement process. Electrical methods can also be used to heat fluids in relatively shallow reservoirs containing high-viscosity oil, but electrical methods are not as common as hot fluid injection methods. The *in situ* combustion method requires compressed air injection after *in situ* oil has been ignited. Steam injection methods require the injection of steam into a reservoir. Steam and hot water injection processes are the most common thermal methods because of the relative ease of generating hot water and steam. The *in situ* combustion process is more difficult to control than steam injection processes, and it requires *in situ* oil that can be set on fire. Hot gases and heat advance through the formation and displace the heated oil to production wells. For more discussion of recent developments in miscible recovery processes, see Jones (2007) and Brigham and Castanier (2007).

Microbial

Microbial enhanced oil recovery uses the injection of microorganisms and nutrients in a carrier medium to increase oil recovery, reduce water production from petroleum reservoirs, or both. Dietrich and colleagues (1996) summarized the results of five successful commercial microbial EOR projects. The projects reflected a diversity of locations, lithologies, depths, porosities, permeabilities, and temperatures. Two of the projects were conducted in the United States, two were done in China, and one was carried out in Argentina. The projects included sandstone, fractured dolomite, siltstone-sandstone, and fractured sandstone reservoirs. Reservoir depths ranged from 4,450 to 6,900 feet, temperatures from 110 to 180°F, porosity from 0.079 to 0.232, and effective permeability from 1.7 to 300 md. Evidence from laboratory research and field case studies shows that microbial EOR processes can result in the incremental recovery of oil and also reduce water production from high-permeability zones. However, research is continuing to maximize the technical and economic potential for microbial EOR. For example, the U.S. Department of Energy has underwritten the development of microbial transport simulators.

16.2 Unconventional Fossil Fuels

Oil and gas fields are considered conventional sources of fossil fuels. Next we discuss several unconventional sources of fossil fuels. Unconventional sources are a more important part of the global energy mix when the price of oil increases.

16.2.1 Coalbed Methane

Coalbeds are an abundant source of methane. The presence of methane gas in coal has been well known to coal miners as a safety hazard, but it is now being viewed as a source of natural gas. The gas is bound in the micropore structure of the coalbed and can diffuse into the natural fracture network when a pressure gradient exists between the matrix and the fracture network. The fracture network in coalbeds consists of microfractures that allow Darcy flow; they are called "cleats."

Gas recovery from coalbeds depends on three processes, as described in Kuuskraa and Brandenburg (1989). Coalbed methane exists as a monomolecular layer on the internal surface of the coal matrix. Its composition is predominately methane, but it can also include other constituents, such as ethane, carbon dioxide, nitrogen, and hydrogen (Mavor et al., 1999). For this reason, coalbed methane is also known as coal gas to highlight the observation that gas from coalbeds is usually a mixture. Gas content can range from approximately 20 standard cubic feet (SCF) gas per ton of coal in the Powder River Basin of Wyoming (Mavor et al., 1999) to 600 SCF per ton in the Appalachian Basin (Gaddy, 1999).

Gas recovery begins with desorption of gas from the internal surface to the coal matrix and micropores. The gas then diffuses into the cleats through the coal matrix and micropores. Finally, gas flows through the cleats to the production well. The flow rate depends, in part, on the pressure gradient in the cleats and on the density and distribution of cleats. The controlling mechanisms for gas production from coalbeds are the rate of desorption from the coal surface to the coal matrix, the rate of diffusion from the coal matrix to the cleats, and the rate of flow of gas through the cleats. The flow rate in the cleats obeys Darcy's law in many systems, but it may also depend on stress-dependent permeability or gas slippage (known as the Klinkenberg effect).

The production performance of a well that is producing gas from a coalbed will typically exhibit three stages. The reservoir dewaters, and methane production increases during the first stage of pressure depletion. Methane production peaks during the second stage. The amount of water produced is relatively small compared to gas production during the second stage because of gas–water relative permeability effects, and desorption of natural gas provides a counterbalance to permeability loss as a result of formation compaction. The third stage of production is similar to conventional gas field production in which the gas rate declines as the reservoir pressure declines.

Coalbed methane recovery can be enhanced by injecting carbon dioxide into the coal seam. Carbon dioxide preferentially displaces methane in the coal matrix, and the displaced methane can then be produced through the cleat system. The resulting adsorption of carbon dioxide by coal can be used to sequester, or store, carbon dioxide in the coal seam. Carbon dioxide sequestration has an environmental benefit that is discussed in more detail in the following subsections. Injection of carbon dioxide into a coalbed methane reservoir is an enhanced coalbed methane process. For more discussion about recent developments in coalbed methane technology, see Jenkins et al. (2007).

16.2.2 Gas Hydrates

The entrapment of natural gas molecules in ice at very low temperatures forms an ice like solid that is a metastable complex called a gas hydrate. Gas hydrates are clathrates, which are chemical complexes that are formed when one type of molecule completely encloses another type of molecule in a lattice. In the case of gas hydrates, hydrogen-bonded water molecules form a cagelike structure in which mobile molecules of gas are absorbed or bound. For more discussion of hydrate properties and technology, see Sloan (2006, 2007) and the references therein.

The presence of gas hydrates can complicate field operations. For example, the existence of hydrates on the ocean floor can affect drilling operations in deep water. The simultaneous flow of natural gas and water in tubing and pipelines can result in the formation of gas hydrates that can impede or completely block the flow of fluids through pipeline networks. Heating the gas or treating the gas–water system with chemical inhibitors can prevent the formation of hydrates, but it increases operating costs.

Gas hydrates are generally considered a problem for oil and gas field operations, but their potential commercial value as a clean energy resource is changing industry perception. The potential as a gas resource is due to the relatively large volume of gas contained in the gas hydrate complex. In particular, Makogon and colleagues (1997) reported that one cubic meter of gas hydrate contains 164.6 m^3 of methane. This is equivalent to one barrel of gas hydrate containing 924 ft^3 of methane and is approximately six times as much gas as the gas contained in an unimpeded gas-filled pore system (Selley, 1998). The gas in gas hydrates occupies approximately 20 percent of the volume of the gas hydrate complex. Water occupies the remaining 80 percent of the gas hydrate complex volume.

Gas hydrates can be found throughout the world (Makogon et al., 1997; Selley, 1998). They exist on land in sub-Arctic sediments and on seabeds where the water is near freezing at depths of at least 600 to 1,500 feet. For instance, favorable conditions for gas hydrate formation exist at seafloor temperatures as low as 39°F in the Gulf of Mexico and as low as 30°F in some sections of the North Sea. According to Makogon and colleagues (1997), over 700 trillion m^3 in explored reserves of methane in the hydrate state exist. Difficulties in cost-effective production have hampered development of the resource.

16.2.3   Tight Gas Sands and Shale Gas

Unconventional gas resources include coalbed methane, tight gas sands, and fractured gas shales. We have already discussed coalbed methane. Both tight gas sands and gas shales are characterized by low permeabilities—that is, permeabilities that are a fraction of a millidarcy. Tight gas sands are low-permeability sands, while gas shales are shales that are typically rich in organic materials and have low permeability. Gas shales are often viewed as source rock for gas fields. The low permeability associated with unconventional gas resources makes it more difficult to produce the gas at economical rates.

Economic production of gas from a gas shale or tight gas sand often requires the creation of fractures by a process known as hydraulic fracturing (Wattenbarger, 2002; Kuuskraa and Bank, 2003; Arthur et al., 2009). In this process, a fluid is injected into the formation at a pressure that exceeds the fracture pressure of the formation. The orientation and length of the induced fracture depend on formation characteristics such as thickness and stress. Once fractures have been created in the formation, a proppant such as manmade pellets or coarse grain sand is injected into the fracture to prevent it from closing, or healing, when the injection pressure is removed. The proppant provides a higher-permeability flow path for gas to flow to the production well.

Unconventional low-permeability gas sands and shales often require more wells per unit area than conventional higher-permeability gas reservoirs. The key to managing an unconventional gas resource is to develop the resource with enough wells to maximize gas recovery without drilling unnecessary wells. The development of hydraulic fracturing techniques has made it possible to economically develop tight gas sands and gas shales. For more discussion of recent developments in tight gas sand technology, see Holditch (2007). Arthur and colleagues (2009) discuss hydraulic fracturing in shale gas.

16.2.4   Shale Oil and Tar Sands

Shale oil is high-API gravity oil contained in porous, low-permeability shale. Sand grains that are cemented together by tar or asphalt are called tar sands. Tar and asphalt are highly viscous, plastic or solid hydrocarbons. Extensive shale oil and tar sand deposits are found throughout the Rocky Mountain region of North America, as well as in other parts of the world. Although difficult to produce, the volume of hydrocarbon in tar sands has stimulated efforts to develop production techniques.

The hydrocarbons in shale oil and tar sands can be extracted by mining when they are close to the surface. Tar pits have been found around the world and have been the source of many fossilized dinosaur bones. In locations where oil shale and tar sands are too deep to mine, it is necessary to increase the hydrocarbon's mobility.

An increase in permeability or a decrease in viscosity can increase mobility. Increasing the temperature of high API gravity oil, tar, or asphalt can significantly reduce viscosity. If there is enough permeability to allow injection, steam or hot water can be used to increase formation temperature and reduce hydrocarbon viscosity. In many cases, however, permeability is too low to allow significant injection of a heated fluid. An alternative to fluid injection is electromagnetic heating. Radio frequency heating has been used in Canada, and electromagnetic heating techniques are being developed for other parts of the world (Callarotti, 2007).

16.3 Geothermal Reservoir Management

Reservoir management principles can be applied to subsurface resources other than oil and gas. The first such application we consider is geothermal energy (Fanchi, 2004; Renner et al., 2007). Heat energy acquired from geological sources is called "geothermal energy." The earth's interior is divided into a solid inner core, molten outer core, mantle, and crust. Drilling into the earth's crust has shown that the temperature of the crust tends to increase linearly with depth. The interior of the earth is much hotter than the crust. The source of heat energy is radioactive decay, and the earth's crust acts as a thermal insulator to prevent heat from escaping into outerspace.

Geothermal energy can be obtained from temperature gradients between the shallow ground and the surface, subsurface hot water, hot rock several kilometers below the earth's surface, and magma. Magma is molten rock in the mantle and crust that is heated by the large heat reservoir in the interior of the earth. In some parts of the crust, magma is close enough to the surface to heat rock or water in the pore spaces of rock. Magma, hot water, and steam are carriers of energy.

The heat carried to the surface from a geothermal reservoir depends on the heat capacity and phase of the produced fluid. We will illustrate this dependence by considering an example. Suppose the pore space of the geothermal reservoir is occupied by hot water. If the temperature of the produced water is at the temperature T_{res} of the geothermal reservoir, the heat produced with the produced water is

$$\Delta H_w = m_w c_w \Delta T \qquad (16.3.1)$$

where

ΔT = the temperature difference
$T_{res} - T_{ref}$, T_{ref} = a reference temperature such as surface temperature
m_w = the mass of produced water
c_w = the specific heat capacity of water

The mass of produced water can be expressed in terms of the volumetric flow rate q_w, the period of flow Δt, and the density of water ρ_w, so

$$m_w = \rho_w q_w \Delta t \tag{16.3.2}$$

Substituting Eq. (16.3.2) into Eq. (16.3.1) gives

$$\Delta H_w = (\rho_w q_w \Delta t) c_w \Delta T = (\rho_w q_w \Delta t) c_w \left(T_{res} - T_{ref}\right) \tag{16.3.3}$$

The heat produced from a geothermal reservoir in time Δt is the geothermal power, or

$$P_{geo} = \frac{\Delta H_w}{\Delta t} = (\rho_w q_w) c_w \Delta T = (\rho_w q_w) c_w \left(T_{res} - T_{ref}\right) \tag{16.3.4}$$

The electrical power that can be generated from geothermal power depends on the efficiency η_{geo} of conversion of geothermal power P_{geo} to electrical power; thus

$$P_{out} = \eta_{geo} P_{geo} = \eta_{geo} \rho_w q_w c_w \left(T_{res} - T_{ref}\right) \tag{16.3.5}$$

If steam is produced instead of hot water or in addition to hot water, the heat produced must account for the latent heat of vaporization.

Some of the largest geothermal production facilities in the world are at the Geysers in California and in Iceland. These areas are determined by the proximity of geothermal energy sources. The technology for converting geothermal energy into useful heat and electricity can be categorized as geothermal heat pumps, direct-use applications, and geothermal power plants. We focus here on geothermal power plants as a reservoir management application.

16.3.1 Geothermal Heating Systems

The geothermal reservoir is an aquifer with hot water or steam. A geothermal heating system is illustrated in Figure 16.1. A production well is used to withdraw hot water from the geothermal reservoir, and an injection well is used to recycle the water. Recycling helps to maintain reservoir pressure. If the geothermal reservoir is relatively small, the recycled, cooler water can lower the temperature of the aquifer. The electric pump in the figure is needed to help withdraw water because the reservoir pressure in this case is not high enough to push the water to the surface. Heat from the geothermal reservoir passes through a heat exchanger and is routed to a distribution network.

Geothermal power plants use steam or hot water from geothermal reservoirs to turn turbines and generate electricity. Dry-steam power plants use steam directly from a geothermal reservoir to turn turbines. Flash-steam power plants allow high-pressure hot water from a geothermal reservoir to flash to steam in lower-pressure tanks. The resulting steam is used to turn turbines. A third type of geothermal power plant, called a binary-cycle plant, uses heat from moderately hot geothermal

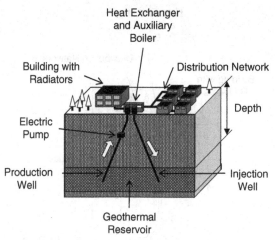

Figure 16.1 A geothermal heating system.
Source: Fanchi (2004).

water to flash a second fluid to the vapor phase. The second fluid must have a lower boiling point than water so it will be vaporized at the lower temperature associated with the moderately hot geothermal water. There must be enough heat in the geothermal water to supply the latent heat of vaporization needed by the secondary fluid to make the phase change from liquid to vapor. The vaporized secondary fluid is then used to turn turbines.

Geothermal power plants can emit toxic gases such as hydrogen sulfide or greenhouse gases such as carbon dioxide. The produced water from a geothermal reservoir will contain dissolved solids that can form solid precipitates when the temperature and pressure of the produced water change.

16.3.2 Managing Geothermal Reservoirs

The hot water or steam in a geothermal reservoir can be depleted by production. The phase of the water in a geothermal reservoir depends on the pressure and temperature of the reservoir. Single-phase steam will be found in low-pressure, high-temperature reservoirs. In high-pressure reservoirs, the water may exist in the liquid phase or in both the liquid and gas phases, depending on the temperature of the reservoir. When water is produced from the geothermal reservoir, both the pressure and temperature in the reservoir can decline. In this sense, geothermal energy is a nonrenewable, finite resource unless the produced hot water or steam is replaced. A new supply of water can be used to replace the produced fluid, or the produced fluid can be recycled after heat transfer at the surface. If the rate of heat transfer from the heat reservoir to the geothermal reservoir is slower than the rate of heat extracted from the geothermal reservoir, the temperature in the geothermal reservoir will decline during production. To optimize the performance of the geothermal reservoir, it must be understood and managed in much the same way that petroleum reservoirs are managed.

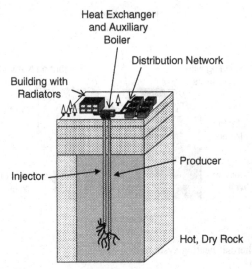

Heat Exchanger
and Auxiliary
Boiler

Distribution Network

Building with
Radiators

Producer

Injector

Hot, Dry Rock

Figure 16.2 Geothermal energy from hot, dry rock.
Source: Fanchi (2004).

16.3.3 Hot, Dry Rock

Another source of geothermal energy is hot, dry rock that is several kilometers
deep inside the earth. These rocks are heated by magma directly below them
and have elevated temperatures, but they do not have a means of transporting
the heat to the surface. In this case, it is technically possible to inject water into
the rock, let it heat up, and then produce the hot water. Figure 16.2 illustrates
a hot, dry rock facility that is designed to recycle the energy-carrying fluid.
Water is injected into fissures in the hot, dry rock through the injector and
then produced through the producer. The power plant at the surface uses the
produced heat energy to drive turbines in a generator. After the hot fluid transfers its heat to the power plant, the cooler fluid can be injected again into the
hot, dry rock.

16.4 Sequestration

A purported cause of adverse global climate change is the greenhouse effect.
The climatic greenhouse effect occurs when carbon dioxide in the atmosphere
absorbs infrared radiation rather than letting it escape into space. The resulting
atmospheric heating is attributed to increasing levels of carbon dioxide in the
atmosphere. One proposed method for reducing the climatic greenhouse effect is
to collect and store greenhouse gases such as carbon dioxide (CO_2) and hydrogen
sulfide (H_2S) in geologic formations. Geologic sequestration is the capture,
separation, and long-term storage of greenhouse gases or other gas pollutants in

a subsurface environment such as a reservoir, an aquifer, or a coal seam. Carbon sequestration is the capture, separation, and long-term storage of CO_2 for environmental purposes but not necessarily in a geologic formation.

A clean energy source is a source that emits negligible amounts of greenhouse gases or other pollutants. Systems that use combustion of fossil fuels—notably gas, oil, and coal—are sources of greenhouse gas emissions, especially CO_2. Mitigation technologies, such as geologic sequestration, can reduce the environmental impact of some systems such as power plants that burn fossil fuels. One goal of CO_2 sequestration is to provide an economically competitive and environmentally safe option to offset projected growth in baseline emissions of greenhouse gases.

Enhanced oil recovery has the capacity to store significant volumes of CO_2. Numerous examples of CO_2 injection for enhanced oil recovery (EOR) have been cited in the literature, with projects beginning as early as the 1970s. Projects designed to sequester gas in oil reservoirs depend on this experience. CO_2 may be stored by injection into an oil reservoir in either an immiscible displacement process or a miscible displacement process. Much more CO_2 can be stored in a miscible process than in an immiscible process.

The usual objectives of miscible gas injection include pressure maintenance, viscosity reduction, and displacement. In cases where CO_2 sequestration is combined with hydrocarbon recovery, two more factors distinguish "sequestration" from being simply an EOR process: The CO_2 will ultimately be left in place when hydrocarbon recovery processes have ended, and the motivation is at least partly environmental—that is, the goal of gas injection is to reduce the release of CO_2 into the atmosphere. Examples of carbon sequestration into coal seams and saline aquifers include a pilot study in the San Juan basin and CO_2 injection into the Sleipner West field, a saline aquifer off Norway's coast.

The sequestration of CO_2 in subsurface formations is a gas storage process that must satisfy the three primary objectives in designing and operating natural gas storage reservoirs: verification of injected gas inventory, monitoring of injected gas migration, and determination of gas injectivity. The ability to predict and verify CO_2 movement is needed to ensure that additional environmental problems are not created by injecting greenhouse gases into the subsurface. If the gas is not contained, it can contaminate other formations, such as a potable aquifer, or be released to the atmosphere through an outcrop.

16.5 Compressed Air Energy Storage

Reservoir management concepts, tools, and principles are applicable to sustainable energy systems. A sustainable energy system is a system that lets us meet present energy needs while preserving the ability of future generations to meet their needs. Renewable energy sources are expected to be key components of sustainable energy systems.

Renewable energy is energy obtained from sources at a rate that is less than or equal to the rate at which the source is replenished. Solar energy and wind energy are two renewable energy sources that hold great promise for replacing fossil fuels as the dominant energy sources in the future energy mix. The amount of wind energy and solar energy that can be converted to useful power depends on location, time of day, season, and weather conditions. Consequently, both solar energy and wind energy have limited reliability; they are not available all day long, every day of the year. This problem of intermittent reliability is currently solved by using a more reliable source of energy, such as nonrenewable fossil fuels or nuclear energy, to provide energy as needed. In this power configuration, solar energy and wind energy are used to supplement other forms of energy. Our ability to rely on solar energy and wind energy as baseline sources of energy depends on our ability to store them in forms that can be easily accessed.

16.5.1 Storing Intermittent Renewable Energy

Wind farms and solar energy plants are both examples of renewable energy systems that may be able to function as sustainable energy systems. Both wind and solar power plants are intermittent sources of energy: they cannot provide power on demand. Large-scale energy storage facilities are needed to optimize the performance of intermittent energy sources. Compressed air energy storage (CAES) is an example of a large-scale energy storage technology that has been implemented in Germany and Alabama. CAES is designed to transfer off-peak energy from primary power plants during peak demand periods.

The Huntorf CAES facility in Germany and the McIntosh CAES facility in Alabama store gas in salt caverns. Off-peak energy is used to pump air underground and compress it in a salt cavern. The compressed air is produced during periods of peak energy demand to drive a turbine and generate additional electrical power. Currently, CAES systems have been limited to areas where wind farms and salt formations coexist. There are many more areas where current or proposed wind farms overlie permeable formations that already have or could contain gas. This could mean a significant expansion for implementation of CAES systems, which are essentially gas reservoir systems.

One way to store solar energy and wind energy is to convert these energy forms to other forms of energy. For example, excess solar energy from a solar power plant could be used to heat and compress a fluid that could be stored in a reservoir. The hot, compressed fluid could be extracted from the reservoir and used to generate electrical power when the amount of energy from the solar power plant declines. Similarly, excess wind energy from a wind farm could be used to compress a fluid that could be stored in a reservoir. A wind energy system that employs compressed air energy storage is shown in Figure 16.3. Neumiller and colleagues (2009) demonstrated how to use reservoir simulators to model a CAES system.

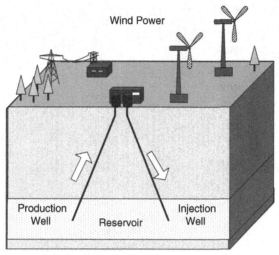

Figure 16.3 CAES for a wind farm.

CS.16 Valley Fill Case Study: Waterflood Prediction

The results of a depletion case were presented in Chapter 14. The results of a waterflood case are presented in Table CS.16A, and cumulative oil recovery results are shown in Figure CS.16A. Production Well 5 was converted to a water injection well, and a second water injection well was placed south of Well 9, shown in Figure CS.11A between production Wells 3 and 4. Both water injection wells are downdip relative to the regional dip, and they are completed in the lowermost layer of the model.

Table CS.16A Field Production Data

Time (Days)	P_{avg} (psia) Depl	WF	Oil Rate (STB/D) Depl	WF	Water Rate (STB/D) Depl	WF	Cumulative Oil (MSTB) Depl	WF
5	3,989	3,989	100	100	0.0	0.0	0.5	0.5
45	3,961	3,961	100	100	0.0	0.0	4.5	4.5
90	3,897	3,897	200	200	0.0	0.0	13.5	13.5
135	3,801	3,801	300	300	0.3	0.3	27.0	27.0
180	3,673	3,673	400	400	0.6	0.6	45.0	45.0
225	3,511	3,511	500	500	1.2	1.2	67.5	67.5
270	3,315	3,315	600	600	1.9	1.9	94.5	94.5
365	2,898	2,898	600	600	3.4	3.4	151.5	151.5
455	2,500	2,765	600	500	5.0	2.6	206.0	196.5
548	2,118	2,634	80	500	3.6	3.3	256.7	242.8
730	2,102	2,370	1	500	0.2	5.2	258.6	334.0
913	2,102	2,162	0	397	0.1	13.7	258.7	416.7
1,095	2,101	2,131	0	312	0.0	22.8	258.7	474.1

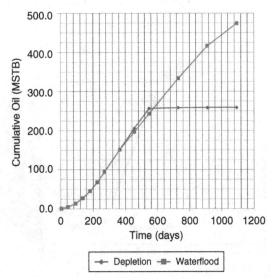

Figure CS.16A CAES for a wind farm.

Exercises

16-1. History match results are presented in file VFILL5_HM.TSS. Find the model calculated oil and water production rates in file VFILL5_HM.TSS and plot them against the historical data shown in Table CS.13B.

16-2. Find the model calculated GOR in VFILL5_HM.TSS and plot it against historical data shown in Table CS.13B.

16-3A. A geothermal power plant was able to provide 2,000 MWe (megawatts electric) power when it began production. Twenty years later, the plant is only able to provide 850 MWe from the geothermal source. Assuming the decline in power production is approximately linear, estimate the average annual decline in power output (in MWe/year).

16.3B. Suppose the plant operator has decided to close the plant when the electric power output declines to 10 MWe. How many more years will the plant operate if the decline in power output calculated in part A continues?

16-4A. A coal seam is 800 feet wide, 1 mile long, and 10 feet thick. The volume occupied by the fracture network is 1 percent. The volume of coal is the bulk volume of the coal seam minus the fracture volume. What is the volume of coal in the coal seam? Express your answer in m^3.

16-4B. If the density of coal is 1.7 lbm/ft^3, how many tonnes of coal are in the coal seam?

16-5. Copy the file VFILL5_WF.DAT to ITEMP.DAT and run IFLO by double clicking on the IFLO.EXE file on your hard drive. Select option "Y" to write the run output to files. When the program ends, it will print "STOP". Close the IFLO window. You do not need to save changes. Copy ITEMP.ROF to VFILL5_WF.ROF, ITEMP.TSS to VFILL5_WF.TSS, and ITEMP.ARR to VFILL5_WF.ARR. Open VFILL5_WF.ROF using a text editor. Search the file for INITIAL FLUID VOLUMES.

(a) How much oil is initially in place?

(b) How much water is initially in place?

(c) How much gas is initially in place?

(d) How much of the gas exists in a free gas phase?

(e) What is the elapsed time of the model run (in days)?

16-6. File VFILL5_WF.DAT is the flow model that matches historical production and makes a prediction of field performance when subjected to a waterflood. Open timestep summary file VFILL5_WF.TSS.

(a) The maximum water injection rate for the field is 400 STB/day. What is the maximum water production rate?

(b) What is the cumulative oil recovery at the end of the run?

(c) What is the oil recovery factor?

16-7. Run the visualization program 3DVIEW and load the file VFILL5_WF.ARR. Select the water saturation attribute at the end of the run. Slice the model to obtain a better view of the interior of the model. Is water injected upstructure or downstructure?

16-8. Plot cumulative oil recovery versus time for the Valley Fill case study depletion prediction and waterflood prediction. The data are available in the timestep summary files VFILL5_DEPL.TSS and VFILL5_WF.TSS.

Appendix A
Unit Conversion Factors

Time

1 hour = 1 hr = 3,600 s
1 day = 8.64×10^4 s
1 year = 1 yr = 3.1536×10^7 s

Length

1 foot = 1 ft = 0.3048 m
1 kilometer = 1 km = 1,000 m
1 mile = 1 mi = 1.609 km
1 cm = 10^{-2} m

Velocity

1 foot per second = 0.3048 m/s
1 kilometer per hour = 1 kph = 1,000 m/hr = 0.278 m/s
1 mile per hour = 1 mph = 1.609 km/hr = 1,609 m/hr = 0.447 m/s

Area

1 square foot = 1 ft^2 = 0.0929 m^2
1 square mile = 1 mi^2 = 2.589 km^2 = 2.589×10^6 m^2
1 square mile = 1 mi^2 = 640 acres
1 acre = 1 ac = 4,047 m^2
1 hectare = 1 ha = 1.0×10^4 m^2
1 millidarcy = 1 md = 0.986923×10^{-15} m^2
1 darcy = 1,000 md = 0.986923×10^{-12} m^2
1 barn = 1.0×10^{-24} cm^2 = 1.0×10^{-28} m^2

Volume

1 liter = 1 L = 0.001 m^3
1 cubic foot = 1 ft^3 = 2.83×10^{-2} m^3
1 standard cubic foot = 1 SCF = 1 ft^3 at standard conditions
1 acre-foot = 1 ac-ft = 1,233.5 m^3
1 barrel = 1 bbl = 0.1589 m^3
1 gallon (U.S. liquid) = 1 gal = 3.785×10^{-3} m^3
1 barrel = 42 gallons = 0.1589 m^3
1 barrel = 5.6148 ft^3

Integrated Reservoir Asset Management. DOI: 10.1016/B978-0-12-382088-4.00022-0

1 gallon = 3.788 liters
1 cm^3 = 1 cc = 10^{-6} m^3
1 cm^3 = 3.534 × 10^{-6} m^3

Mass

1 gram = 1 g = 0.001 kg
1 pound (avoirdupois) = 1 lb (avdp) = 1 lbm = 0.453592 kg
1 tonne = 1,000 kg

Mass Density

1 g/cm^3 = 1,000 kg/m^3

Force

1 pound-force = 1 lbf = 4.4482 N

Pressure

1 pascal = 1 Pa = 1 N/m^2 = 1 kg/m·s^2
1 megapascal = 1 MPa = 10^6 Pa
1 gigapascal = 1 GPa = 10^9 Pa
1 pound-force per square inch = 1 psi = 6,894.8 Pa
1 atmosphere = 1 atm = 1.01325 × 10^5 Pa
1 atmosphere = 1 atm = 14.7 psi
1 bar = 14.5 psi = 0.1 MPa
1 MPa = 145 psi

Energy

1 megajoule = 1 MJ = 1.0 × 10^6 J
1 gigajoule = 1 GJ = 1.0 × 10^9 J
1 exajoule = 1 EJ = 1.0 × 10^{18} J
1 eV = 1.6022 × 10^{-19} J
1 MeV = 10^6 eV = 1.6022 × 10^{-13} J
1 erg = 10^{-7} J
1 BTU = 1,055 J
1 calorie (thermochemical) = 1 cal = 4.184 J
1 kilocalorie = 1 kcal = 1,000 calories = 4.184 × 10^3 J
1 calorie = 1,000 calories = 4.184 × 10^3 J
1 kilowatt-hour = 1 kWh = 1 kW · 1 hr = 3.6 × 10^6 J
1 quad = 1 quadrillion BTU = 1.0 × 10^{15} BTU = 1.055 × 10^{18} J
1 quad = 2.93 × 10^{11} kWh = 1.055 × 10^{12} MJ
1 quad = 1.055 exajoule = 1.055 EJ
1 barrel of oil equivalent = 1 BOE = 5.8 × 10^6 BTU = 6.12 × 10^9 J
1 quad = 1.72 × 10^8 BOE = 172 × 10^6 BOE

Energy Density

1 BTU/lbm = 2,326 J/kg
1 BTU/SCF = 3.73×10^4 J/m^3

Power

1 watt = 1 W = 1 J/s
1 megawatt = 10^6 W = 10^6 J/s
1 kilowatt-hour per year = 1 kWh/yr = 0.114 W = 0.114 J/s
1 horsepower = 1 hp = 745.7 W

Viscosity

1 centipoise = 1 cp = 0.001 Pa·s
1 mPa·s = 0.001 Pa·s = 1 cp = 10^{-3} Pa·s
1 poise = 100 cp = 0.1 Pa·s

Radioactivity

1 curie = 1 Ci = 3.7×10^{10} decays/s
1 roentgen = 1 R = 2.58×10^{-4} C/kg
1 radiation absorbed dose = 1 rad = 100 erg/g = 0.01 J/kg
1 gray = 1 Gy = 1 J/kg
100 rems = 1 sievert = 1 Sv

Temperature

Kelvin to Centigrade: °C = °K − 273.15
Centigrade to Fahrenheit: °F = (9/5) °C + 32
Rankine to Fahrenheit: °F = °R − 460

Volumetric Flow Rate

1 cm^3/s = 1 cc/s = 10^{-6} m^3/s
1 cm^3/s = 3.053 ft^3/day = 0.5437 bbl/day
1 bbl/day = 1.839 cm^3/s

Source: Adapted from Cohen (1996).

Appendix B
IFLO User's Manual
John R. Fanchi, Ph.D.

Integrated Reservoir Asset Management. DOI: 10.1016/B978-0-12-382088-4.00023-2
Copyright © 2010 Elsevier Inc. All rights reserved.

Contents

Chapter	Section		Page
B1	**Introduction to IFLO**		**301**
	B1.1	Program Overview	301
	B1.2	Program Configuration	302
	B1.3	Input Data File	302
B2	**Initialization Data**		**303**
	B2.1	Model Dimensions and Geometry	303
	B2.2	Porosity and Permeability Distributions	307
	B2.3	Rock Region Information	310
	B2.4	Modifications to Pore Volumes and Transmissibilities	312
	B2.5	Reservoir Geophysical Parameters	314
	B2.6	Fluid PVT Tables	319
	B2.7	Miscible Solvent Data	322
	B2.8	Pressure and Saturation Initialization	325
	B2.9	Run Control Parameters	327
	B2.10	Analytic Aquifer Models	330
	B2.11	Coal Gas Model	330
B3	**Recurrent Data**		**332**
	B3.1	Timestep and Output Control	332
	B3.2	Well Information	334
B4	**Program Output**		**341**
	B4.1	Initialization Data	341
	B4.2	Recurrent Data	341

B1 Introduction to IFLO

The program IFLO is a pseudomiscible, multicomponent, multidimensional fluid flow simulator. IFLO models isothermal Darcy flow in up to three dimensions. It assumes reservoir fluids can be described by up to three fluid phases (oil, gas, and water) with physical properties that depend on pressure and composition. Natural gas and injected solvent are allowed to dissolve in both the oil and water phases. A coal gas desorption option is available for modeling coalbed methane production. The integrated flow model IFLO includes a petrophysical algorithm that allows the calculation of reservoir geophysical attributes that make it possible to track changes in seismic variables as a function of time and to perform geomechanical calculations. The features of IFLO are outlined below and discussed in more detail in later manual sections. For more information about flow simulation and integrated flow modeling, see Fanchi (2000, 2006a).

B1.1 Program Overview

IFLO was designed to run on personal computers with Intel Pentium or equivalent processors. This size simulator is well suited for learning how to use a reservoir simulator, developing an understanding of reservoir management concepts, and solving many types of reservoir engineering problems.

IFLO is a modified and expanded version of MASTER, a black oil simulator with multicomponent, pseudomiscible options for modeling carbon dioxide or nitrogen flooding (Ammer et al., 1991). MASTER is an improved version of BOAST, an implicit pressure, explicit saturation (IMPES) simulator published by the U.S. Department of Energy in 1982 (Fanchi et al., 1982). IFLO includes several enhancements to MASTER, including algorithms from earlier versions of BOAST. A variety of useful geoscience, geomechanical, and reservoir engineering features are available in IFLO, including the representation of horizontal or deviated wells, improvements for modeling heterogeneous reservoir characteristics, the calculation of reservoir geophysical information that can be used to model seismic data, estimates of geomechanical parameters, a coal gas production model, and a stress-dependent permeability model for improving the calculation of well and reservoir flow performance.

IFLO has been tested under a variety of conditions, including oil and gas reservoir depletion, waterflooding, gas injection into an undersaturated oil reservoir, aquifer influx into a gas reservoir, and carbon dioxide injection. Favorable comparisons with other simulators have been obtained. IFLO is based on a mass conserving Newton-Raphson solution procedure. It includes a material balance correction that reduces cumulative material balance error to the magnitude of material balance error associated with a single timestep. Coding and algorithm enhancements increase program robustness.

B1.2   Program Configuration

IFLO is designed to run under Windows 95/98/NT/XP or Vista. Dynamic memory management is used to control the size of grids. Memory allocation controls, such as the number of blocks in a given direction or the number of wells in the model, are entered by the user as part of the input data file described in Section B2.

IFLO must be copied to a folder on your hard drive before running. The following procedure is recommended for a CD drive and hard drive running Windows 95/98/NT/XP or Vista.

Open **Windows Explorer** and select your CD drive.

Use a **Windows**-based Unzip program to extract all of the files from the **IFLO** file on the CD to a folder on your hard drive.

Run **IFLO** by double clicking on the **IFLO.EXE** file on your hard drive.

B1.3   Input Data File

IFLO reads a file called ITEMP.DAT and outputs to files with the prefix ITEMP. The output files are described in Section B4. You should copy and rename any files you wish to save because IFLO overwrites the ITEMP.* files each time it runs.

The easiest way to prepare a new data file is to edit an old one. This will give you an example of the formats needed for most options. If you start with an old data set, make sure that you check all applicable data entries and make changes where appropriate.

IFLO input data are divided into two parts: initialization data and recurrent data. Initialization data are described in Section B2. They include data that are set at the beginning of the study and not expected to change during a model run. Such data include the reservoir description and fluid properties. Recurrent data are described in Section B3 and refer to data that are expected to change during the course of a simulation. They include well schedules and timestep control information.

B2 Initialization Data

Initialization data records are read once at the beginning of the simulation. They must be read in the order presented below. Title or heading records are read before each major and many minor sections. These records are designed to make the input data file easier to read and edit.

In many cases, codes are read that will specify the type of input to follow and the number of values that will be read. These codes increase the efficiency and flexibility of entering input data. All input data, with the exception of well names, are entered by free format. Data entered on the same line must be separated by a comma or a space.

Tabular data entered by the user should cover the entire range of values expected to occur during a simulation. The table interpolation algorithms in IFLO will return table end point values if the independent variable goes outside the range of the input tabular values. No message will be printed if this occurs.

1. Title   Up to 80 characters; this record will appear as the run title.

B2.1   Model Dimensions and Geometry

B2.1.1   Model Dimensions

1. Heading   Up to 80 characters.

2. II, JJ, KK, NWELL, NWCON

Code	Meaning
II	Number of gridblocks in the x-direction
JJ	Number of gridblocks in the y-direction
KK	Number of gridblocks in the z-direction
NWELL	Number of wells
NWCON	Number of connections per well

3. Heading   Up to 80 characters.

4. KDX, KDY, KDZ, KDZNET
- KDX   Control code for input of x-direction grid size
- KDY   Control code for input of y-direction grid size
- KDZ   Control code for input of z-direction gross thickness
- KDZNET   Control code for input of z-direction net thickness

Code	Value	Meaning
KDX	−1	The x-direction grid dimensions are the same for all blocks. Read only one value.
	0	The x-direction dimensions are read for each block in the first row (J = 1) of layer 1 (K = 1). These values are assigned to all other rows and layers. Read II values.
	1	The x-direction dimensions are read for each block in layer 1 (K = 1). These values are assigned to all other layers. Read II × JJ values.
KDY	−1	The y-direction grid dimensions are the same for all blocks. Read only one value.
	0	The y-direction dimensions are read for each block in the first column (I = 1) of layer 1 (K = 1). These values are assigned to all other columns and layers. Read JJ values.
	1	The y-direction dimensions are read for each block in layer 1 (K = 1). These values are assigned to all other layers. Read II × JJ values.
KDZ	−1	The z-direction gross thickness is the same for all blocks. Read only one value.
	0	A constant gross thickness is read for each layer; each layer may have a different value. Read KK values.
	1	The z-direction gross thickness is read for each block in the grid. Read II × JJ × KK values.
KDZNET	−1	The z-direction net thickness is the same for all blocks. Read only one value.
	0	A constant net thickness is read for each layer; each layer may have a different value. Read KK values.
	1	The z-direction net thickness is read for each block in the grid. Read II × JJ × KK values.

5. **DX**
   DX   Gridblock size in x-direction (ft)
   If KDX = −1, read one constant value
   If KDX = 0, read II values (one for each row)
   If KDX = +1, read II × JJ values (one for each K = 1 block)

6. **DY**
   DY   Gridblock size in y-direction (ft)
   If KDY = −1, read one constant value
   If KDY = 0, read JJ values (one for each column)
   If KDY = +1, read II × JJ values (one for each K = 1 block)

7. DZ

DZ   Gross gridblock thickness in z-direction (ft)

If KDZ $= -1$, read one constant value

If KDZ $= 0$, read KK values (one for each layer)

If KDZ $= +1$, read II $\times$ JJ $\times$ KK values (one for each block)

8. DZNET

DZNET   Net gridblock thickness in z-direction (ft)

If KDZ $= -1$, read one constant value

If KDZ $= 0$, read KK values (one for each layer)

If KDZ $= +1$, read II $\times$ JJ $\times$ KK values (one for each block)

NOTE: Those gridblocks with zero pore volume should be defined by setting DZNET $= 0$ or porosity $= 0$. Bulk volume (DX $\times$ DY $\times$ DZ) should be a nonzero positive value for every gridblock. The IFLO calculation assumes that all gridblocks have a nonzero pore volume. A block with zero pore volume is treated as a water-filled block with (porosity) $\times$ (net-to-gross ratio) $= 0.0001$. Transmissibilities for these blocks are set to zero to prevent flow into or out of the block.

B2.1.2 *Modifications to Grid Dimensions*

1. **Heading**   Up to 80 characters.

2. **NUMDX, NUMDY, NUMDZ, NUMDZN, IDCODE**

NUMDX   Number of regions where x-direction length (DX) is changed

NUMDY   Number of regions where y-direction length (DY) is changed

NUMDZ   Number of regions where z-direction gross thickness (DZ) is changed

NUMDZN Number of regions where z-direction net thickness (DZN) is changed

IDCODE $= 0$ means do not print the modified distributions

                      $= 1$ means print the modified distributions

3. **I1, I2, J1, J2, K1, K2, DX**

Omit this record if NUMDX $= 0$.

I1   Coordinate of first region block in I direction

I2   Coordinate of last region block in I direction

J1   Coordinate of first region block in J direction

J2   Coordinate of last region block in J direction

K1   Coordinate of first region block in K direction

K2   Coordinate of last region block in K direction

DX New value of x-direction grid size for region (ft)

NOTE: NUMDX records must be read.

4. **I1, I2, J1, J2, K1, K2, DY**

Omit this record if NUMDY $= 0$.

I1   Coordinate of first region block in I direction

I2   Coordinate of last region block in I direction

J1   Coordinate of first region block in J direction

J2   Coordinate of last region block in J direction

K1   Coordinate of first region block in K direction
K2   Coordinate of last region block in K direction
DY   New value of y-direction grid size for region (ft)

NOTE: NUMDY records must be read.

5. I1, I2, J1, J2, K1, K2, DZ
Omit this record if NUMDZ = 0.
I1   Coordinate of first region block in I direction
I2   Coordinate of last region block in I direction
J1   Coordinate of first region block in J direction
J2   Coordinate of last region block in J direction
K1   Coordinate of first region block in K direction
K2   Coordinate of last region block in K direction
DZ   New value of z-direction gross thickness for region (ft)

NOTE: NUMDZ records must be read.

6. I1, I2, J1, J2, K1, K2, DZNET
Omit this record if NUMDZN = 0.
I1   Coordinate of first region block in I direction
I2   Coordinate of last region block in I direction
J1   Coordinate of first region block in J direction
J2   Coordinate of last region block in J direction
K1   Coordinate of first region block in K direction
K2   Coordinate of last region block in K direction
DZNET   New value of z-direction net thickness for region (ft)

NOTE: NUMDZN records must be read.

B2.1.3   Depths to Top of Gridblocks

The coordinate system used in IFLO is defined so values in the z (vertical) direction increase as the layer gets deeper. Negative values will be read as heights above the datum.

1. Heading   Up to 80 characters.

2. KEL
KEL   Control code for input of depth values

KEL	Meaning
0	A single constant value is read for the depth to the top of all gridblocks in layer 1 (horizontal plane). Each layer is contiguous in this option. Depths to the top of gridblocks in layers below layer 1 are calculated by adding the layer thickness to the preceding layer top; thus, top (I, J, K + 1) = top (I, J, K) + DZ (I, J, K).
1	A separate depth value must be read for each gridblock in layer 1. Read II × JJ values. Each layer is contiguous in this option. Depths to the top of gridblocks in layers below layer 1 are calculated by adding the

KEL	Meaning
	layer thickness to the preceding layer top; thus, top (I, J, K + 1) = top (I, J, K) + DZ (I, J, K).
2	A separate depth value is read for each layer. Read KK values. Each layer is horizontal (layer cake) in this option.
3	A separate depth value is read for each gridblock. Read II × JJ × KK values.

3. **ELEV**
  ELEV　　Depth to top of gridblock (ft)
  If KEL = 0, read one constant value
  If KEL = 1, read II × JJ values (one for each block in layer 1)
  If KEL = 2, read KK values (one for each layer)
  If KEL = 3, read II × JJ × KK values (one for each block)

B2.2　Porosity and Permeability Distributions

B2.2.1　Porosity and Permeability

1. **Heading**　Up to 80 characters.

2. **KPH, KKX, KKY, KKZ**
  KPH　　Control code for input of porosity
  KKX　　Control code for input of x-direction permeability
  KKY　　Control code for input of y-direction permeability
  KKZ　　Control code for input of z-direction permeability

Code	Value	Meaning
KPH	−1	The porosity is constant for all gridblocks. Read only one value.
	0	A constant value is read for each layer. Read KK values.
	1	A value is read for each block. Read II × JJ × KK values.
KKX	−1	The x-direction permeability is constant for all gridblocks. Read only one value.
	0	A constant value is read for each layer. Read KK values.
	1	A value is read for each block. Read II × JJ × KK values.
KKY	−1	The y-direction permeability is constant for all gridblocks. Read only one value.
	0	A constant value is read for each layer. Read KK values.
	1	A value is read for each block. Read II × JJ × KK values.
KKZ	−1	The z-direction permeability is constant for all gridblocks. Read only one value.
	0	A constant value is read for each layer. Read KK values.
	1	A value is read for each block. Read II × JJ × KK values.

3. PHI
PHI   Porosity (fraction)

If KPH $= -1$, read one constant value

If KPH $= 0$, read KK values (one for each layer)

If KPH $= +1$, read II $\times$ JJ $\times$ KK values (one for each block)

4. PERMX
PERMX   Permeability in x-direction (md)

If KKX $= -1$, read one constant value

If KKX $= 0$, read KK values (one for each layer)

If KKX $= +1$, read II $\times$ JJ $\times$ KK values (one for each block)

5. PERMY
PERMY   Permeability in y-direction (md)

If KKY $= -1$, read one constant value

If KKY $= 0$, read KK values (one for each layer)

If KKY $= +1$, read II $\times$ JJ $\times$ KK values (one for each block)

6. PERMZ
PERMZ   Permeability in z-direction (md)

If KKZ $= -1$, read one constant value

If KKZ $= 0$, read KK values (one for each layer)

If KKZ $= +1$, read II $\times$ JJ $\times$ KK values (one for each block)

B2.2.2   Modifications to Porosities and Permeabilities

1. Heading   Up to 80 characters.

2. NUMP, NUMKX, NUMKY, NUMKZ, IPCODE
NUMP   Number of regions where porosity (PHI) is changed

NUMKX   Number of regions where x-direction permeability (PERMX) is changed

NUMKY   Number of regions where y-direction permeability (PERMY) is changed

NUMKZ   Number of regions where z-direction permeability (PERMZ) is changed

IPCODE $= 0$ means do not print the modified distributions

         $= 1$ means print the modified distributions

3. I1, I2, J1, J2, K1, K2, VALPHI
Omit this record if NUMP $= 0$.

I1   Coordinate of first region block in I direction

I2   Coordinate of last region block in I direction

J1   Coordinate of first region block in J direction

J2   Coordinate of last region block in J direction

K1   Coordinate of first region block in K direction

K2   Coordinate of last region block in K direction

Code	Value	Meaning
NUMP	>0	New value of porosity (fr).
	<0	Multiply porosity by VALPHI.

NOTE: NUMP records must be read.

4. **I1, I2, J1, J2, K1, K2, VALKX**
   Omit this record if NUMKX = 0.
   I1 Coordinate of first region block in I direction
   I2 Coordinate of last region block in I direction
   J1 Coordinate of first region block in J direction
   J2 Coordinate of last region block in J direction
   K1 Coordinate of first region block in K direction
   K2 Coordinate of last region block in K direction

Code	Value	Meaning
NUMKX	>0	New value of x-direction permeability (md).
	<0	Multiply x-direction permeability by VALKX.

NOTE: NUMKX records must be read.

5. **I1, I2, J1, J2, K1, K2, VALKY**
   Omit this record if NUMKY = 0.
   I1 Coordinate of first region block in I direction
   I2 Coordinate of last region block in I direction
   J1 Coordinate of first region block in J direction
   J2 Coordinate of last region block in J direction
   K1 Coordinate of first region block in K direction
   K2 Coordinate of last region block in K direction

Code	Value	Meaning
NUMKY	>0	New value of y-direction permeability (md).
	<0	Multiply y-direction permeability by VALKY.

NOTE: NUMKY records must be read.

6. **I1, I2, J1, J2, K1, K2, VALKZ**
   Omit this record if NUMKZ = 0.
   I1 Coordinate of first region block in I direction
   I2 Coordinate of last region block in I direction
   J1 Coordinate of first region block in J direction
   J2 Coordinate of last region block in J direction
   K1 Coordinate of first region block in K direction
   K2 Coordinate of last region block in K direction

Code	Value	Meaning
NUMKZ	>0	New value of z-direction permeability (md).
	<0	Multiply z-direction permeability by VALKZ.

NOTE: NUMKZ records must be read.

B2.3   Rock Region Information

B2.3.1   Definition of Rock Regions

1. **Heading**   Up to 80 characters.

2. **KR3P, NROCK, KPHIMOD**

   KR3P   Code specifying desired relative permeability option

   NROCK   Number of distinct rock regions. A separate set of saturation-dependent
               tables must be entered for each rock region

   KPHIMOD   Code specifying desired φ-*K* model for initial permeability calculation and
               transmissibility updates

Code	Value	Meaning
KR3P	0	Oil relative permeability calculated from the relative permeability data for the two-phase water–oil system.
	1	Oil relative permeability calculated from the relative permeability data for the two-phase gas–oil system.
	2	Three-phase oil relative permeability based on modified Stone equation.
KPHIMOD	0	Do not use φ-*K* model.
	1	Use φ-*K* model to calculate initial permeability. Do not update transmissibility.
	2	Use φ-*K* model to calculate initial permeability and update transmissibility.
	3	Use φ-*K* model to update transmissibility. Do not calculate initial permeability.

3. **Heading**   Up to 80 characters.
   Omit this record if NROCK = 1.

4. **NUMROK**
   Omit this record if NROCK = 1.

   NUMROK = 0   Enter rock region value for each block

   NUMROK > 0   Number of regions where the rock region default value of 1 is
                 changed

5. **IVAL**
   Omit this record if NROCK = 1 or NUMROK > 0.

   IVAL   Array of rock region values. Read II × JJ × KK values

6. **I1, I2, J1, J2, K1, K2, IVAL**
   Omit this record if NROCK = 1 or NUMROK = 0.

   I1   Coordinate of first region block in I direction

   I2   Coordinate of last region block in I direction

   J1   Coordinate of first region block in J direction

J2   Coordinate of last region block in J direction
K1   Coordinate of first region block in K direction
K2   Coordinate of last region block in K direction
IVAL   Number of the saturation-dependent data set to be assigned to this rock region and
        IVAL ≤ NROCK

NOTE: NUMROK records must be read.

B2.3.2 Porosity-Permeability Model for Transmissibility Calculation

1. **Heading** Up to 80 characters.
Include this record if KPHIMOD > 0.

2. **XKBASE, YKBASE, ZKBASE, PHIBASE**
Include this record if KPHIMOD = 1 or 2.
XKBASE   Base permeability in x-direction (md)
YKBASE   Base permeability in y-direction (md)
ZKBASE   Base permeability in z-direction (md)
PHIBASE   Base porosity (fr)

NOTE: The x-direction ϕ-K model is

$$K_x = K_{x,base}\left\{ a_1 \left(\frac{\phi}{\phi_{base}}\right)^{b_1} + a_2 e^{[b_2(\phi-\phi_{base})]} \right\}$$

Similar models apply to y-direction and z-direction permeabilities. Coefficients are defined as follows:

3. **XKPHIA1, XKPHIB1, XKPHIA2, XKPHIB2**
Include this record if KPHIMOD > 0.
XKPHIA1   Coefficient a_1 for ϕ-K model in x-direction
XKPHIB1   Coefficient b_1 for ϕ-K model in x-direction
XKPHIA2   Coefficient a_2 for ϕ-K model in x-direction
XKPHIB2   Coefficient b_2 for ϕ-K model in x-direction

4. **YKPHIA1, YKPHIB1, YKPHIA2, YKPHIB2**
Include this record if KPHIMOD > 0.
YKPHIA1   Coefficient a_1 for ϕ-K model in y-direction
YKPHIB1   Coefficient b_1 for ϕ-K model in y-direction
YKPHIA2   Coefficient a_2 for ϕ-K model in y-direction
YKPHIB2   Coefficient b_2 for ϕ-K model in y-direction

5. **ZKPHIA1, ZKPHIB1, ZKPHIA2, ZKPHIB2**
Include this record if KPHIMOD > 0.
ZKPHIA1   Coefficient a_1 for ϕ-K model in z-direction
ZKPHIB1   Coefficient b_1 for ϕ-K model in z-direction
ZKPHIA2   Coefficient a_2 for ϕ-K model in z-direction
ZKPHIB2   Coefficient b_2 for ϕ-K model in z-direction

NOTE: Repeat records 1 through 5 a total of NROCK times (one set of records for each rock region defined in Section B2.3.1).

B2.3.3 Relative Permeability and Capillary Pressure Tables

1. **Heading** Up to 80 characters.

2. **SAT1 KROW1 KRW1 PCOW1**
   !
   SATn KROWn KRWn PCOWn
   SAT   Water phase saturation (fr). Set SATn = 1.0
   KROW   Oil relative permeability for oil–water system (fr)
   KRW   Water relative permeability for oil–water system (fr)
   PCOW   Oil–water capillary pressure (psi)

   NOTE: There must be table entries for irreducible water saturation (S_{wr}) and residual oil saturation (S_{orw}). Capillary pressure is defined as PCOW = $P_o - P_w$, where P_o and P_w are the oil and water phase pressures, respectively.
   NOTE: Repeat records 1 and 2 a total of NROCK times (one set of records for each rock region defined in Section B2.3).

3. **Heading** Up to 80 characters.

4. **SAT1 KROG1 KRG1 PCGO1**
   !
   SATn KROGn KRGn PCGOn
   SAT   Gas phase saturation (fr). Set SAT1 = 0.0 and SATn = 1.0
   KROG   Oil relative permeability for gas–oil system (fr)
   KRG   Gas relative permeability for gas–oil system (fr)
   PCGO   Gas–oil capillary pressure (psi)

   NOTE: The gas–oil table assumes that irreducible water saturation (S_{wr}) is present. As a matter of consistency, KROG at SAT1 = 0 must equal KROW at S_{wr}. There must be table entries for residual gas saturation (S_{gr}) and residual oil saturation (S_{orgw}). Capillary pressure is defined as PCGO = $P_g - P_o$, where P_o and P_g are the oil and gas phase pressures, respectively. If solvent is included in the model, gas–oil capillary pressure will only be used at gridblocks that have pressures below the miscibility pressure.
   NOTE: Repeat records 3 and 4 a total of NROCK times (one set of records for each rock region defined in Section B2.3.1).

B2.4 Modifications to Pore Volumes and Transmissibilities

1. **Heading** Up to 80 characters.

2. **NUMPV, NUMTX, NUMTY, NUMTZ, ITCODE**
   NUMPV   Number of regions where pore volume is changed
   NUMTX   Number of regions where x-direction transmissibility (TX) is changed
   NUMTY   Number of regions where y-direction transmissibility (TY) is changed
   NUMTZ   Number of regions where z-direction transmissibility (TZ) is changed
   ITCODE   = 0 means do not print the modified distributions
               = 1 means print the modified distributions

NOTE: The conventions for gridblock (I, J, K) transmissibility follow:

TX(I, J, K) refers to flow between blocks I-1 and I
TY(I, J, K) refers to flow between blocks J-1 and J
TZ(I, J, K) refers to flow between blocks K-1 and K

3. I1, I2, J1, J2, K1, K2, VALPV
Omit this record if NUMPV = 0.

I1	Coordinate of first region block in I direction
I2	Coordinate of last region block in I direction
J1	Coordinate of first region block in J direction
J2	Coordinate of last region block in J direction
K1	Coordinate of first region block in K direction
K2	Coordinate of last region block in K direction
VALPV	Multiplier of pore volume for region

NOTE: NUMPV records must be read.

4. I1, I2, J1, J2, K1, K2, VALTX
Omit this record if NUMTX = 0.

I1	Coordinate of first region block in I direction
I2	Coordinate of last region block in I direction
J1	Coordinate of first region block in J direction
J2	Coordinate of last region block in J direction
K1	Coordinate of first region block in K direction
K2	Coordinate of last region block in K direction
VALTX	Multiplier of x-direction transmissibility for region

NOTE: NUMTX records must be read.

5. I1, I2, J1, J2, K1, K2, VALTY
Omit this record if NUMTY = 0.

I1	Coordinate of first region block in I direction
I2	Coordinate of last region block in I direction
J1	Coordinate of first region block in J direction
J2	Coordinate of last region block in J direction
K1	Coordinate of first region block in K direction
K2	Coordinate of last region block in K direction
VALTY	Multiplier of y-direction transmissibility for region

NOTE: NUMTY records must be read.

6. I1, I2, J1, J2, K1, K2, VALTZ
Omit this record if NUMTZ = 0.

I1	Coordinate of first region block in I direction
I2	Coordinate of last region block in I direction
J1	Coordinate of first region block in J direction
J2	Coordinate of last region block in J direction
K1	Coordinate of first region block in K direction
K2	Coordinate of last region block in K direction
VALTZ	Multiplier of z-direction transmissibility for region

NOTE: NUMTZ records must be read.

B2.5 Reservoir Geophysical Parameters

B2.5.1 Moduli and Grain Densities

1. **Heading** Up to 80 characters.

2. **KGPMOD, KDSMOD**
   KGPMOD   Control code for reservoir geophysical model
   KDSMOD   Control code for dynamic-to-static conversion model

KGPMOD	KDSMOD	Meaning
-1	0	No reservoir geophysical model.
0	0	Constant moduli model: enter moduli as arrays of constant values; moduli do not depend on effective pressure, porosity, or clay content.
1	0	IFM model: enter moduli as functions of porosity, effective pressure, and clay content; enter model parameters by rock region (NROCK values).
1	1	IFM model plus conversion of Young's modulus and Poisson's ratio from dynamic to static conditions; enter model parameters by rock region (NROCK values).

3. **Heading** Up to 80 characters.
   Enter this record if KGPMOD = 0.

4. **KKB, KKG, KMU, KRHO**
   Enter this record if KGPMOD = 0.
   KKB   Control code for input of the dry frame bulk modulus (evacuated porous rock)
   KKG   Control code for input of the grain bulk modulus (solid matrix material)
   KMU   Control code for input of the shear modulus (evacuated porous rock)
   KRHO   Control code for input of the grain density (solid matrix material)

Code	Value	Meaning
KKB	-1	Dry frame bulk moduli are the same for all blocks. Read only one value.
	0	A constant value of dry frame bulk modulus is read for each layer; each layer may have a different value. Read KK values.
	1	Dry frame bulk moduli are read for each block. Read II × JJ × KK values.
KKG	-1	Grain bulk moduli are the same for all blocks. Read only one value.
	0	A constant value of grain bulk modulus is read for each layer; each layer may have a different value. Read KK values.
	1	Grain bulk moduli are read for each block. Read II × JJ × KK values.

Code	Value	Meaning
KMU	−1	Shear moduli are the same for all blocks. Read only one value.
	0	A constant value of shear modulus is read for each layer; each layer may have a different value. Read KK values.
	1	Shear moduli are read for each block. Read II × JJ × KK values.
KRHO	−1	Grain densities are the same for all blocks. Read only one value.
	0	A constant value of grain density is read for each layer; each layer may have a different value. Read KK values.
	1	Grain densities are read for each block. Read II × JJ × KK values.

5. KB

Enter this record if KGPMOD = 0.

KB   Dry frame bulk modulus (psia)

If KKB = −1, read one constant value

If KKB = 0, read KK values (one for each layer)

If KKB = +1, read II × JJ × KK values (one for each block)

NOTE: In the absence of data, a value of 3×10^6 psia is reasonable.

6. KG

Enter this record if KGPMOD = 0.

KG   Grain bulk modulus (psia)

If KKG = −1, read one constant value

If KKG = 0, read JJ values (one for each layer)

If KKG = +1, read II × JJ values (one for each block)

NOTE: In the absence of data, a value of 3×10^6 psia is reasonable.

7. MU

Enter this record if KGPMOD = 0.

MU   Effective shear modulus (psia)

If KMU = −1, read one constant value

If KMU = 0, read KK values (one for each layer)

If KMU = +1, read II × JJ × KK values (one for each block)

NOTE: In the absence of data, a value of 3×10^6 psia is reasonable.

8. RHOMA

Enter this record if KGPMOD = 0.

RHOMA   Grain density (lbf/ft^3)

If KRHO = −1, read one constant value

If KRHO = 0, read KK values (one for each layer)

If KRHO = +1, read II × JJ × KK values (one for each block)

NOTE: In the absence of data, a value of 168 lbf/ft^3 (corresponding to 2.7 g/cm^3) is reasonable.

B2.5.2   IFM Model

1. Heading   Up to 80 characters.
Enter this record if KGPMOD = 1.

2. AIKMA, AIKMB, AIKMC, AIKMD, AIKME, AIKMF
Enter this record if KGPMOD = 1.
AIKMA   Dry frame bulk modulus parameter a_0
AIKMB   Dry frame bulk modulus parameter a_1
AIKMC   Dry frame bulk modulus parameter a_2
AIKMD   Dry frame bulk modulus parameter a_3
AIKME   Dry frame bulk modulus parameter a_4
AIKMF   Dry frame bulk modulus parameter a_5

3. EXK1, EXK2
Enter this record if KGPMOD = 1.
EXK1   Dry frame bulk modulus exponent e_1
EXK2   Dry frame bulk modulus exponent e_2

4. AIMUA, AIMUB, AIMUC, AIMUD, AIMUE, AIMUF
Enter this record if KGPMOD = 1.
AIMUA   Shear modulus parameter α_0
AIMUB   Shear modulus parameter α_1
AIMUC   Shear modulus parameter α_2
AIMUD   Shear modulus parameter α_3
AIMUE   Shear modulus parameter α_4
AIMUF   Shear modulus parameter α_5

5. EXM1, EXM2
Enter this record if KGPMOD = 1.
EXM1   Dry frame bulk modulus exponent ε_1
EXM2   Dry frame bulk modulus exponent ε_1

6. AIRHOA, AIRHOB, AIRHOC
Enter this record if KGPMOD = 1.
AIRHOA   Rock matrix grain density parameter b_0
AIRHOB   Rock matrix grain density parameter b_1
AIRHOC   Rock matrix grain density parameter b_2

NOTE: Repeat records 1 through 6 a total of NROCK times (one set of records for each rock region defined in Section B2.3).

B2.5.3   Confining Pressure and Clay Content for IFM Model

1. KPCON, KCLAY
Enter this record if KGPMOD = 1.
KPCON   Control code for input of confining pressure
KCLAY   Control code for input of clay content

Code	Value	Meaning
KPCON	−1	Confining pressure is the same for all blocks. Read only one value.
	0	A constant value of confining pressure is read for each layer; each layer may have a different value. Read KK values.
	1	Confining pressures are read for each block. Read II × JJ × KK values.
	11	Calculate confining pressures from block elevations and overburden pressure gradient.
KCLAY	−1	Clay content is the same for all blocks. Read only one value.
	0	A constant value of clay content is read for each layer; each layer may have a different value. Read KK values.
	1	Clay content is read for each block. Read II × JJ × KK values.

2. PCON

Enter this record if KGPMOD = 1.

PCON   Confining pressure (psia)

If KPCON = −1, read one constant value

If KPCON = 0, read KK values (one for each layer)

If KPCON = +1, read II × JJ × KK values (one for each block)

If KPCON = +11, read constant values for **OBGRAD, OBDAT**

OBGRAD   Overburden pressure gradient (psia/ft)

OBDAT   Overburden datum (ft)

NOTE: In the absence of data, values of OBGRAD = 1.0 psia/ft and OBDAT = 0.0 ft are reasonable.

3. CLAY

Enter this record if KGPMOD = 1.

CLAY   Clay content (volume fraction)

If KCLAY = −1, read one constant value

If KCLAY = 0, read JJ values (one for each layer)

If KCLAY = +1, read II × JJ values (one for each block)

NOTE: In the absence of data, a value of 0.0 is reasonable.

B2.5.4   Modifications to Confining Pressure and Clay Content

1. Heading   Up to 80 characters.

2. NUMCON, NUMCLA, IDCODE

NUMCON   Number of regions where confining pressure (PCON) is changed

NUMCLA   Number of regions where clay content (CLAY) is changed

IDCODE   = 0 means do not print the modified distributions

              = 1 means print the modified distributions

3. I1, I2, J1, J2, K1, K2, PCON
Omit this record if NUMCON = 0.

I1   Coordinate of first region block in I direction
I2   Coordinate of last region block in I direction
J1   Coordinate of first region block in J direction
J2   Coordinate of last region block in J direction
K1   Coordinate of first region block in K direction
K2   Coordinate of last region block in K direction
PCON   New value of confining pressure (psia)

NOTE: NUMCON records must be read.

4. I1, I2, J1, J2, K1, K2, CLAY
Omit this record if NUMCLA = 0.

I1   Coordinate of first region block in I direction
I2   Coordinate of last region block in I direction
J1   Coordinate of first region block in J direction
J2   Coordinate of last region block in J direction
K1   Coordinate of first region block in K direction
K2   Coordinate of last region block in K direction
CLAY   New value of clay content (volume fraction)

NOTE: NUMCLA records must be read.

B2.5.5   Dynamic to Static Conversion of Young's Modulus and Poisson's Ratio

1. Heading   Up to 80 characters.
Enter this record if KGPMOD = 1 and KDSMOD = 1.

2. YDSA1, YDSA2, YDSB1, YDSB2, YDSC
Enter this record if KGPMOD = 1 and KDSMOD = 1.

YDSA1   Coefficient a_1 for dynamic to static Young's modulus conversion
YDSA2   Coefficient a_2 for dynamic to static Young's modulus conversion
YDSB1   Coefficient b_1 for dynamic to static Young's modulus conversion
YDSB2   Coefficient b_2 for dynamic to static Young's modulus conversion
YDSC   Coefficient c for dynamic to static Young's modulus conversion

NOTE: The dynamic to static conversion algorithm for Young's modulus E is

$$E_s = aE_d^b + c$$
$$a = a_1 + a_2 \log(P_e)$$
$$b = b_1 + b_2 \log(P_e)$$

where subscript s denotes static and subscript d denotes dynamic. The coefficients $\{a, a_1, a_2, b, b_1, b_2, c\}$ are empirical fit parameters, and P_e is effective pressure. An analogous dynamic to static conversion algorithm may be specified for Poisson's ratio.

3. **PDSA1, PDSA2, PDSB1, PDSB2, PDSC**
   Enter this record if KGPMOD = 1 and KDSMOD = 1.
   PDSA1   Coefficient a_1 for dynamic to static Poisson's ratio conversion
   PDSA2   Coefficient a_2 for dynamic to static Poisson's ratio conversion
   PDSB1   Coefficient b_1 for dynamic to static Poisson's ratio conversion
   PDSB2   Coefficient b_2 for dynamic to static Poisson's ratio conversion
   PDSC   Coefficient c for dynamic to static Poisson's ratio conversion

NOTE: Repeat records 1 through 3 a total of NROCK times (one set of records for each rock region defined in Section B2.3).

B2.6   Fluid PVT Tables

1. **Heading**   Up to 80 characters.

2. **PBO, VOSLP, BOSLP, BWSLP, PMAX**
   PBO   Initial reservoir oil bubble point pressure (psia). If no oil or natural gas exists, set PBO = 14.7 psia.
   VOSLP   Slope of the oil viscosity versus pressure curve for undersaturated oil—that is, for pressures above PBO. The slope ($\Delta\mu_o/\Delta P_o$) should be in cp/psia.
   BOSLP   Slope of the oil formation volume factor versus pressure curve for undersaturated oil. The slope ($\Delta B_o/\Delta P_o$) should be in RB/STB/psia and should be negative or zero.
   BWSLP   Slope of the water formation volume factor versus pressure curve for under-saturated water—that is, for pressures above PBO. The slope ($\Delta B_w/\Delta P_o$) should be in RB/STB/psia and should be negative or zero.
   PMAX   Maximum pressure entry for all PVT tables (psia)

NOTE: VOSLP, BOSLP, and BWSLP are used only for undersaturated oil and water. The slope ($\Delta R_{so}/\Delta P_o$) of the solution natural gas–oil ratio versus the pressure curve for undersaturated oil is assumed to be zero.

3. **Heading**   Up to 80 characters; oil table follows.

4. **P1   MUO1   BO1   RSO1**
   !

   PMAX   MUO(PMAX)   BO(PMAX)   RSO(PMAX)
   P   Pressure (psia). Pressures must be in ascending order from P1 (normally 14.7 psia) to PMAX. The last table entry must be PMAX
   MUO   Oil viscosity (cp)
   BO   Oil formation volume factor (RB/STB)
   RSO   Solution natural-gas–oil ratio (SCF/STB)

NOTE: Oil properties must be entered as saturated data over the entire pressure range. Saturated oil data are required because of the bubble point pressure tracking algorithm.

5. **Heading**   Up to 80 characters; water table follows.

6. **P1   MUW1   BW1   RSW1**
   !
   PMAX   MUW(PMAX)   BW(PMAX)   RSW(PMAX)
   P   Pressure (psia). Pressures must be in ascending order from P1 (normally 14.7
            psia) to PMAX. The last table entry must be PMAX
   MUW   Water viscosity (cp)
   BW   Water formation volume factor (RB/STB)
   RSW   Solution natural-gas–water ratio (SCF/STB). Water properties must be entered
            as saturated data over the entire pressure range if RSW is nonzero

NOTE: It is usually assumed in black oil simulations that the solubility of gas in
water can be neglected. In this case, set RSW = 0.0 for all pressures. IFLO includes
gas solubility in the water phase to account for CO_2 solubility in water, gas production from geopressured aquifers, or any other case where gas solubility in water can
be significant.

7. **Heading**   Up to 80 characters.

8. **KGCOR**

Code	Value	Meaning
KGCOR	0	Read gas and rock properties table.
	1	Activate gas correlation option and read rock compressibility versus pressure table.

9. **Heading**   Up to 80 characters; gas table follows.
   Omit this record if KGCOR = 1.

10. **P1   MUG1   BG1   PSI1   CR1**
    !
    PMAX   MUG(PMAX)   BG(PMAX)   PSI(PMAX)   CR(PMAX)
    Omit this record if KGCOR = 1.
    P   Pressure (psia). Pressures must be in ascending order from P1 (normally 14.7
             psia) to PMAX. The last table entry must be PMAX
    MUG   Natural gas viscosity (cp)
    BG   Natural gas formation volume factor (RCF/SCF)
    PSI   Real gas pseudo-pressure ($psia^2$/cp)
    CR   Rock compressibility (1/psia)

11. **KODEA, MPGT, TEM, SPG**
    Omit this record if KGCOR = 0.
    KODEA   Gas composition option
    MPGT   Number of gas PVT table entries (1 < MPGT ≤ 25)
    TEM   Reservoir temperature (°F)
    SPG   Gas specific gravity (air = 1.0)

KODEA	Gas Description
1	Sweet gas: input 12 component mole fractions as 0. 0. 0. 1. 0. 0. 0. 0. 0. 0. 0. 0.
2	Sour gas: input 12 component mole fractions in the order y_1 y_2 y_3 y_4 0. 0. 0. 0. 0. 0. 0. 0, where y_1 = mole fraction of H_2S, y_2 = mole fraction of CO_2, y_3 = mole fraction of N_2, and $y_4 = 1 - (y_1 + y_2 + y_3)$.
3	Sweet or sour gas with the following 12 component mole fractions read in the following order: H_2S, CO_2, N_2, C_1, C_2, C_3, iC_4, nC_4, iC_5, nC_5, C_6, C_{7+}. The sum of the mole fractions should equal 1.
4	Same as KODEA = 3 but also read critical pressure, critical temperature, and molecular weight of C_{7+}.

12. FRCI

Omit this record if KGCOR = 0.

FRCI   Component mole fraction of gas. Read 12 entries in the following order:

FRCI(I)	Component I	FRCI(I)	Component I
1	H_2S	7	iC_4
2	CO_2	8	nC_4
3	N_2	9	iC_5
4	C_1	10	nC_5
5	C_2	11	C_6
6	C_3	12	C_{7+}

13. PRSCI, TEMCI, RMWTI

Omit this record if KGCOR = 0 or if KODEA $\neq$ 4.

PRSCI   Critical pressure (psia)
TEMCI   Critical temperature (°R)
RMWTI   Molecular weight

14. Heading   Up to 80 characters; rock compressibility table follows.
Omit this record if KGCOR = 0.

15. P1   CR1
!
PMAX   CR(PMAX)
Omit this record if KGCOR = 0.

Option	Code	Meaning
Constant rock compressibility NOTE: Enter 1 record.	PMAX	Maximum table pressure (psia) from record 4.
	CR	Rock compressibility (1/psia).

Option	Code	Meaning
Pressure-dependent rock compressibility	P	Pressure (psia). Pressures must be in ascending order from P1 (normally 14.7 psia) to PMAX. The last table entry must be PMAX.
NOTE: Enter MPGT records.	CR	Rock compressibility (1/psia).

16. Heading   Up to 80 characters.

17. RHOSCO, RHOSCW, RHOSCG
RHOSCO   Stock tank oil density (lbm/cu ft)
RHOSCW   Stock tank water density (lbm/cu ft)
RHOSCG   Gas density at standard conditions (lbm/cu ft). If no natural gas exists, set RHOSCG = 0

NOTE: At stock tank conditions (14.7 psia and 60°F for oilfield units), pure water has a density of 62.4 lbm/cu ft and air has a density of 0.0765 lbm/cu ft.

B2.7   Miscible Solvent Data

1. Heading   Up to 80 characters.

2. NSLUGS, NSREAD
NSLUGS   Number of solvents
NSREAD   Number of solvent PVT tables to be read (up to 4). NSREAD must be equal to or greater than NSLUGS

NOTE: NSREAD is provided as a convenience. PVT data for one to four solvents may be left in the input data set for an oil-water-natural gas run by setting NSREAD = 1 to 4 and NSLUGS = 0.
If NSREAD = 0, omit the data in the remainder of this section and proceed to Section B2.8.

3. Heading   Up to 80 characters.

4. PBO1, VO1OPE, BO1OPE
PBO1   Initial base solvent-oil bubble point pressure (psia)
VO1OPE   Undersaturated slope of oil viscosity (cp/psi)
BO1OPE   Undersaturated slope of oil formation volume factor (RB/STB/psi)

5. Heading   Up to 80 characters.

6. PBW1, VW1OPE, BW1OPE
PBW1   Initial base solvent-water bubble point pressure (psia)
VW1OPE   Undersaturated slope of water viscosity (cp/psi)
BW1OPE   Undersaturated slope of water formation volume factor (RB/STB/psi)

7. Heading   Up to 80 characters.

8. PMISC, FPMISC, SOMIN, REDK, BETA, SORM, VSMISC

PMISC	Miscibility pressure (psia)
FPMISC	Fraction of PMISC (fr) for calculating multicontact miscibility pressure PMCM (psia). PMISC and PMCM are related by PMCM = FPMISC × PMISC
SOMIN	Minimum oil saturation for solid precipitation (fr). SOMIN > 0 only if SORM = 0
REDK	Relative permeability reduction factor for solid precipitation (fr)
BETA	Parameter for water blocking function
SORM	Miscible region residual oil saturation (fr). SORM > 0 only if SOMIN = 0
VSMISC	Total solvent volume fraction required to obtain full miscibility (fr)

Code	Value	Meaning
SOMIN	0	No solid precipitation.
	>0	Low solid precipitation.
BETA	<0	No water blocking.
	≥0	Water blocking on.

NOTE: If the automatic timestep control is on, saturation convergence requires that SOMIN < DSMAX (Section B2.9).

9. Heading Up to 80 characters.

10. OM1, OM2

OM1 Mixing parameter ω_1 for natural gas solvent miscibility
OM2 Mixing parameter ω_2 for oil-gas-solvent miscibility

NOTE: Only OM1 is used if the gridblock pressure P < PMCM. Only OM2 is used if P > PMISC. Both OM1 and OM2 are used if P is in the multicontact miscibility pressure range PMCM < P < PMISC.

11. Heading Up to 80 characters.

12. RHOSC1, RHOSC2, RHOSC3, RHOSC4

RHOSC1 Stock tank density of base solvent (lbm/cu ft)
RHOSC2 Stock tank density of solvent 2 (lbm/cu ft)
RHOSC3 Stock tank density of solvent 3 (lbm/cu ft)
RHOSC4 Stock tank density of solvent 4 (lbm/cu ft)

13. Heading Up to 80 characters.

14. Heading Up to 80 characters.

15.

P1	MUS1	BS1	RSOS1	RSWS1	BO1	MUO1	BW1	MUW1
!	!	!	!	!	!	!	!	!
PMAX	MUS1 @ PMAX	BS1 @ PMAX	RSOS1 @ PMAX	RSWS1 @ PMAX	BO1 @ PMAX	MUO1 @ PMAX	BW1 @ PMAX	MUW1 @ PMAX

P	Pressure (psia). Pressures must be in ascending order from P1 (normally 14.7 psia) to PMAX. The last table entry must be PMAX
MUS1	Viscosity of base solvent (cp)
BS1	Formation volume factor of base solvent (RB/STB)
RSOS1	Solubility of base solvent in oil (SCF/STB)
RSWS1	Solubility of base solvent in water (SCF/STB)
BO1	Formation volume factor of oil with base solvent (RB/STB)
MUO1	Viscosity of oil with base solvent (cp)
BW1	Formation volume factor of water with base solvent (RB/STB)
MUW1	Viscosity of water with base solvent (cp)

NOTE: Base solvent PVT data are required if NSREAD > 0. Base solvent PVT data are used only if NSLUGS > 0. Oil and water properties must be entered as base solvent saturated data over the entire pressure range. Saturated oil and water data are required because of the bubble point pressure tracking algorithm. Oil-base solvent properties should be determined with dead oil that is fully saturated with base solvent at each pressure.

16. Heading   Up to 80 characters.

17. Heading   Up to 80 characters.

18. P1   **MUS2**   **BS2**   **RSOS2**
  !

PMAX MUS2(PMAX) BS2(PMAX)   RSOS2(PMAX)

P	Pressure (psia). Pressures must be in ascending order from P1 (normally 14.7 psia) to PMAX. The last table entry must be PMAX
MUS2	Viscosity of solvent 2 (cp)
BS2	Formation volume factor of solvent 2 (RB/STB)
RSOS2	Solubility of solvent 2 in oil (SCF/STB)

NOTE: Solvent 2 PVT data are required if NSREAD > 1. Solvent 2 PVT data are used only if NSLUGS > 1.

19. Heading   Up to 80 characters.

20. Heading   Up to 80 characters.

21. P1   **MUS3**   **BS3**   **RSOS3**
  !

PMAX MUS3(PMAX) BS3(PMAX)   RSOS3(PMAX)

P	Pressure (psia). Pressures must be in ascending order from P1 (normally 14.7 psia) to PMAX. The last table entry must be PMAX
MUS3	Viscosity of solvent 3 (cp)
BS3	Formation volume factor of solvent 3 (RB/STB)
RSOS3	Solubility of solvent 3 in oil (SCF/STB)

NOTE: Solvent 3 PVT data are required if NSREAD > 2. Solvent 3 PVT data are used only if NSLUGS > 2.

22. Heading   Up to 80 characters.

23. Heading   Up to 80 characters.

24. P1   **MUS4**   **BS4**   **RSOS4**
    !
    PMAX MUS4(PMAX)   BS4(PMAX)   RSOS4(PMAX)
    P   Pressure (psia). Pressures must be in ascending order from P1 (normally 14.7
                psia) to PMAX. The last table entry must be PMAX
    MUS4   Viscosity of solvent 4 (cp)
    BS4   Formation volume factor of solvent 4 (RB/STB)
    RSOS4   Solubility of solvent 4 in oil (SCF/STB)

   NOTE: Solvent 4 PVT data are required if NSREAD > 3. Solvent 4 PVT data
are used only if NSLUGS > 3.

25. Heading   Up to 80 characters.

26. NOMOB, MOBCTL, SCI
    NOMOB   Number of entries in the mobility control table
    MOBCTL   Mobility control switch
    SCI   Surfactant concentration index. SCI multiplies the mobility reduction
                  values FRCO2 defined as follows:

Code	Value	Meaning
MOBCTL	0	No mobility control.
	1	Apply mobility control.

27. Heading   Up to 80 characters.
    Omit this record if MOBCTL = 0.

28. NSC, FRCO2
    Omit this record if MOBCTL = 0.
    NSC   Normalized surfactant concentration (fr)
    FRCO2   Reduction of base solvent mobility (fr)

   NOTE: NOMOB records must be read.

B2.8   Pressure and Saturation Initialization

1. Heading   Up to 80 characters.

2. KPI, KSI
    KPI   Pressure initialization code
    KSI   Saturation initialization code

Code Values		Meaning
KPI	**KSI**	
0	0	Equilibrium pressure and saturation initialization. Enter pressures and depths at the OWC and GOC. This option assumes no solvent present at initialization. Saturations are calculated from capillary pressures.
1		Specify pressure throughout grid. Read II × JJ × KK values of P.
	1	Specify constant initial oil, water, and gas saturations; specify constant initial solvent volume fractions.
	2	Specify variable saturations throughout grid. Read II × JJ × KK values of SO, SW, and solvent volume fractions. IFLO sets SG = 1 − SO − SW internally.
0	3	Gravity segregated oil, water, and gas saturations. This option assumes no solvent present at initialization.

NOTE: Option {KPI = 1, KSI = 2} may be used to prepare a restart data file.

3. WOC, PWOC, GOC, PGOC
Enter this record if KPI = 0.
WOC   Depth to the water–oil contact (ft below datum)
PWOC   Pressure at the water–oil contact (psia)
GOC   Depth to the gas–oil contact (ft below datum)
PGOC   Pressure at the gas–oil contact (psia)

NOTE: Repeat this record a total of NROCK times: one record for each rock region defined in Section B2.3.

4. PO
Enter this record if KPI = 1.
PO   Oil phase pressure (psia). Read II × JJ × KK values

5. SOI, SWI, SGI, VS1I, VS2I, VS3I, VS4I
Enter this record if KSI = 1.
SOI   Initial oil saturation (fr)
SWI   Initial water saturation (fr)
SGI   Initial gas saturation (fr)

Omit the following values if NSLUGS = 0.
VS1I   Initial base solvent volume fraction in the gaseous phase (fr). Enter this value if NSLUGS ≥ 1
VS2I   Initial solvent 2 volume fraction in the gaseous phase (fr). Enter this value if NSLUGS ≥ 2
VS3I   Initial solvent 3 volume fraction in the gaseous phase (fr). Enter this value if NSLUGS ≥ 3
VS4I   Initial solvent 4 volume fraction in the gaseous phase (fr). Enter this value if NSLUGS ≥ 4

NOTE: The sum of the saturations must satisfy SOI + SWI + SGI = 1, and the sum of the volume fractions must satisfy the constraint VGG + VS1 + VS2 + VS3 + VS4 = 1.0, where VGG is the fraction of natural gas in the gaseous phase.

6. **SO, SW, VS1, VS2, VS3, VS4**

Enter this record if KSI = 2.

SO   Oil saturation (fr). Read II × JJ × KK values

SW   Water saturation (fr). Read II × JJ × KK values

Omit the following arrays if NSLUGS = 0.

VS1   Base solvent volume fraction in the gaseous phase (fr). Read II × JJ × KK values. Enter this array if NSLUGS ≥ 1

VS2   Solvent 2 volume fraction in the gaseous phase (fr). Read II × JJ × KK values. Enter this array if NSLUGS ≥ 2

VS3   Solvent 3 volume fraction in the gaseous phase (fr). Read II × JJ × KK values. Enter this array if NSLUGS ≥ 3

VS4   Solvent 4 volume fraction in the gaseous phase (fr). Read II × JJ × KK values. Enter this array if NSLUGS ≥ 4

NOTE: If NSLUGS > 0, then the sum of the volume fractions must satisfy the constraint VGG + VS1 + VS2 + VS3 + VS4 = 1.0, where VGG is the fraction of natural gas in the gaseous phase.

7. **SOI, SGI, SOR**

Enter this record if KSI = 3.

SOI   Initial oil saturation (fr) for the oil–water zone. Initial water saturation in the oil–water zone is $1 - SOI$

SGI   Initial gas saturation (fr) for the gas–water zone. Initial water saturation in the gas–water zone is $1 - SGI$

SOR   Irreducible oil saturation (fr). If SOR > 0, set $S_o = 0$ when $S_o < SOR$. Water and gas saturations are then renormalized

NOTE: Repeat this record a total of NROCK times B, one record for each rock region defined in Section B2.3.

B2.9   Run Control Parameters

1. **Heading**   Up to 80 characters.

2. **KSW1, KSW2, KSW3, KSW4**

KSW1   Control code for printing material balance information. Information includes the gridblock location with the largest component material balance error, the magnitude of the error, and the elapsed time

KSW2   Control code for graphical image orientation

KSW3   Control code for printing the number of iterations required for convergence of the iterative solution techniques (SSOR, LSOR, ORTHOMIN)

KSW4   Control code for printing the timestep summary to the terminal

Code	Value	Meaning
KSW1	0	Do not print the information.
	1	Print the information to file ITEMP.MBE.
KSW2	0	Image aligned with grid.
	1	Image inverted relative to y-axis.
KSW3	0	Do not print the information.
	1	Print the information to file ITEMP.MBE.
KSW4	0	Print summary at each timestep.
	1	Print summary at FTIO times (Section B3.1).

3. **Heading** Up to 80 characters.

4. **NMAX, FACT1, FACT2, TMAX, WORMAX, GORMAX, PAMIN, PAMAX**
   - NMAX Maximum number of timesteps per simulation run
   - FACT1 Factor for increasing timestep size using automatic timestep control. FACT1 = 1.0 for fixed timestep size. A common value for FACT1 is 1.25
   - FACT2 Factor for decreasing timestep size using automatic timestep control. FACT2 = 1.0 for fixed timestep size. A common value for FACT2 is 0.5
   - TMAX Maximum elapsed time to be simulated (days); the run will be terminated when the time exceeds TMAX
   - WORMAX Maximum allowed water–oil ratio for a producing oil well (STB/STB)
   - GORMAX Maximum allowed gas–oil ratio for a producing oil well (SCF/STB)
   - PAMIN Minimum field average pressure (psia)
   - PAMAX Maximum field average pressure (psia)

NOTE: The run will be terminated if producing WOR > WORMAX or producing GOR > GORMAX. GORMAX is the total natural gas plus solvent–oil ratio. PAMIN and PAMAX should be within the range of pressures covered by the fluid PVT tables discussed in Section B2.6. The run will be terminated when the pore volume weighted average reservoir pressure P_{avg} < PAMIN or P_{avg} > PAMAX. Each of the controls {WORMAX, GORMAX, PAMIN, PAMAX} will be ignored if it is set to zero.

5. **Heading** Up to 80 characters.

6. **KSOL, MITR, OMEGA, TOL, NCYCLE, DSMAX, DPMAX, ITMAX, RTOL, NERR**
   - KSOL Solution method code
   - MITR For KSOL = 1 or 2: maximum number of SOR iterations for convergence with a typical value of 100. For KSOL = 4: maximum number of conjugate gradient iterations for convergence with a typical value of 50
   - OMEGA For KSOL = 1 or 2: initial SOR acceleration parameter. Initial value of OMEGA should be between 1.0 and 2.0. A typical initial value is 1.2. The model will attempt to optimize OMEGA if NCYCLE $\neq$ 0
   - TOL For KSOL = 1 or 2: maximum acceptable SOR pressure convergence tolerance with a typical value of 0.001 psia. For KSOL = 4: pressure convergence tolerance with a typical value of 0.001 psia to 0.0001 psia

NCYCLE   For KSOL = 1 or 2: number of SOR iteration cycles for determining when to change (optimize) OMEGA. A typical value is 12. If NCYCLE = 0, the initial value of OMEGA will be used for the entire run

DSMAX   Maximum saturation change (fraction) allowed per timestep. The timestep size DT will be reduced by FACT2 if the saturation change of any phase or any component in any gridblock exceeds DSMAX and DT > DTMIN (the user-specified minimum timestep size defined in Section B3.1). If the resulting step size is less than DTMIN, the timestep will be repeated with DT = DTMIN. A typical value of DSMAX is 0.05

DPMAX   Maximum pressure change (psia) allowed per timestep. The timestep size will be reduced by FACT2 if the pressure change in any gridblock exceeds DPMAX and DT > DTMIN. If the resulting step size is less than DTMIN, the timestep will be repeated with DT = DTMIN. A typical value of DPMAX is 100 psia

ITMAX   Maximum number of Newton-Raphson iterations per timestep for convergence. A typical value is 5

RTOL   Maximum allowed residual for Newton-Raphson convergence. A typical value is 0.001. ITMAX overrides RTOL if RTOL is not reached

NERR   Code for controlling the material balance error technique. NERR = 1 is recommended

Code	Value	Meaning
KSOL	1	SSOR: iterative, slice (planar) successive overrelaxation method for 2-D and 3-D models.
	2	LSOR: iterative, line successive overrelaxation method for 0-D, 1-D, 2-D, and 3-D models.
	3	D4: direct solution method for 1-D, 2-D, and moderate-sized 3-D models.
	4	ORTHOMIN: iterative, preconditioned conjugate gradient algorithm for large 2-D and 3-D models.
NERR	0	Material balance error control technique is off.
	1	Material balance error control technique is on.

7. Heading   Up to 80 characters.

8. WEIGHT

WEIGHT   Fluid property weighting factor

Code Value	Meaning
0.5	Average properties are used.
1.0	Upstream properties are used.

NOTE: The weighting factor applies to formation volume factor and viscosity of oil, water, natural gas, and solvents; the solubility of natural gas and solvents in oil; and the solubility of natural gas and base solvent in water.

B2.10   Analytic Aquifer Models

1. **Heading**   Up to 80 characters.

2. **IAQOPT**
   IAQOPT   Analytic aquifer model code

Code	Value	Meaning
IAQOPT	0	No analytic aquifer model.
	1	Steady-state aquifer model (constant aquifer pressure).

NOTE: Different aquifer influx strengths may be specified for a given aquifer.

3. **NAQEN**
   Omit this record if IAQOPT $\neq$ 1.
   NAQEN   Number of regions containing a steady-state aquifer

4. **I1, I2, J1, J2, K1, K2, SSAQ**
   Omit this record if IAQOPT $\neq$ 1.
   I1   Coordinate of first region block in I direction
   I2   Coordinate of last region block in I direction
   J1   Coordinate of first region block in J direction
   J2   Coordinate of last region block in J direction
   K1   Coordinate of first region block in K direction
   K2   Coordinate of last region block in K direction
   SSAQ   Steady-state aquifer strength (SCF/day/psia)

NOTE: NAQEN records must be read.

B2.11   Coal Gas Model

1. **Heading**   Up to 80 characters.

2. **ICGOPT**
   ICGOPT   Coal gas model code

Code	Value	Meaning
ICGOPT	0	No coal gas model.
	1	Coal gas model with diffusive desorption.

3. **NCGREG**
   Omit this record if ICQOPT $\neq$ 1.
   NCGREG   Number of regions containing a coalbed

4. ITMPCG, ITMPMOD, NCGSUB

Omit this record if ICGOPT $\neq$ 1.

ITMPCG   Coal gas region number
ITMPMOD   Coal gas model number
NCGSUB   Number of subregions within coal gas region

Code	Value	Meaning
ITMPMOD	1	Saturated coal gas model with diffusive desorption.
ITMPMOD	2	Undersaturated coal gas model with diffusive desorption and critical desorption pressure.

NOTE: ITMPCG and NCGSUB must be greater than zero.

5. CGDIFF, CGRAD, CGDEN, CGVL, CGPL, CGASH, CGWC, CGPCD

Omit this record if ICGOPT = 0.

CGDIFF   Coal diffusion (ft^2/day)
CGRAD   Spherical radius of coal (ft)
CGDEN   Coal density (g/cc)
CGVL   Langmuir isotherm volume (SCF gas/ton coal)
CGPL   Langmuir isotherm pressure (psia)
CGASH   Ash content of coal (wt fr)
CGWC   Moisture content of coal (wt fr)
CGPCD   Critical desorption pressure (psia)

NOTE: Set CGPCD = 0 psia if ITMPMOD = 1.

6. I1, I2, J1, J2, K1, K2

Omit this record if ICGOPT = 0.

I1   Coordinate of first region block in I direction
I2   Coordinate of last region block in I direction
J1   Coordinate of first region block in J direction
J2   Coordinate of last region block in J direction
K1   Coordinate of first region block in K direction
K2   Coordinate of last region block in K direction

NOTE: NCGSUB records must be read.
NOTE: Records 4 and 5 should be repeated NCGREG times.

B3 Recurrent Data

Recurrent data records are read periodically during the course of the simulation run. These data include the location and the specification of wells in the model, the changes in well completions and field operations over time, a schedule of well rate and/or pressure performance over time, the timestep control information for advancing the simulation through time, and the controls on the type and frequency of printout information provided by the simulator.

1. **Major Heading** Up to 80 characters.

   NOTE: This record signifies the start of the recurrent data section.

B3.1 Timestep and Output Control

1. **Heading** Up to 80 characters.

2. **IWREAD, IOMETH, IWLREP, ISUMRY**
   IWREAD   Controls input of well information
   IOMETH   Controls scheduling of well input and array print controls
   IWLREP   Controls output of well report
   ISUMRY   Controls output of summary report

Code	Value	Meaning
IWREAD	0	Do not read well information.
	1	Read well information.
IOMETH	≥1	Number of elapsed time values to be read on record 3. The program will print results to output files at these elapsed times and allow you to change well characteristics after the last elapsed time entered during this recurrent data period.
IWLREP	0	Do not print well report.
	1	Print well report.
ISUMRY	0	Do not print summary report.
	1	Print summary report.
	2	Write ITEMP.ARR file.

3. **FTIO**
   FTIO   Array containing total elapsed times at which output will occur (days). Up to 50 monotonically increasing values may be entered. The first entry must be greater than 0, and each succeeding entry must be greater than any previous entry

NOTE: When the elapsed time of a run equals an FTIO value, the well and basic summary reports will be printed. Maps will also be printed according to the instructions given below. When the elapsed time of a run equals the last FTIO value, the program will allow the user to enter a new set of recurrent data records (repeat Sections B3.1 and B3.2).

4. **IPMAP, ISOMAP, ISWMAP, ISGMAP, IPBMAP, IRSMAP**

   IPMAP   Control code for printing pressure array
   ISOMAP   Control code for printing oil saturation array
   ISWMAP   Control code for printing water saturation array
   ISGMAP   Control code for printing gas saturation array
   IPBMAP   Control code for printing bubble point pressure array
   IRSMAP   Control code for printing natural gas solubility array

Code Value	Meaning
0	Do not print the array.
1	Print the array.

5. **IS1MAP, IS2MAP, IS3MAP, IS4MAP, IAQMAP**

   IS1MAP   Control code for printing base solvent volume fraction array
   IS2MAP   Control code for printing solvent 2 volume fraction array
   IS3MAP   Control code for printing solvent 3 volume fraction array
   IS4MAP   Control code for printing solvent 4 volume fraction array
   IAQMAP   Control code for printing aquifer influx array

Code Value	Meaning
0	Do not print the array.
1	Print the array.

6. **IVPMAP, IZMAP, IRCMAP, IVSMAP, IVRMAP**

   IVPMAP   Control code for printing seismic compressional velocity (V_p) array
   IZMAP   Control code for printing seismic acoustic impedance array
   IRCMAP   Control code for printing seismic reflection coefficient array
   IVSMAP   Control code for printing seismic shear velocity (V_s) array
   IVRMAP   Control code for printing seismic velocity ratio V_p/V_s array

Code Value	Meaning
0	Do not print the array.
1	Print the array.

7. **INUMAP, IYMMAP, IUNMAP, ISVMAP, ISHMAP**

   INUMAP   Control code for printing Poisson's ratio array
   IYMMAP   Control code for printing Young's modulus array
   IUNMAP   Control code for printing uniaxial compaction array
   ISVMAP   Control code for printing vertical stress (confining pressure) array
   ISHMAP   Control code for printing horizontal stress array

Code Value	Meaning
0	Do not print the array.
1	Print the array.

8. DT, DTMIN, DTMAX

DT   Starting timestep size (days)

DTMIN   Minimum timestep size (days). A typical value is 1 day

DTMAX   Maximum timestep size (days). A typical value is 3–5 days

B3.2   Well Information

Omit this section if IWREAD = 0.

1. Heading   Up to 80 characters.

2. NWELLN, NWELLO, KSIS

NWELLN   Number of new wells for which complete well information is entered

NWELLO   Number of previously defined wells for which new rates and/or rate controls are entered

KSIS   Control code for surfactant-water injection

Code	Value	Meaning
NWELLN	0	Do not read new well information.
	≥1	Read new well information.
NWELLO	0	Do not change data for previously defined wells.
	≥1	Change data for previously defined wells.
KSIS	0	Do not inject surfactant.
	2, −2, or −12	Inject surfactant in the water phase as a gas phase mobility control agent.

3. Heading   Up to 80 characters.

Include this record if NWELLN > 0.

4. WELLID

Include this record if NWELLN > 0.

WELLID   Well name with up to five characters

5. IDWELL, KONECT, KWCNTL, KWPID

Include this record if NWELLN > 0.

IDWELL   Well identification number. Each well should have a unique IDWELL number. If two or more wells have the same IDWELL number, the characteristics of the last well entered will be used

KONECT   Total number of gridblocks connected to well IDWELL

KWCNTL   Control code for well limits applied to well IDWELL

KWPID   Control code for calculating well PID

Code	Value	Meaning
KWCNTL	0	Do not read well rate constraints and workovers.
	1	Read well rate constraints and workovers.
KWPID	0	User enters PID.
	1	Model calculates PID.

6. I, J, K, PID, PWF
Include this record if NWELLN > 0 and KWPID = 0.

I x-coordinate of gridblock containing well
J y-coordinate of gridblock containing well
K z-coordinate of gridblock containing well
PID   Layer flow index for gridblock
PWF   Flowing bottomhole pressure for block (psia)

NOTE: KONECT records must be read.
Deviated (slanted) and horizontal wells may be represented by calculating an appropriate PID and specifying gridblock locations that model the expected well trajectory. For example, a horizontal well that is aligned in the x-direction will have constant J and K indices, and index I will vary if there is more than one connection.

To shut in a connection, set that connection PID to 0. To shut in a well, set all of its connection PID values to zero.

7A. I, J, K, IWDIR, KHMOD, PIDRW, PIDS, PWF
Include this record if NWELLN > 0 and KWPID = 1.

I x-coordinate of gridblock containing well
J y-coordinate of gridblock containing well
K z-coordinate of gridblock containing well
IWDIR   Well orientation
KHMOD   Flow capacity model for PID calculation
PIDRW   Wellbore radius (ft)
PIDS   Well skin (fr)
PWF   Flowing bottomhole pressure for block (psia)

Code	Value	Meaning
IWDIR	1	Well aligned in x-direction.
	2	Well aligned in y-direction.
	3	Well aligned in z-direction.
KHMOD	0	PID calculated with constant KH.
	1	PID calculated with pressure dependent KH.

NOTE: The x-direction ϕ-K model is

$$K_x = K_{x0}\left\{a_1\left(\frac{\phi}{\phi_0}\right)^{b_1} + a_2 e^{[b_2(\phi-\phi_0)]}\right\}$$

where K_{x0} is the initial permeability and ϕ_0 is initial porosity. Similar models apply to y-direction and z-direction permeabilities. Coefficients for the directional ϕ-K models are specified below. Values of net thickness and permeability in well PID are calculated as functions of pressure and saturation when KPHIMOD > 0.

7B. XKA1, XKB1, XKA2, XKB2
Include this record if NWELLN > 0, KWPID = 1, and KHMOD = 1.
XKA1   Coefficient a_1 for ϕ-K model in x-direction
XKB1   Coefficient b_1 for ϕ-K model in x-direction
XKA2   Coefficient a_2 for ϕ-K model in x-direction
XKB2   Coefficient b_2 for ϕ-K model in x-direction

7C. YKA1, YKB1, YKA2, YKB2
Include this record if NWELLN > 0, KWPID = 1, and KHMOD = 1.
YKA1   Coefficient a_1 for ϕ-K model in y-direction
YKB1   Coefficient b_1 for ϕ-K model in y-direction
YKA2   Coefficient a_2 for ϕ-K model in y-direction
YKB2   Coefficient b_2 for ϕ-K model in y-direction

7D. ZKA1, ZKB1, ZKA2, ZKB2
Include this record if NWELLN > 0, KWPID = 1, and KHMOD = 1.
ZKA1   Coefficient a_1 for ϕ-K model in z-direction
ZKB1   Coefficient b_1 for ϕ-K model in z-direction
ZKA2   Coefficient a_2 for ϕ-K model in z-direction
ZKB2   Coefficient b_2 for ϕ-K model in z-direction

NOTE: Repeat records 7A through 7D a total of KONECT times (one set of records for each connection).

8. KIP, QO, QW, QG, QT, QS
Include this record if NWELLN > 0.
KIP   Code for specifying well operating characteristics
QO   Oil rate (STB/D)
QW   Water rate (STB/D)
QG   Natural gas rate (MSCF/D)
QT   Total fluid voidage rate (RB/D). QT includes oil, water, natural gas, and solvent
QS   Solvent rate (MSCF/D)

NOTE: Sign conventions for rates: negative rates indicate fluid injection and positive rates indicate fluid production. To impose a maximum target rate on an explicit pressure-controlled well, set KWCNTL = 1, and set the primary phase rate (QO, QW, QG, or QT) to the maximum target rate.

9. ALIT, BLIT
Include this record if NWELLN > 0 and KIP = 10.
ALIT a coefficient of LIT gas well analysis
BLIT b coefficient of LIT gas well analysis

NOTE: Gas rate QG > 0 in record 8 will be used as a target rate if KWCNTL = 1; if KWCNTL = 0, the value of QG will be ignored.

10. **WQMAX, WQMIN, WWOR, WGOR**
   Include this record if NWELLN > 0 and KWCNTL = 1.
   WQMAX   Maximum allowed rate for primary phase (QO, QW, QG, or QT)
   WQMIN   Minimum allowed rate for primary phase (QO, QW, QG, or QT)
   WWOR   Maximum allowed WOR (SCF/STB); shut-in worst offending connection.
           Set WWOR = 0 to ignore
   WGOR   Maximum allowed GOR (SCF/STB); shut-in worst offending connection.
           Set WGOR = 0 to ignore

NOTE: Rates are expressed in the same units as the rates in record 8. WOR and GOR constraints apply to production wells only. If a maximum target rate is set in record 8 for an explicit pressure-controlled well, the value of WQMAX will take precedence.

Records 4 through 10 should be repeated NWELLN times.

11. **Heading**   Up to 80 characters.
   Include this record if NWELLO > 0.

12. **WELLID**
   Include this record if NWELLO > 0.
   WELLID   Well name with up to five characters

13. **IDWELL, KONECT, KWCNTL**
   Include this record if NWELLO > 0.
   IDWELL   Well identification number
   KONECT   Total number of gridblocks connected to well IDWELL
   KWCNTL   Control code for well limits applied to well IDWELL

14. **I, J, K, PID, PWF**
   Include this record if NWELLO > 0 and KWPID = 0.
   I x-coordinate of gridblock containing well
   J y-coordinate of gridblock containing well
   K z-coordinate of gridblock containing well
   PID   Layer flow index for gridblock
   PWF   Flowing bottomhole pressure for gridblock (psia)

   NOTE: KONECT records must be read.

15A. **I, J, K, IWDIR, KHMOD, PIDRW, PIDS, PWF**
   Include this record if NWELLO > 0 and KWPID = 1.
   I x-coordinate of gridblock containing well
   J y-coordinate of gridblock containing well
   K z-coordinate of gridblock containing well
   IWDIR   Well orientation
   KHMOD   Flow capacity model for PID calculation
   PIDRW   Wellbore radius (ft)
   PIDS   Well skin (ft)
   PWF   Flowing bottomhole pressure for gridblock (psia)

Code	Value	Meaning
IWDIR	1	Well aligned in x-direction.
	2	Well aligned in y-direction.
	3	Well aligned in z-direction.
KHMOD	0	PID calculated with constant KH.
	1	PID calculated with pressure-dependent KH.

NOTE: The x-direction ϕ-K model is

$$K_x = K_{x0}\left\{ a_1\left(\frac{\phi}{\phi_0}\right)^{b_1} + a_2 e^{[b_2(\phi-\phi_0)]} \right\}$$

where K_{x0} is the initial permeability and ϕ_0 is initial porosity. Similar models apply to y-direction and z-direction permeabilities. Coefficients for the directional ϕ-K models are specified below. Values of net thickness and permeability in well PID are calculated as functions of pressure and saturation when KPHIMOD > 0.

15B. XKA1, XKB1, XKA2, XKB2
Include this record if NWELLO > 0, KWPID = 1, and KHMOD = 1.
XKA1   Coefficient a_1 for ϕ-K model in x-direction
XKB1   Coefficient b_1 for ϕ-K model in x-direction
XKA2   Coefficient a_2 for ϕ-K model in x-direction
XKB2   Coefficient b_2 for ϕ-K model in x-direction

15C. YKA1, YKB1, YKA2, YKB2
Include this record if NWELLO > 0, KWPID = 1, and KHMOD = 1.
YKA1   Coefficient a_1 for ϕ-K model in y-direction
YKB1   Coefficient b_1 for ϕ-K model in y-direction
YKA2   Coefficient a_2 for ϕ-K model in y-direction
YKB2   Coefficient b_2 for ϕ-K model in y-direction

15D. ZKA1, ZKB1, ZKA2, ZKB2
Include this record if NWELLO > 0, KWPID = 1, and KHMOD = 1.
ZKA1   Coefficient a_1 for ϕ-K model in z-direction
ZKB1   Coefficient b_1 for ϕ-K model in z-direction
ZKA2   Coefficient a_2 for ϕ-K model in z-direction
ZKB2   Coefficient b_2 for ϕ-K model in z-direction

NOTE: Repeat records 15A through 15D a total of KONECT times (one set of records for each connection).

16. KIP, QO, QW, QG, QT, QS
Include this record if NWELLO > 0.
KIP   Code for specifying well operating characteristics
QO   Oil rate (STB/D)
QW   Water rate (STB/D)

QG   Natural gas rate (MSCF/D)
QT   Total fluid voidage rate (RB/D)
QS   Solvent rate (MSCF/D)

NOTE: Sign conventions for rates: negative rates indicate fluid injection, and positive rates indicate fluid production. To impose a maximum target rate on an explicit pressure-controlled well, set KWCNTL = 1, and set the primary phase rate (QO, QW, QG, or QT) to the maximum target rate.

17. ALIT, BLIT
Include this record if NWELLO > 0 and KIP = 10.
ALIT   *a* coefficient of LIT gas well analysis
BLIT   *b* coefficient of LIT gas well analysis

NOTE: Gas rate QG > 0 in record 16 will be used as a target rate if KWCNTL = 1; if KWCNTL = 0, the value of QG will be ignored.

18. WQMAX, WQMIN, WWOR, WGOR
Include this record if NWELLO > 0 and KWCNTL = 1.
WQMAX   Maximum allowed rate for primary phase (QO, QW, QG, or QT)
WQMIN   Minimum allowed rate for primary phase (QO, QW, QG, or QT)
WWOR   Maximum allowed WOR (SCF/STB); shut-in worst offending connection.
        Set WWOR = 0 to ignore
WGOR   Maximum allowed GOR (SCF/STB); shut-in worst offending connection.
        Set WGOR = 0 to ignore

NOTE: Rates are expressed in the same units as the rates in record 16. WOR and GOR constraints apply to production wells only. If a maximum target rate is set in record 16 for an explicit pressure-controlled well, the value of WQMAX will take precedence.

Records 12 through 18 should be repeated NWELLO times.

Table B3.2-1 Options for Controlling Production Wells

Primary Phase	Well Control	KIP	Nonzero Rates	Well Controls?
	Rate	1	QO > 0	Yes
Oil	Explicit P	−1	QO > 0	Yes
	Implicit P	−11		No
	Rate	1	QG > 0	Yes
Gas	Explicit P	−1	QG > 0	Yes
	Implicit P			No
	LIT	10	QG > 0	Yes
	Rate	1	QW > 0	Yes
Water	Explicit P	−1	QW > 0	Yes
	Implicit P			No
	Rate	1	QT > 0	Yes
Total OWG	Explicit P	−1	QT > 0	Yes
	Implicit P			No

Table B3.2-2 Options for Controlling Injection Wells

Primary Phase	Well Control	KIP	Nonzero Rates	Well Controls?
Water	Rate	2	QW < 0	Yes
	Explicit P	−2	QW < 0	Yes
	Implicit P	−12		No
Gas	Rate	3	QG < 0	Yes
	Explicit P	−3	QG < 0	Yes
	Implicit P	−13		No
Base Solvent	Rate	100	QS < 0	No
(Solvent 1)	Explicit P	−4		No
	Implicit P	−14		No
Solvent 2	Rate	200	QS < 0	No
	Explicit P	−5		No
	Implicit P	−15		No
Solvent 3	Rate	300	QS < 0	No
	Explicit P	−6		No
	Implicit P	−16		No
Solvent 4	Rate	400	QS < 0	No
	Explicit P	−7		No
	Implicit P	−17		No

B4 Program Output

You are given the option at the start of an IFLO run to direct output to either the screen or a set of files. It is often worthwhile to send output to the screen when first building and debugging a data set. IFLO will abort at the point in the data set where it encounters improperly entered data. For evaluating run results, it is preferable to send output to files. In this case, a one-line timestep summary is sent to the screen each timestep so that you can monitor the progress of a run. All output files are in text format.

B4.1 Initialization Data

IFLO outputs the following initialization data in the text file ITEMP.ROF:

- Gridblock sizes
- Node midpoint elevations
- Porosity distributions
- Permeability distributions
- Rock and PVT region distributions
- Relative permeability and capillary pressure tables
- Petrophysical distributions
- PVT tables
- Slopes calculated from PVT data
- Timestep control parameters
- Analytic aquifer model selection
- Coal gas model selection
- Initial fluid volumes in place
- Initial pressure and saturation arrays
- Initial reservoir geophysical attribute arrays
- Initial well information

Other output can be obtained at your request. For example, if a modification option is invoked, you may print out the altered array. It is worthwhile to do this as a check on the input changes.

B4.2 Recurrent Data

All output files are text files so that they may be read by a variety of commercially available spreadsheets. IFLO output may then be manipulated using spreadsheet options. This is especially useful for making plots or displaying array data. Different output files are defined so simulator output file sizes are more manageable. The output files are designed to contain information that is logically connected—for example, well data in one file, timestep information in another file, and so on. The different output files are described next.

B4.2.1   Timestep Summary File—ITEMP.TSS

A one-line timestep summary is automatically printed to the terminal as a record of the progress of the run. This summary provides you with necessary information for evaluating the stability of the solution as a function of time. Significant oscillations in GOR or WOR or large material balance errors are indicative of simulation problems and should be corrected. A smaller timestep through the difficult period is often sufficient to correct IMPES instabilities.

The timestep summary is written to the file ITEMP.TSS. The output quantities include cumulative production of oil, water, and gas; pore volume weighted average pressure; aquifer influx rate and cumulative aquifer influx; and fieldwide WOR and GOR values. The WOR and the GOR are ratios of total producing fluid rates. Consequently, these ratios are comparable to observed fieldwide ratios. These quantities are output as functions of time and timestep number.

B4.2.2   Run Output File—ITEMP.ROF

Model initialization data and run output information, including well performance, are in the file ITEMP.ROF. You may output the following arrays whenever desired: pressure, saturations, bubble point pressure, cumulative aquifer influx, acoustic velocities, acoustic impedances, seismic reflection coefficient, Poisson's ratio, Young's modulus, and uniaxial compaction. Output arrays may be used as input pressure and saturation distributions for restarting a run. It is usually unnecessary to print all of the arrays. To avoid excessive output and correspondingly large output files, you should be judicious in deciding which arrays are printed.

B4.2.3   Well Output File—ITEMP.WEL

Well performance is in the file ITEMP.WEL. The information is provided for easy access and includes production (injection) for each well completion and total well production (injection) for all production (injection) wells.

B4.2.4   Array File—ITEMP.ARR

Selected parameter arrays are tabulated in the file ITEMP.ARR. The arrays are displayed as functions of the Cartesian (x, y, z) coordinate locations of each gridblock midpoint. The parameter arrays include pressure, saturations, and acoustic velocity information.

B4.2.5   Material Balance Error File—ITEMP.MBE

Material balance errors as a function of time are summarized in the file ITEMP. MBE.

References

Ahmed, T., 2000. Reservoir Engineering Handbook, Gulf Publishing, Houston.

Ammer, J.R., Brummert, A.C., Sams, W.N., 1991. Miscible Applied Simulation Techniques for Energy Recovery—Version 2.0, U.S. Department of Energy Report DOE/BC-91/2/SP, Morgantown Energy Technology Center, West Virginia.

Amorim, T., Mocyzdlower, B., 2007. Validating the Use of Experimental Design Techniques in Exploratory Evaluations, Paper SPE 107441. Society of Petroleum Engineers, Richardson, Texas.

Amudo, C., Graf, T., Dandekar, R., Randle, J.M., 2008. The Pains and Gains of Experimental Design and Response Surface Applications in Reservoir Simulation Studies, Paper SPE 107441. Society of Petroleum Engineers, Richardson, Texas.

Amyx, J.W., Bass, D.H., Whiting, R.L., 1960. Petroleum Reservoir Engineering, McGraw-Hill, New York.

Arps, J.J., 1945. Analysis of Decline Curves. Transactions of AIME 160, 228–247.

Arthur, J.D., Bohm, B., Coughlin, B.J., Layne, M., 2009. Evaluating Implications of Hydraulic Fracturing in Shale Gas Reservoirs, Paper SPE 121038. Society of Petroleum Engineers, Richardson, Texas.

Badri, M.A., Sayers, C.M., Awad, R., Graziano, A., 2000. Feasibility study for pore-pressure prediction using seismic velocities in the offshore Nile Delta, Egypt. The Leading Edge, 1103–1108.

Barree, R.D., Conway, M.W., 2005. Beyond Beta Factors: A Model for Darcy, Forchheimer, and Trans-Forchheimer Flow in Porous Media. Journal of Petroleum Technology, 43–45.

Bassiouni, Z., 1994. Theory, Measurement, and Interpretation of Well Logs, SPE Textbook Series 4, Society of Petroleum Engineers, Richardson, Texas.

Batycky, R.P., Thiele, M.R., Coats, K.H., Grindheim, A., Ponting, D., Killough, J.E., et al., 2007. Chapter 17: Reservoir Simulation. In: Holstein, E.D. (Ed.); Lake, L.W. (Editor-in-Chief), Petroleum Engineering Handbook, Volume 5: Reservoir Engineering and Petrophysics. Society of Petroleum Engineers, Richardson, Texas.

Batzle, M., 2006. Chapter 13. Rock Properties. In: Fanchi, J.R. (Ed.); Lake, L.W. (Editor-in-Chief), Petroleum Engineering Handbook, Volume 1: General Engineering. Society of Petroleum Engineers, Richardson, Texas.

Bear, J., 1972. Dynamics of Fluids in Porous Media, Elsevier, New York.

Beggs, H.D., 1984. Gas Production Operations, OGCI Publications, Tulsa.

Biondi, D.L., 2006. 3-D Seismic Imaging. In: Investigations in Geophysics, Number 14, Society of Exploration Geophysicists, Tulsa.

Bowen, D.W., Weimer, P., Scott, A.J., 1993. Application of Siliciclastic Sequence Stratigraphy to Exploration: Examples from Frontier and Mature Basins, Paper SPE 26438. Society of Petroleum Engineers, Richardson, Texas.

Brigham, W.E., Castanier, L., 2007. Chapter 16: In-Situ Combustion. In: Holstein, E.D. (Ed.); Lake, L.W. (Editor-in-Chief), Petroleum Engineering Handbook, Volume 5: Reservoir Engineering and Petrophysics. Society of Petroleum Engineers, Richardson, Texas.

344 References

Brundtland, G., 1987. Our Common Future. Oxford University Press, Oxford.

de Buyl, M., Guidish, T., Bell, F., 1988. Reservoir Description from Seismic Lithologic Parameter Estimation. Journal of Petroleum Technology, 475–482.

Buckley, S.E., Leverett, M.C., 1942. Mechanisms of Fluid Displacement in Sands. Transactions of the AIME 146, 107–116.

Byrnes, A.P., Bhattacharya, S., 2006. Influence of Initial and Residual Oil Saturation and Relative Permeability on Recovery from Transition Zone Reservoirs in Shallow Shelf Carbonates, Paper SPE 99736. Society of Petroleum Engineers, Richardson, Texas.

Callarotti, R.C., 2007. Chapter 12: Electromagnetic Heating of Oil. In: Warner Jr., H.R. (Ed.); Lake, L.W. (Editor-in-Chief), Petroleum Engineering Handbook, Volume 6: Emerging and Peripheral Technologies. Society of Petroleum Engineers, Richardson, Texas.

Calvert, R., 2005. Insights and Methods for 4D Reservoir Monitoring and Characterization, 2005 Distinguished Instructor Short Course. Society of Exploration Geophysicists, Tulsa.

Canadian Energy Resources Conservation Board, 1975. Theory and Practice of the Testing of Gas Wells, third ed. Energy Resources Conservation Board, Calgary.

Carlson, M.R., 2003. Practical Reservoir Simulation, PennWell, Tulsa.

Castagna, J.P., Batzle, M.L., Eastwood, R.L., 1985. Relationships between Compressional-Wave and Shear-Wave Velocities in Clastic Silicate Rocks. Geophysics 50, 571–581.

Chambers, R.L., Yarus, J.M., Hird, K.B., 2000. Petroleum Geostatistics for Nongeostatisticians. The Leading Edge, Part 1, 474–479, May; and Part 2, 592–599, June.

Chawathé, A., Taggart, I., 2004. Insights into Upscaling using 3D Streamlines. SPE Reservoir Evaluation & Engineering, 285–296, August.

Christie, M.A., 1996. Upscaling for Reservoir Simulation. Journal of Petroleum Technology, 1004–1010, November.

Christie, M.A., Blunt, M.J., 2001. Tenth SPE Comparative Solution Project: A Comparison of Upscaling Techniques, Paper SPE 66599. Society of Petroleum Engineers, Richardson, Texas.

Christiansen, R.L., 2006. Chapter 15: Relative Permeability and Capillary Pressure. In: Fanchi, J.R. (Ed.); Lake, L.W. (Editor-in-Chief), Petroleum Engineering Handbook, Volume 1: General Engineering. Society of Petroleum Engineers, Richardson, Texas.

Clark, I., Harper, W., 2000. Practical Geostatistics 2000 Book and CD. Ecosse North America, LLC., Columbus, Ohio.

Coe, A.L. (Ed.), 2003. The Sedimentary Record of Sea-Level Change. Cambridge University Press, Cambridge.

Cohen, E.R., 1996. The Physics Quick Reference Guide, American Institute of Physics, Woodbury, New York.

Collins, R.E., 1961. Flow of Fluids Through Porous Materials, Petroleum Publishing-Reinhold, Tulsa.

Craft, B.C., Hawkins, M.F., 1991. Applied Petroleum Reservoir Engineering, second ed. Prentice-Hall, Englewood Cliffs, NJ, revised by R.E. Terry.

Cramer, R., Eken, A., Dunham, C., der Kinderen, W., Rudenko, P., Adaji, N., et al., 2000. Shell Tests Low-Cost, Real-Time Oil Well Surveillance. Oil & Gas Journal, 53–57, February.

Dahlberg, E.C., 1975. Relative Effectiveness of Geologists and Computers in Mapping Potential Hydrocarbon Exploration Targets. Mathematical Geology 7, 373–394.

Dake, L.P., 2001. The Practice of Reservoir Engineering, revised edition, Elsevier, Amsterdam.

Davis, J.C., 2002. Statistics and Data Analysis in Geology, third ed., J. Wiley, New York.

Deutsch, C.V., Journel, A.G., 1998. GSLIB: Geostatistical Software Library and User's Guide, Oxford University Press, New York.

Dietrich, F.L., Brown, F.G., Zhou, Z.H., Maure, M.A., 1996. Microbial EOR Technology Advancement: Case Studies of Successful Projects, Paper SPE 36746. Society of Petroleum Engineers, Richardson, Texas.

Dietz, R.N., 1987. Tracer Technology. In: Regional and Long-Range Transport of Air Pollution. Elsevier, Amsterdam.

Dogru, A.H., 2000. Megacell Reservoir Simulation. Journal of Petroleum Technology, 54–60, May.

Dogru, A.H., Fung, L.S.K., Middya, U., Al-Shaalan, T.M., Pita, J.A., Kumar, K.H., et al., 2009. A Next-Generation Parallel Reservoir Simulator for Giant Reservoirs, Paper SPE 119272. Society of Petroleum Engineers, Richardson, Texas.

Dorn, G.A., 1998. Modern 3-D Seismic Interpretation. The Leading Edge, 1262–1283, September.

Dugstad, Ø, 2007. Chapter 6: Well-to-Well Tracer Tests. In: Holstein, E.D. (Ed.); Lake, L.W. (Editor-in-Chief), Petroleum Engineering Handbook, Volume 5: Reservoir Engineering and Petrophysics. Society of Petroleum Engineers, Richardson, Texas.

van Dyke, K., 1997. Fundamentals of Petroleum, fourth ed. Petroleum Extension Service, University of Texas.

Earlougher Jr., R.C., 1997. Advances in Well Test Analysis, Society of Petroleum Engineers, Richardson, Texas.

Ebanks Jr., W.J., 1987. Flow Unit Concept—Integrated Approach to Reservoir Description for Engineering Projects, Paper Presented at the AAPG Annual Meeting, Los Angeles.

Ebanks Jr., W.J., Scheihing, M.H., Atkinson, C.D., 1993. Flow Units for Reservoir Characterization. In: Morton-Thompson, D., Woods, A.M. (Eds.), Development Geology Reference Manual, AAPG Methods in Exploration Series, Number 10, 282–285.

Eberhart-Phillips, D.M., 1989. Investigation of Crustal Structure and Active Tectonic Processes in the Coast Ranges, Central California, Ph.D. Dissertation, Stanford University, Stanford, California.

Economides, M.J., Hill, A.D., Ehlig-Economides, C., 1994. Petroleum Production Systems, Prentice-Hall, Upper Saddle River, NJ.

Emerick, A.A., de Moraes, R.J., Rodrigues, J.R.P., 2007. Calculating Seismic Attributes within a Reservoir Flow Simulator, Paper SPE 107001. Society of Petroleum Engineers, Richardson, Texas.

Ertekin, T., Abou-Kassem, J.H., King, G.R., 2001. Basic Applied Reservoir Simulation, Society of Petroleum Engineers, Richardson, Texas.

Fancher, G.H., Lewis, J.A., 1933. Flow of Simple Fluids through Porous Materials. Ind. Eng. Chem. 25, 1139–1147.

Fanchi, J.R., 1983. Multidimensional Numerical Dispersion. Society of Petroleum Engineering Journal, 143–151.

Fanchi, J.R., 1990. Calculation of Parachors for Compositional Simulation: An Update. Society of Petroleum Engineers Reservoir Engineering, 433–436.

Fanchi, J.R., 2000. Integrated Flow Modeling, Elsevier, Amsterdam.

Fanchi, J.R., 2002. Shared Earth Modeling, Elsevier–Butterworth-Heinemann, Amsterdam.

Fanchi, J.R., 2003. The Han-Eberhart-Phillips Model and Integrated Flow Modeling. Geophysics 68, 574–576.

Fanchi, J.R., 2004. Energy: Technology and Directions for the Future, Elsevier-Academic Press, Boston.

Fanchi, J.R., 2006. Principles of Applied Reservoir Simulation, third ed., Elsevier-Gulf Professional Publishing, Boston.

Fanchi, J.R. (Ed.), 2006b. Petroleum Engineering Handbook, Volume 1: General Engineering, Society of Petroleum Engineers, Richardson, Texas, Editor-in-Chief L.W. Lake.

Fanchi, J.R., 2006. Math Refresher for Scientists and Engineers, third ed., J. Wiley, New York.

Fanchi, J.R., 2008. Directional Permeability. SPE Reservoir Evaluation and Engineering Journal, 565–568.

Fanchi, J.R., 2009. Embedding a Petroelastic Model in a Multipurpose Flow Simulator to Enhance the Value of 4D Seismic. SPE Reservoir Evaluation and Engineering Journal, to be published; First Presented as Paper SPE 118839.

Fanchi, J.R., Harpole, K.J., Bujnowski, S.W., 1982. BOAST: A Three- Dimensional, Three-Phase Black Oil Applied Simulation Tool, 2 Volumes, U.S. Department of Energy, Bartlesville Energy Technology Center, Tulsa.

Fanchi, J.R., Pagano, T.A., Davis, T.L., 1999. State of the Art of 4-D Seismic Monitoring. Oil & Gas Journal, 38–43, May.

Fayers, F.J., Hewett, T.A., 1992. A Review of Current Trends in Petroleum Reservoir Description and Assessing the Impacts on Oil Recovery. In: Proceedings of Ninth International Conference on Computational Methods in Water Resources, June.

Gaddy, D.E., 1999. Coalbed Methane Production Shows Wide Range of Variability. Oil & Gas Journal, 41–42, April.

Gassmann, F., 1951. Elastic Waves through a Packing of Spheres. Geophysics 16, 673–685.

Gluyas, J., Swarbrick, R., 2004. Petroleum Geoscience, Blackwell Publishing, Malden, Massachusetts.

Gorell, S., Bassett, R., 2001. Trends in Reservoir Simulation: Big Models, Scalable Models? Will You Please Make Up Your Mind? Paper SPE 71596. Society of Petroleum Engineers, Richardson, Texas.

Govier, G.W. (Ed.), 1978. Theory and Practice of the Testing of Gas Wells. Energy Resources Conservation Board, Calgary.

Green, D.W., Wilhite, G.P., 1998. Enhanced Oil Recovery, Society of Petroleum Engineers, Richardson, Texas.

Greenkorn, R.A., 1962. Experimental Study of Waterflood Tracers. Journal of Petroleum Technology, 87–92.

Gunter, G.W., Finneran, J.M., Hartmann, D.J., Miller, J.D., 1997. Early Determination of Reservoir Flow Units Using an Integrated Petrophysical Method, Paper SPE 38679. Society of Petroleum Engineers, Richardson, Texas.

Haldorsen, H.H., Damsleth, E., 1993. Challenges in Reservoir Characterization. American Association of Petroleum Geologists Bulletin 77 (4), 541–551.

Haldorsen, H.H., Lake, L.W., 1989. A New Approach to Shale Management in Field-Scale Models, Reservoir Characterization 2, SPE Reprint Series #27. Society of Petroleum Engineers, Richardson, Texas.

Han, D.-H., 1986. Effects of Porosity and Clay Content on Acoustic Properties of Sandstones and Unconsolidated Sandstones, Ph.D. Dissertation, Stanford University, Stanford, California.

Harpole, K.J., 1985. Reservoir Environments and Their Characterization, International Human Resources Development Corporation, Boston.

Harris, D.G., 1975. The Role of Geology in Reservoir Simulation Studies. Journal of Petroleum Technology, 625–632, May.

Heymans, M.J., 1997. A Method for Developing 3-D Hydrocarbon Saturation Distributions in Old and New Reservoirs. In: Coalson, E.B., Osmond, J.C., Williams, E.T. (Eds.), Innovative Applications of Petroleum Technology in the Rocky Mountain Area. Rocky Mountain Association of Geologists, Denver, pp. 171–182.

Hirsche, K., Porter-Hirsche, J., Mewhort, L., Davis, R., 1997. The Use and Abuse of Geostatistics, The Leading Edge, 253–260, March.

Holditch, S.A., 2007. Chapter 7: Tight Gas Reservoirs. In: Warner Jr., H.R. (Ed.); Lake, L.W. (Editor-in-Chief), Petroleum Engineering Handbook, Volume 6: Emerging and Peripheral Technologies. Society of Petroleum Engineers, Richardson, Texas.

Holstein, E.D. (Ed.); Lake, L.W. (Editor-in-Chief), 2007. Petroleum Engineering Handbook, Volume 5: Reservoir Engineering and Petrophysics. Society of Petroleum Engineers, Richardson, Texas.

Holstein, E.D., Stalkup, F.I., 2007. Chapter 14: Miscible Processes. In: Holstein, E.D. (Ed.); Lake, L.W. (Editor-in-Chief), Petroleum Engineering Handbook, Volume 5: Reservoir Engineering and Petrophysics. Society of Petroleum Engineers, Richardson, Texas.

Honarpour, M.M., Koederitz, L.F., Harvey, A.H., 1982. Empirical Equations for Estimating Two-Phase Relative Permeability in Consolidated Rock. Journal of Petroleum Technology, 2905–2908.

Honarpour, M.M., Nagarajan, N.R., Sampath, K., 2006. Rock/Fluid Characterization and Their Integration—Implications on Reservoir Management. Journal of Petroleum Technology 120–131.

Horne, R.N., 1995. Modern Well Test Analysis, Petroway, Palo Alto, California.

Hubbard, R.J., Pape, J., Roberts, D.G., 1985. Depositional Sequence Mapping as a Technique to Establish Tectonic and Stratigraphic Framework, and Evaluate Hydrocarbon Potential on Passive Continental Margin. AAPG Memoirs 39, 107–135.

Isaaks, E.H., Srivastava, R.M., 1989. Applied Geostatistics, Oxford University Press, New York.

Jack, I., 1998. Time-Lapse Seismic in Reservoir Management, 1998 Distinguished Instructor Short Course. Society of Exploration Geophysicists, Tulsa.

Jenkins, C., Freyder, D., Smith, J., Starley, G., 2007. Chapter 6: Coalbed Methane. In: Warner Jr., H.R. (Ed.); Lake, L.W. (Editor-in-Chief), Petroleum Engineering Handbook, Volume 6: Emerging and Peripheral Technologies. Society of Petroleum Engineers, Richardson, Texas.

Jones, J., 2007. Chapter 15: Thermal Recovery by Steam Injection. In: Holstein, E.D. (Ed.); Lake, L.W. (Editor-in-Chief), Petroleum Engineering Handbook, Volume 5: Reservoir Engineering and Petrophysics. Society of Petroleum Engineers, Richardson, Texas.

Kabir, C.S., Chawathe, A., Jenkins, S.D., Olayomi, A.J., Aigbe, C., Faparusi, D.B., 2004. Developing New Fields Using Probabilistic Reservoir Forecasting. SPE Reservoir Evaluation and Engineering, 15–23.

Kalla, S., White, C.D., 2005. Efficient Design of Reservoir Simulation Studies for Development and Optimization, Paper SPE 95456. Society of Petroleum Engineers, Richardson, Texas.

Kamal, M.M., Freyder, D.G., Murray, M.A., 1995. Use of Transient Testing in Reservoir Management. Journal of Petroleum Technology, 992–999.

Kelamis, P.G., Uden, R.C., Dunderdale, I., 1997. 4-D Seismic Aspects of Reservoir Management, Paper OTC 8293. In: Proceedings of the 1997 Offshore Technology Conference, Houston.

Kelkar, M., 2000. Application of Geostatistics for Reservoir Characterization—Accomplishments and Challenges. Journal of Canadian Petroleum Technology 59, 25–29.

King, G.R., David, W., Tokar, T., Pope, W., Newton, S.K., Wadowsky, J., et al., 2002. Takula Field: Data Acquisition, Interpretation, and Integration for Improved Simulation and Reservoir Management. SPE Reservoir Evaluation & Engineering, pp. 135–145, April.

King, M.J., Mansfield, M., 1999. Flow Simulation of Geologic Models. SPE Reservoir Evaluation & Engineering, 351–367, August.

Kuuskraa, V.A., Bank, G.C., 2003. Gas from Tight Sands, Shales a Growing Share of U.S. Supply. Oil & Gas Journal, 34–43, December.

Kuuskraa, V.A., Brandenburg, C.F., 1989. Coalbed Methane Sparks a New Energy Industry. Oil & Gas Journal, 49, October.

Lake, L.W., 1989. Enhanced Oil Recovery, Prentice Hall, Englewood Cliffs.

Lake, L.W., Jensen, J.L., 1989. A Review of Heterogeneity Measures Used in Reservoir Characterization, Paper SPE 20156. Society of Petroleum Engineers, Richardson, Texas.

Lantz, R.B., 1971. Quantitative Evaluation of Numerical Diffusion. Society of Petroleum Engineering Journal, 315–320.

Lasseter, T.J., Jackson, S.A., 2004. Improving Integrated Interpretation Accuracy and Efficiency using a Single Consistent Reservoir Model from Seismic to Simulation. The Leading Edge, 1118–1121, November.

Lee, J., 2007. Chapter 8: Fluid Flow through Permeable Media. In: Holstein, E.D. (Ed.); Lake, L.W. (Editor-in-Chief), Petroleum Engineering Handbook, Volume 5: Reservoir Engineering and Petrophysics. Society of Petroleum Engineers, Richardson, Texas.

Levin, H.L., 1991. The Earth Through Time, fourth ed., Saunders College Publishing, New York.

Loder Jr., W.R., 2000. Tracer Technologies International. Exhibitor's Technology Update Series, SPE/DOE Eleventh Symposium on Improved Oil Recovery, Tulsa.

Lumley, D.E., 2001. Time-lapse seismic reservoir characterization. Geophysics 66, 50–53.

Lumley, D.E., 2004. Business and Technology Challenges for 4D Seismic Reservoir Monitoring. The Leading Edge, 1166–1168.

Lumley, D.E., Behrens, R., 1997. Practical Issues for 4-D Reservoir Modeling. Journal of Petroleum Technology, 998–999, September.

Makogon, Y.F., Dunlap, W.A., Holditch, S.A., 1997. Recent Research on Properties of Gas Hydrates, Paper OTC 8299, presented at the 1997 Offshore Technology Conference. Society of Petroleum Engineers, Richardson, Texas.

Martin, F.D., 1992. Enhanced Oil Recovery for Independent Producers, Paper SPE/DOE 24142. Society of Petroleum Engineers, Richardson, Texas.

Maschio, C., Schiozer, D.J., Moura Filho, M.A.B., 2005. A Methodology to Quantify the Impact of Uncertainties in the History Matching Process and in the Production Forecast, Paper SPE 96613. Society of Petroleum Engineers, Richardson, Texas.

Mattax, C.C., Dalton, R.L., 1990. Reservoir Simulation, SPE Monograph #13. Society of Petroleum Engineers, Richardson, Texas.

Matthews, C.S., Russell, D.G., 1967. Pressure Buildup and Flow Tests in Wells, Society of Petroleum Engineers, Richardson, Texas.

Mavko, G., Mukerji, T., Dvorkin, J., 1998. The Rock Physics Handbook, Cambridge University Press, Cambridge.

Mavor, M., Pratt, T., DeBruyn, R., 1999. Study Quantifies Powder River Coal Seam Properties. Oil & Gas Journal, 35–40, April 26.

McCain Jr., W.D., 1990. The Properties of Petroleum Fluids, second ed., PennWell Publishing, Tulsa.

McCain Jr., W.D., 1991. Reservoir-Fluid Property Correlations—State of the Art. Society of Petroleum Engineers Reservoir Engineering, 266–272.

Miller, M.A., Holstein, E.D., 2007. Chapter 9: Gas Reservoirs. In: Holstein, E.D. (Ed.); Lake, L.W. (Editor-in-Chief), Petroleum Engineering Handbook, Volume 5: Reservoir Engineering and Petrophysics. Society of Petroleum Engineers, Richardson, Texas.

Mitchell, R.F. (Ed.); Lake, L.W. (Editor-in-Chief), 2007. Petroleum Engineering Handbook, Volume 2: Drilling Engineering. Society of Petroleum Engineers, Richardson, Texas.

Montgomery, C.W., 1990. Physical Geology, second ed., Wm. C. Brown, Publishers, Dubuque.

Moses, P.L., 1986. Engineering Applications of Phase Behavior of Crude Oil and Condensate Systems. Journal of Petroleum Technology, 715–723, July.

Mulholland, J.W., 1998. Sequence stratigraphy: Basic elements, concepts and terminology. The Leading Edge, 37–40.

Murphy, W., Reisher, A., Hsu, K., 1993. Modulus Decomposition of Compressional and Shear Velocities in Sand Bodies. Geophysics 58, 227–239.

Murtha, J.A., 1997. Monte Carlo Simulation: Its Status and Future. Journal of Petroleum Technology, 361–373.

Neumiller, J., Graves, R., Fanchi, J., 2009. Feasibility of Using Wind Energy and CAES Systems in a Variety of Geologic Settings, Paper SPE 118839, Society of Petroleum Engineers, Richardson, Texas, first presented in proceedings of 2009 SPE EUROPEC/ EAGE Annual Conference and Exhibition, Amsterdam, June.

Nørgård, J.-P., 2006. Revolutionizing History Matching and Uncertainty Assessment. GEO ExPro, 34–35.

Offshore staff, 1998. Refurbishment/Abandonment. Offshore Magazine 148, 186, September.

Passone, S., McRae, G.J., 2007. Probabilistic Field Development in Presence of Uncertainty, Paper IPTC 11294, presented at the 2007 International Petroleum Technology Conference, Dubai.

Peaceman, D.W., 1977. Fundamentals of Numerical Reservoir Simulation, Elsevier, New York.

Peaceman, D.W., 1978. Interpretation of Well-Block Pressures in Numerical Reservoir Simulation. Society of Petroleum Engineering Journal, 183–194, June.

Peaceman, D.W., 1983. Interpretation of Well-Block Pressures in Numerical Reservoir Simulation with Non-Square Grid Blocks and Anisotropic Permeability. Society of Petroleum Engineering Journal, 531–543, June.

Pedersen, K.S., Fredenslund, A., Thomassen, P., 1989. Properties of Oil and Natural Gases, Gulf Publishing, Houston.

Peijs-van Hilten, M., Good, T.R., Zaitlin, B.A., 1998. Heterogeneity Modeling and Geopseudo Upscaling Applied to Waterflood Performance Prediction of an Incised Valley Reservoir: Countess YY Pool, Southern Alberta. AAPG Bulletin 82, 2220–2245.

Pennington, W.D., 2001. Reservoir Geophysics. Geophysics 66, 25–30.

Pennington, W.D., 2007. Chapter 1: Reservoir Geophysics. In: Warner Jr., H.R. (Ed.); Lake, L.W. (Editor-in-Chief), Petroleum Engineering Handbook, Volume 6: Emerging and Peripheral Technologies. Society of Petroleum Engineers, Richardson, Texas.

Pletcher, J.L., 2002. Improvements to Reservoir Material Balance Methods. SPE Reservoir Evaluation & Engineering, 49–59, February.

Press, F., Siever, R., 2001. Understanding Earth, third ed., W.H. Freeman and Company, San Francisco.

Rao, N.D., Girard, M.G., 1997. A New Technique for Reservoir Wettability Characterization. Journal of Canadian Petroleum Technology 35, 31–39, January.

Raza, S.H., 1992. Data Acquisition and Analysis for Efficient Reservoir Management. Journal of Petroleum Technology, 466–468.

Renner, J.L., Shook, G.M., Garg, S., Finger, J.T., Kasameyer, P.W., Bloomfield, K.K., et al., 2007. Chapter 9: Geothermal Engineering. In: Warner Jr., H.R. (Ed.); Lake, L.W. (Editor-in-Chief), Petroleum Engineering Handbook, Volume 6: Emerging and Peripheral Technologies. Society of Petroleum Engineers, Richardson, Texas.

Reynolds, S.J., Johnson, J.K., Kelly, M.M., Morin, P.J., Carter, C.M., 2008. Exploring Geology, McGraw-Hill, Boston.

Richardson, J.G., Sangreem, J.B., Sneider, R.M., 1987a. Applications of Geophysics to Geologic Models and to Reservoir Descriptions. Journal of Petroleum Technology, 753–755.

Richardson, J.G., Sangree, J.B., Sneider, R.M., 1987b. Introduction to Geologic Models. Journal of Petroleum Technology (first of series), 401–403.

Ridley, M., 1996. Evolution, second ed. Blackwell Science, Cambridge.

Robinson, M., 2007. Chapter 3: Intelligent-Well Completions. In: Warner Jr., H.R. (Ed.); Lake, L.W. (Editor-in-Chief), Petroleum Engineering Handbook, Volume 6: Emerging and Peripheral Technologies. Society of Petroleum Engineers, Richardson, Texas.

Sabet, M.A., 1991. Well Test Analysis, Gulf Publishing, Houston.

Saleri, N.G., 2002. Learning' Reservoirs: Adapting to Disruptive Technologies. Journal of Petroleum Technology, 57–60.

Saleri, N.G., 2005. Reservoir Management Tenets: Why They Matter to Sustainable Supplies. Journal of Petroleum Technology, 28–30.

Scheidegger, A.E., 1954. Statistical Hydrodynamics in Porous Media. Journal of Applied Physics 25, 91 ff.

Schilthuis, R.D., 1936. Active Oil and Reservoir Energy. Transactions of the AIME 118, 33 ff.

Schneider, F.N, 1987. Three Procedures Enhance Relative Permeability Data. Oil and Gas Journal, 45–51, May.

Schön, J.H., 1996. Physical Properties of Rocks: Fundamentals and Principles of Petrophysics, volume 18. Elsevier, New York.

Selley, R.C., 1998. Elements of Petroleum Geology, Academic Press, San Diego.

Senum, G.I., Fajer, R., 1992. Petroleum Reservoir Characterization by Perfluorocarbon Tracers. In: Proceedings of the SPE/DOE Eighth Symposium on EOR, Tulsa, April, pp. 22–24, 337–345.

Sheriff, R.E., 1992. Reservoir Geophysics, Investigations in Geophysics, Number 7, Society of Exploration Geophysicists, Tulsa.

Slatt, R.M., Hopkins, G.L., 1990. Scaling Geologic Reservoir Description to Engineering Needs. Journal of Petroleum Technology, 202–210.

Sloan, E.D., 2006. Chapter 11: Phase Behavior of $H2O$ + Hydrocarbon Systems. In: Fanchi, J.R. (Ed.); Lake, L.W. (Editor-in-Chief), Petroleum Engineering Handbook, Volume 1: General Engineering. Society of Petroleum Engineers, Richardson, Texas.

Sloan, E.D., 2007. Chapter 11: Hydrate Emerging Technologies. In: Warner Jr., H.R. (Ed.); Lake, L.W. (Editor-in-Chief), Petroleum Engineering Handbook, Volume 6: Emerging and Peripheral Technologies. Society of Petroleum Engineers, Richardson, Texas.

SPE-PRMS, 2007. SPE/WPC/AAPG/SPEE Petroleum Resources Management System, Society of Petroleum Engineers, Richardson, Texas. Accessed on June 1, 2009, at http://www.spe.org/spe-app/spe/industry/reserves/.

Stern, D., 2005. Practical Aspects of Scaleup of Simulation Models. Journal of Petroleum Technology, 74–82, September.

Stone, H.L., 1973. Estimation of Three-Phase Relative Permeability and Residual Oil Data. Journal of Canadian Petroleum Technology, 53ff.

Sutton, R.P., 2006. Chapter 6. Oil System Correlations. In: Fanchi, J.R. (Ed.); Lake, L.W. (Editor-in-Chief), Petroleum Engineering Handbook, Volume 1: General Engineering. Society of Petroleum Engineers, Richardson, Texas.

Taber, J.J., Martin, F.D., 1983. Technical Screening Guides for the Enhanced Recovery of Oil. Paper SPE 12069. Society of Petroleum Engineers, Richardson, Texas.

Taber, J.J., Martin, F.D., Seright, R.S., 1997a. EOR Screening Criteria Revisited: Part 1— Introduction to Screening Criteria and Enhanced Recovery Field Projects. SPE Reservoir Evaluation and Engineering, 189–198.

Taber, J.J., Martin, F.D., Seright, R.S., 1997b. EOR Screening Criteria Revisited: Part 2— Applications and Implications of Oil Prices. SPE Reservoir Evaluation and Engineering, 199–205.

Tatham, R.H., McCormack, M.D., 1991. Multicomponent Seismology in Petroleum Exploration, Society of Exploration Geophysicists, Tulsa.

Tearpock, D.J., Bischke, R.E., 1991. Applied Subsurface Geological Mapping, Prentice-Hall, Englewood Cliffs, NJ.

Tearpock, D.J., Brenneke, J.C., 2001. Multidisciplinary Teams, Integrated Software for Shared Earth Modeling Key E&P Success. Oil and Gas Journal, 84–88, December.

Tek, M.R., 1996. Natural Gas Storage Underground: Inventory and Deliverability, PennWell Publishing, Tulsa.

Telford, W.M., Geldart, L.P., Sheriff, R.E., 1990. Applied Geophysics, second ed. Cambridge University Press, New York.

Terry, R.E., Michnick, M.J., Vossoughi, S., 1981. Manual for Tracer Test Design and Evaluation, Institute of Mineral Resources Research. University of Kansas, Lawrence.

Tiab, D., Donaldson, E.C., 2003. Petrophysics, second ed., Elsevier, Boston.

Toronyi, R.M., Saleri, N.G., 1989. Engineering Control in Reservoir Simulation: Part II, Paper SPE 17937. Society of Petroleum Engineers, Richardson, Texas.

Towler, B.F., 2002. Fundamental Principles of Reservoir Engineering. Society of Petroleum Engineers, Richardson, Texas, SPE Textbook Series Volume 8.

Towler, B.F., 2006. Chapter 5. Gas Properties. In: Fanchi, J.R. (Ed.); Lake, L.W. (Editor-in-Chief), Petroleum Engineering Handbook, Vol. 1, General Engineering, Society of Petroleum Engineers, Richardson, Texas.

Turcotte, D.L., Schubert, G., 2002. Geodynamics, second ed., Cambridge University Press, New York.

Turner, R.G., Hubbard, M.G., Dukler, A.E., 1969. Analysis and Prediction of Minimum Flow Rate for the Continuous Removal of Liquids from Gas Wells. Journal of Petroleum Technology, 1475–1482, November.

Uland, M.J., Tinker, S.W., Caldwell, D.H., 1997. 3-D Reservoir Characterization for Improved Reservoir Management. Paper Presented at the 1997 Society of Petroleum Engineers' 10th Middle East Oil Show and Conference, Bahrain, March.

Vail, P.R., Mitchum, R.M., Todd, R.G., Widmeir, J.M., Thompson, S., Sangree, J.B., et al., 1977. Seismic Stratigraphy and Global Changes of Sea Level. Memoirs—American Association of Petroleum Geologists 26, 49–212.

Van Dyke, K., 1997. Fundamentals of Petroleum, fourth ed. University of Texas–Austin, Petroleum Extension Service.

Van Wagoner, J.C., Posamentier, H.W., Mitchum, R.M., Vail, P.R., Sarg, J.F., Loutit, T.S., et al., 1988. An Overview of the Fundamentals of Sequence Stratigraphy and Key Definitions. Special Publication—Society of Economic Paleontology and Mineralogy 42, 39–46.

Walker, G.J., Lane, H.S., 2007. Assessing the Accuracy of History Match Predictions and the Impact of Time-Lapse Seismic Data: A Case Study for the Harding Reservoir. Paper SPE 106019. Society of Petroleum Engineers, Richardson, Texas.

Walsh, M.P., Lake, L.W., 2003. A Generalized Approach to Primary Hydrocarbon Recovery, Elsevier, Amsterdam.

Wang, Z., 1997. Feasibility of Time-Lapse Seismic Reservoir Monitoring: the Physical Basis. The Leading Edge 16, 1327–1329, September.

Wang, Z., 2000. Dynamic versus Static Elastic Properties of Reservoir Rocks. In: Wang, Z., Nur, A. (Eds.), Seismic and Acoustic Velocities in Reservoir Rocks, Volume 3, Recent Developments. Society of Exploration Geophysicists, Tulsa.

Warner Jr., H.R., 2007. Chapter 11: Waterflooding. In: Holstein, E.D. (Ed.); Lake, L.W. (Editor-in-Chief), Petroleum Engineering Handbook, Volume 5: Reservoir Engineering and Petrophysics. Society of Petroleum Engineers, Richardson, Texas.

Warner Jr., H.R. (Ed.); Lake, L.W. (Editor-in-Chief), 2007. Petroleum Engineering Handbook, Volume 6: Emerging and Peripheral Technologies. Richardson, Texas, Society of Petroleum Engineers.

Wattenbarger, R.A., 2002. Trends in Tight Gas Sand Production. Journal of Canadian Petroleum Technology, 17–20, July.

Weimer, P., Davis, T.L. (Eds.), Applications of 3-D Seismic Data to Exploration and Production, Studies in Geology, no. 42. American Association of Petroleum Geologists, Tulsa.

Weimer, R.J., Sonnenberg, S.A., 1989. Sequence Stratigraphic Analysis of Muddy (J) Sandstone Reservoir, Wattenberg Field, Denver Basin. In: Petrogenesis and Petrophysics of Selected Sandstone Reservoirs of the Rocky Mountain Region. Rocky Mountain Association of Geologists, Denver.

Welge, H., 1952. A Simplified Method for Computing Oil Recoveries by Gas or Water Drive. Transactions of the American Institute of Mechanical Engineers 195, 91–98.

Westlake, D.W.S., 1999. Bioremediation, Regulatory Agencies, and Public Acceptance of this Technology. Journal of Canadian Petroleum Technology 38, 48–50.

Whitson, C.H., Brulé, M.R., 2000. Phase Behavior, SPE Monograph Series. Society of Petroleum Engineers, Richardson, Texas.

Wigley, T.M.L., Richels, R., Edmonds, J.A., 1996. Economic and Environmental Choices in the Stabilization of Atmospheric CO_2 Concentrations. Nature, 240–243, January.

Wilhite, G.P., 1986. Waterflooding, SPE Textbook Series, vol. 3. Society of Petroleum Engineers, Richardson, Texas.

Wilkinson, A.J., 1997. Improving Risk-Based Communications and Decision Making. Journal of Petroleum Technology, 936–943, September.

Williams, M.A., Keating, J.F., Barghouty, M.F., 1998. The Stratigraphic Method: A Structured Approach to History-Matching Complex Simulation Models. SPE Reservoir Evaluation and Engineering Journal, 169–176.

Williams, M.C., Leighton, V.L., Vassilou, A.A., Tan, H., Nemeth, T., Cox, V.D., et al., 1997. Crosswell Seismic Imaging: A Technology Whose Time Has Come? The Leading Edge, 285–291.

Wilson, M.J., Frederick, J.D., 1999. Environmental Engineering for Exploration and Production Activities, SPE Monograph Series, Society of Petroleum Engineers, Richardson, Texas.

Wittick, T., 2000. Exploration vs. Development Geophysics: Why Is Development Geophysics So Much More Quantitative? Oil and Gas Journal, 29–32, July.

Yang, C., Nghiem, L., Card, C., Bremeier, M., 2007. Reservoir Model Uncertainty Quantification Through Computer Assisted History Matching, Paper SPE 109825. Society of Petroleum Engineers, Richardson, Texas.

Yergin, D., 1992. The Prize, Simon and Schuster, New York.

Yeten, B., Castellini, A., Guyaguler, B., Chen, W.H., 2005. A Comparison Study on Experimental Design and Response Surface Methodologies, Paper SPE 93347. Society of Petroleum Engineers, Richardson, Texas.

Index

A

accumulation, 17, 36, 42, 86, 178, 223, 224, 235
acoustic
 impedance, 78, 80, 81, 82, 88, 97, 105, 238
 log, 115, 116, 124
algorithm, 102, 105, 181, 191, 192, 193, 195, 201, 202, 238, 239
API, 22, 32, 284, 285
aquifer, 23, 128, 151, 153, 154, 160, 224, 265t, 268, 281, 286, 288, 289
Archie, 119, 122
areal model, 221

B

base case, 240, 266, 276
basin modeling, 42
bin size, 82, 200
black oil
 model, 28, 31
 simulator, 27, 28, 229, 232
block size, 227, 241
borehole imaging, 120
boundaries, 1, 34, 36, 47, 71, 86, 87, 128, 141, 142t, 155, 192, 234, 254t
boundary conditions, 127, 128, 214, 215, 226
brown field, 243, 257, 263, 267, 268, 275
bubble point, 24, 25, 28, 31, 32t, 165, 236, 240, 245, 246, 248, 250t, 256, 276
Buckley-Leverett, 209, 210, 211, 214
buildup, 126, 130, 132, 138, 139, 142t, 143
bulk
 density, 78, 79, 88, 97, 104, 116, 124, 190
 modulus, 89, 90, 92, 96, 97, 98, 99, 100, 101, 104, 105, 107, 108, 187
 volume, 43, 45, 49, 50, 89, 100, 116, 223

C

C–D equation, 161, 225, 226
CAES, 290, 291f, 292f
calibration, 84, 85, 266
caliper log, 113

CAPEX, 11
capillary pressure, 167, 169, 170, 171, 172f, 173, 174t, 175, 183, 184, 205, 206, 207, 208, 209, 213, 214, 235, 254
Cartesian, 49, 161, 187, 203, 225
cash flow, 6, 8, 9f
casing, 109, 110f, 268
cementation exponent, 118
checkshot, 81
climate change, 13, 288
coalbed methane, 1, 58, 282, 283, 284
completion, 4f, 110f, 111, 249, 254t, 280
compositional model, 29
compressed air energy storage, 290
compressibility
 fluid, defined, 20, 99
 porosity, defined, 51
compressional velocity, 78, 79, 88, 97, 100, 104, 105, 107, 108, 238, 239
computer
 mapping, 192, 200, 202
 program, 27, 192, 193, 223, 232, 264
concentration, 2, 14, 19, 117, 119, 159, 161, 163t, 164, 224, 225, 226, 227, 229, 230
condensate, 24f, 25, 27t, 28, 229, 245, 246, 248
conditional simulation, 199t, 201
conservation of mass, 149, 209, 223
constant
 composition expansion, 245, 246f
 volume depletion, 245, 246, 247f
constraint, 76, 77, 197, 201, 202, 230, 266, 268, 271, 273f, 274f, 277
contact angle, 168, 169f, 171, 172, 173
continuity equation, 214, 224, 225, 226, 229
convection–dispersion (C-D), 161
core analysis, 44, 46, 72, 187, 252, 254t
cricondentherm, 24f, 25
critical point, 24, 25
cross-section, 33, 61, 91, 117, 118, 164, 171, 206, 208, 212, 222, 223, 237
cylindrical, 52, 69, 112, 113

D

Darcy's law, 23
data
  acquisition, 80, 85, 106, 243, 249, 251,
    252, 255, 280
  management, 243
  preparation, 253
datum, 65, 96, 143, 188, 252
decline curve analysis, 145, 146
deliverability, 125, 126, 137, 145, 147,
    148, 156, 249
density
  gradient, 171, 175
  log, 96, 115, 120, 122, 123, 124
depth
  migration, 81
  of investigation, 112, 119
design of experiments, 258, 263
deterministic, 194, 195, 201, 202, 263,
    264, 275, 276
development geophysics, 84, 243
dew point, 25, 28, 245, 246
differential
  equations, 215, 216
  liberation, 28, 245, 246, 247*f*
diffusivity equation, 126, 127, 130, 131,
    133, 137, 138
digitize, 190, 204
dimensional analysis, 52
dipmeter log, 120
dipping, 65, 154, 155, 207, 213, 215
disciplines, 5
discount rate, 8, 9*t*, 10, 11
displacement efficiency, 8, 15, 48, 155, 157,
    218, 280
dominant frequency, 82, 83, 84, 88
drawdown, 126, 139, 140, 142*t*, 144, 250*t*
drill
  bit, 1, 2, 3, 4, 15, 109, 113
  pipe, 109, 120
  stem test, 142*t*, 143, 144, 245
driller's log, 113
drilling mud, 2, 109, 111, 113, 124, 244
drive mechanisms, 149, 151, 153,
    154, 156
dry gas, 25, 26
Dykstra-Parsons, 63, 64, 69
dynamic model, 74, 254, 257, 259*t*, 260*f*,
    263, 264, 269, 270*f*, 271, 273, 275, 276

E

economics, 8, 52, 174, 218, 250
elastic constant, 89
elasticity theory, 93
electrode log, 119, 120
elevation, 170, 175
enhanced oil recovery, 41, 157, 279, 289
environmental impact, 3, 5, 11, 289
equation of state, 18, 21, 29, 30, 55
  cubic, 30*t*, 30
equilibration, 19, 20, 29, 30, 248
equilibrium, 66, 151, 248
equivalent height, 170, 175
era, 35, 36*t*, 109
expenses, 1, 9, 10, 11
exploration geophysics, 84
extended reach drilling, 4

F

facies, 38, 47, 82, 87, 120, 201
facilities, 3, 4*f*, 5, 11, 31, 245, 249, 267,
    286, 290
falloff, 126, 142*t*
fine grid, 82, 191
finite difference, 227, 234
flash, 25, 27, 28, 31, 248, 286
flow
  capacity, defined, 49
  regime, 53, 54*t*, 128, 129, 133, 145, 147
  unit, 187, 188*t*, 189, 244, 254
fluid
  classification, 23
  contacts, 82, 191, 254*t*
  movement, 39, 40, 104, 109, 128, 129
  properties defined, 18
  sampling, 244, 245
  type, 25*t*, 26*f*, 27*t*, 113, 120, 218, 253
flux, 223, 224, 225, 229, 230
Forcheimer, 54
formation volume factor, 21, 28, 32, 46,
    48, 54, 55, 61, 128, 135, 150*t*, 229, 230,
    233*t*, 234*t*
forward modeling, 105
fracture, 39, 40, 44, 58, 77, 83, 86, 120,
    123, 157, 158, 159, 218, 224, 281, 282,
    284, 292
Fresnel zone, 83, 84
frontal advance, 162, 164, 205, 209, 210,
    211, 213*f*, 215, 216, 227, 241

fully implicit, 227, 235, 241
gamma ray, 114, 115, 116, 117, 120, 121*t*, 122, 123, 252

G
gas
  cap, 80, 106, 150*t*, 151, 152*t*, 153, 154, 155, 156*f*, 240, 250*t*
  hydrates, 283
gas-oil, 2
Gassmann, 98, 100, 108
geochemistry, 40, 42
geologic model, 74, 81, 82, 195
geology, 33, 81, 125, 195
geomechanical model, 101
geophysics, 80, 83, 84, 85, 194
geostatistics, 194, 195, 197, 199, 200, 201, 202
geothermal, 1, 5, 285, 286, 287*f*, 288, 292
giga scale, 72*f*, 109, 117
grain
  density, 78, 96, 97, 101, 105, 107
  modulus, 100, 107, 108
  volume, 43, 50
gravity drainage, 281
green field, 243, 257, 258*t*, 259, 260*f*, 261*f*, 262*f*, 263, 268, 269, 273, 275, 277
greenhouse gas, 41, 104, 287, 288, 289
gridblock, 227, 234, 235, 236, 241
gross thickness, defined, 45

H
heating value, 22
heavy oil, 22, 25*t*
heterogeneity, 34, 42, 63, 65, 67, 73, 195, 202, 221
historical data, 259, 263, 264, 277
history matching
  deterministic, 266, 268, 276
Honarpour, 167, 179, 183, 253
Hooke's law, 89, 94, 96
Horner
  analysis, 132, 136, 139
  plot, 132, 133, 143, 144
hysteresis, 170*f*, 173, 179

I
IFLO, 15, 106, 232, 241
igneous, 37, 39

immiscible, 26, 157, 167, 168, 169, 171, 175, 176, 205, 208, 209, 210, 214, 279, 280, 281, 289
IMPES, 227, 235, 236, 241
implicit, 227, 235, 241
improved oil recovery, 157
incompressible, 209, 210, 214, 220, 226
induction log, 119, 120
infill, 157, 220
influx, 150*t*, 151*t*, 153, 156*f*, 224, 281
interfacial tension, 157, 167, 171, 172, 176, 245, 280, 281
inverse distance weighting, 193, 194*f*, 200, 204
irrotational, 76, 77
isothermal, 151
intelligent
  field, 280
  well, 280

J
Jacobian, 227

K
K value, 20, 29
Klinkenberg, 57, 282
kriging, 197, 198, 199*t*, 200, 201

L
laboratory measurements, 105, 133, 167, 171, 172, 178, 179, 181
lag, 196, 197, 199, 200*f*, 204, 213
Lagrange multiplier, 197, 198
Lamé, 76, 95
lateral resolution, 83
life cycle, 1, 5, 156
log
  cutoff, 191
  lithology, 114
  suite, 122
Lorenz coefficient, 63, 64, 69
LSE, 47
LWD, 120

M
Macleod-Sugden, 167, 172
macro scale, 71, 72*f*
mapping, 80, 87, 187, 190, 192, 200, 202
mass conservation, 224, 229, 230

material balance equation, 149, 150, 151, 152*t*, 235
matrix, 58, 68
mega scale, 71, 72*f*, 109, 117, 141
metamorphic, 37, 39
micro scale, 71, 72*f*
microbial, 279, 281
minimum miscibility pressure (MMP), 26, 248, 281
miscibility, 26, 248, 280
miscible, 26, 279, 280, 281, 289
mobility
   defined, 181
   ratio, defined, 182
model calibration, 266
modified Lorenz plot, 188
mole fraction, defined, 19
molecular weight, 21, 22, 25, 26, 32, 167, 223, 245
momentum, 53, 228
Monte Carlo analysis, 263, 273, 276
mud cake, 111, 112*f*
multidisciplinary, 1
multilateral wells, 2, 3*f*
multiple contact miscibility, 248
MWD, 4, 120

N
naturally fractured, 77, 79, 141, 224
Navier-Stokes equation, 228
net present value (NPV), 8, 9*t*, 10, 11
neutron log, 115, 117, 120, 122, 252
Newton-Raphson, 227, 236
normal distribution, 6, 7, 254, 255
NOx, 13
nugget, 196, 197*f*, 204
numerical dispersion, 227, 241

O
objective function, 268, 269, 271, 272*f*, 273
objectives, 6, 7, 14, 145, 250, 253, 264, 266, 280, 289
oil productive capacity, 80
oil–water, 45, 78, 123
OPEX, 11

P
P–T diagram, 24
parachor, 167
paraffin, 24

partial differential equations, 226, 227
performance predictions, 73, 266, 267
permeability
   absolute, 57, 61, 176, 177, 182, 208, 233*t*, 234*t*, 237, 240
   anisotropic, 67
   defined, 40, 45
   directional, 49, 65
   effective, 125, 133, 147, 176, 177, 237, 240, 281
   homogeneous
   horizontal, 68, 142*t*, 143
   inhomogeneous
   isotropic, 67–68, 237
   tensor, 67, 93
   vertical, 68, 142*t*, 143
petroelastic model, 96, 101, 105, 106, 239, 265
petroleum, 3, 6*t*, 17, 19, 21, 23, 24, 25, 27*t*, 31, 39, 41, 42, 109, 281, 287
petrophysics, 89, 190
phase
   behavior, 24, 25, 26, 106, 135, 181, 248
   envelope, 24, 25
   potential, 69, 214, 235
photoelectric log, 120
pipe, 2, 15, 110*f*
pipeline, 5, 11, 12, 283
plate tectonics, 34, 72
Poisson's ratio, 91, 92, 95, 101, 102, 103, 106, 108
pore
   pressure, 20, 95, 96, 100, 103, 104, 239
   radius, 171
   volume defined, 43
porosity
   compressibility, 49, 68, 103, 239, 240
   defined, 40, 43
   log, 115, 122, 123
porous medium, 37, 43, 49, 50, 51, 52, 54, 57, 67
prediction, 7, 11, 85, 263, 264, 266, 276
pressure
   buildup test, 126, 130, 138, 139, 143
   derivative, 135
   drawdown, 139, 140
   effective, 99, 100, 102, 238
   maintenance, 156, 276, 289
   three-phase relative permeability, 180

transient testing, 71, 125, 126, 129, 130, 135, 136, 138, 140, 141, 144, 187, 243, 254*t*
primary production, 2, 154, 155, 156
prime mover, 2
probability distribution, 6, 11, 195, 258, 259, 263, 273, 275, 276
productivity index, 236, 240, 241, 266
proxy, 261, 262*f*, 263, 273, 275, 276, 277
pseudocomponent, 29, 30, 31, 232
pulse, 75, 126, 142*t*
PVT
    data, 28, 32
    property, 254

R
radial
    coordinates, 55
    flow, 54, 55, 59, 60, 125, 127, 129, 133, 135*t*, 136, 139, 164, 236, 240
radioactive tracer, 160
radius of investigation, 125, 136, 140, 141, 144
real gas pseudopressure, 56, 57, 137, 138, 139
realizations, 145, 195, 198, 201, 202, 257, 263
recovery efficiency, 4, 7, 15, 149, 156, 157
reflection coefficient, 78, 88, 105, 123
regression, 30, 31, 85, 167, 194, 199, 261, 263
relative
    mobility, 182, 184
    permeability, defined, 177
reliability, 71, 202, 290
repeat formation test, 142*t*, 143
reserves, defined, 6*t*
reservoir
    architecture, 72, 86
    characterization, defined, 187
    description, 264
    engineering, 80, 106, 111, 181, 245, 250
    geophysics, 83, 84, 85, 194
    limits test, 140, 142*t*
    management, 1, 4, 5, 7, 8, 11, 14, 25, 73, 74, 83, 84, 104, 106, 109, 125, 145, 156, 160, 167, 201, 205, 240, 250, 251, 252, 253, 257, 266, 275, 279, 280, 285, 286, 289
    scale, 71
    simulation, 192, 223
    structure, 16, 74, 79, 125, 135

reservoir forcasting
    deterministic, 263, 264, 275
    probabilistic, 257, 263, 264, 267, 275
resistivity
    factor, 118, 119, 122, 124
    log, 118, 119, 120, 123, 124
resolution
    lateral, 83
    vertical, 82
revenue, 9, 10, 11
Reynolds number, 53
risk, 11, 257, 267
rock
    quality, 57, 80
    region, 254
rotary drilling, 1, 2, 109, 243

S
saturation
    constraint, 45
    defined, 45
    pressure, 24, 28, 32*t*, 245, 246, 248
secondary recovery, 157, 279
sedimentary rock, 37, 39, 40, 42, 45, 85, 114
seismic
    data interpretation, 80
    data processing, 80
    inversion, 80
    methods, 80, 83, 192
    stratigraphy, 85
    velocity, 190, 238, 239, 276
    waves, 74, 75, 76, 78, 104
semivariance, 196, 198, 199, 200
semivariogram, 196, 197, 198, 199, 200, 201, 204
sensitivity analysis, 266
separator test, 31, 32*t*, 245, 248
sequence stratigraphy, 86
sequestration, 1, 5, 14, 104, 283, 288
shale
    gas, 52, 284
    oil, 284
shear
    modulus, defined, 91
    velocity, defined, 97
sill, 196, 197*f*, 204
sinusoidal, 77
skin, 111, 128, 133, 138, 139, 140, 142*t*, 164, 165*t*, 237, 240, 266
slippage, 57, 282

solenoidal, 77
sonic log, 116, 122, 123
source/sink, 224, 233*t*, 234*t*
SP log, 114, 115, 123
spacing, 119, 218, 220, 254*t*, 279
SPE/WPC Reserves, 6*t*
specific gravity, 21, 32, 208
spontaneous potential, 114, 123
stability, 41, 213, 214, 216, 217, 218, 227
stabilization time, 140, 141, 144, 147
stabilized rate, 130, 138
standard deviation, 6, 7, 196, 254, 259
static model, 74, 257, 264, 268
stiffness, 97, 104, 105
stochastic, 195, 201, 202
storage capacity, 49, 189
strain defined, 89
stratigraphy, 85, 243
stress
  defined, 89
  effective vertical, 95
structures, 80, 194
subsidence, 38, 42, 86, 91
superficial velocity, 53
superposition principle, 130, 131, 138
sustainable development, 12, 279
sweep efficiency
  areal, 8, 15
  vertical, 8, 15
  volumetric, 8, 221
swelling test, 245, 248
symmetry, 33

T
tar sands, 284
temperature scales, 18
tertiary production, 157
thermal, 96, 117, 157, 279, 281, 285
tight gas, 52, 58, 141, 284
time-lapse, 89, 104, 203, 265
tortuosity, 44, 118
tracer test, 157, 160, 166, 254*t*
transgression, 47, 85
transient tests, 68, 125, 126, 129, 130, 134,
  135, 136, 138, 140, 141, 142*t*, 254*t*
transition zone, 167, 170, 174, 175, 176*f*,
  185, 214
transmissibility, 49, 61, 235
traps, 39, 40

trend surface analysis, 194, 198
triangle distribution, 255, 256*f*
TSE, 47
tubing, 4, 109, 110*f*, 111, 164, 245, 283

U
uncertainty analysis, 11
unconventional, 1, 52, 282, 284
undersaturated, 28, 29*f*, 31, 32*t*, 165
uniaxial compaction, 101, 103*f*, 238, 239
United Nations' World Commission on
  Environment and Development, 12

V
validity, 104, 245, 276
valley fill geology, 47
variance, 71, 196, 198
variogram, 200
vertical
  conformance, 158
  resolution, 82, 84, 88
viscosity
  defined, 23
  dynamic, 23
  kinematic, 23
visualization technology, 202
volatile oil, 24, 25*t*, 26*f*, 27*t*
volume element, 62, 188*f*, 209
volumetrics, 160, 174, 221, 265*t*
VSP, 81, 83

W
water drive, 153, 154, 155, 156, 174
waterflood, 41, 157, 163, 178, 181, 182, 185,
  208, 211, 212, 279, 280, 281, 291, 293
wave
  equation, 75, 76, 77
  propagation, 76, 77, 79
Weinaug-Katz, 167, 172
Welge, 211, 212*f*, 214
well log, defined, 111
well
  density, 4, 157, 279
  horizontal, 125, 141, 218
  pattern, 4, 158, 205, 218, 219*f*, 220, 279
  spacing, 157, 220
  test, 57, 58, 72, 109, 125, 126, 132, 133,
    136, 138, 139, 141, 143, 201, 249,
    251*t*, 252, 257, 263, 265*t*

wellbore
  diagram, 110*f*
  storage, 132, 133, 134, 135
wet gas, 246
wettability, 157, 168, 171, 172, 178, 179
workflow, 105, 252, 253, 255, 257, 258*t*,
    263, 264, 266, 267, 268, 275, 276

workover, 254*t*
Wyllie, 116, 117

Y
Young's modulus, 89, 91, 92, 95, 101,
    102, 108

Printed in the United States
By Bookmasters